Radiation Tolerant Electronics

Radiation Tolerant Electronics

Special Issue Editor
Paul Leroux

MDPI • Basel • Beijing • Wuhan • Barcelona • Belgrade

Special Issue Editor
Paul Leroux
KU Leuven ESAT-ADVISE
Belgium

Editorial Office
MDPI
St. Alban-Anlage 66
4052 Basel, Switzerland

This is a reprint of articles from the Special Issue published online in the open access journal *Electronics* (ISSN 2079-9292) from 2018 to 2019 (available at: https://www.mdpi.com/journal/electronics/special_issues/Radiation_Tolerant_Electronics).

For citation purposes, cite each article independently as indicated on the article page online and as indicated below:

LastName, A.A.; LastName, B.B.; LastName, C.C. Article Title. *Journal Name* **Year**, *Article Number*, Page Range.

ISBN 978-3-03921-279-8 (Pbk)
ISBN 978-3-03921-280-4 (PDF)

© 2019 by the authors. Articles in this book are Open Access and distributed under the Creative Commons Attribution (CC BY) license, which allows users to download, copy and build upon published articles, as long as the author and publisher are properly credited, which ensures maximum dissemination and a wider impact of our publications.

The book as a whole is distributed by MDPI under the terms and conditions of the Creative Commons license CC BY-NC-ND.

Contents

About the Special Issue Editor vii

Paul Leroux
Radiation Tolerant Electronics
Reprinted from: *Electronics* **2019**, *8*, 730, doi:10.3390/electronics8070730 . . . . **1**

Teng Wang, Xin Wan, Hu Jin, Hao Li, Yabin Sun, Renrong Liang, Jun Xu and Lirong Zheng
Optimization of the Cell Structure for Radiation-Hardened Power MOSFETs
Reprinted from: *Electronics* **2019**, *8*, 598, doi:10.3390/electronics8060598 . . . . **4**

Mohan Liu, Wu Lu, Xin Yu, Xin Wang, Xiaolong Li, Shuai Yao and Qi Guo
Mechanism of Degradation Rate on the Irradiated Double-Polysilicon Self-Aligned Bipolar Transistor
Reprinted from: *Electronics* **2019**, *8*, 657, doi:10.3390/electronics8060657 . . . . **15**

Kyungsoo Jeong, Duckhoon Ro, Gwanho Lee, Myounggon Kang and Hyung-Min Lee
A Radiation-Hardened Instrumentation Amplifier for Sensor Readout Integrated Circuits in Nuclear Fusion Applications
Reprinted from: *Electronics* **2018**, *7*, 429, doi:10.3390/electronics7120429 . . . . **23**

Jan Budroweit, Mattis Paul Jaksch Maciej Sznajder
Proton Induced Single Event Effect Characterization on a Highly Integrated RF-Transceiver
Reprinted from: *Electronics* **2019**, *8*, 519, doi:10.3390/electronics8050519 . . . . **33**

Bjorn Van Bockel, Jeffrey Prinzie and Paul Leroux
Radiation Assessment of a 15.6 ps Single-Shot Time-to-Digital Converter in Terms of TID
Reprinted from: *Electronics* **2019**, *8*, 558, doi:10.3390/electronics8050558 . . . . **53**

Charalambos M. Andreou, Diego M. González-Castaño, Simone Gerardin, Marta Bagatin, Faustino Gómez, Alessandro Paccagnella, Alexander V. Prokofiev, Arto Javanainen, Ari Virtanen, Valentino Liberali, Cristiano Calligaro, Daniel Nahmad and Julius Georgiou
Low-Power, Subthreshold Reference Circuits for the Space Environment: Evaluated with γ-rays, X-rays, Protons and Heavy Ions
Reprinted from: *Electronics* **2019**, *8*, 562, doi:10.3390/electronics8050562 . . . . **64**

Jeffrey Prinzie and Valentijn De Smedt
Single Event Transients in CMOS Ring Oscillators
Reprinted from: *Electronics* **2019**, *8*, 618, doi:10.3390/electronics8060618 . . . . **88**

V. Díez-Acereda, Sunil L. Khemchandani, J. del Pino and S. Mateos-Angulo
RHBD Techniques to Mitigate SEU and SET in CMOS Frequency Synthesizers
Reprinted from: *Electronics* **2019**, *8*, 690, doi:10.3390/electronics8060690 . . . . **100**

Maria Munoz-Quijada, Samuel Sanchez-Barea, Daniel Vela-Calderon, Hipolito Guzman-Miranda
Fine-Grain Circuit Hardening Through VHDL Datatype Substitution
Reprinted from: *Electronics* **2019**, *8*, 24, doi:10.3390/electronics8010024 . . . . **114**

Jeffrey Prinzie, Karel Appels, Szymon Kulis
Optimal Physical Implementation of Radiation Tolerant High-Speed Digital Integrated Circuits in Deep-Submicron Technologies
Reprinted from: *Electronics* **2019**, *8*, 432, doi:10.3390/electronics8040432 . . . . **132**

Chang Cai, Xue Fan, Jie Liu, Dongqing Li, Tianqi Liu, Lingyun Ke, Peixiong Zhao and Ze He
Heavy-Ion Induced Single Event Upsets in Advanced 65 nm Radiation Hardened FPGAs
Reprinted from: *Electronics* **2019**, *8*, 323, doi:10.3390/electronics8030323 . . . . **142**

Solomon Banteywalu, Baseem Khan, Valentijn De Smedt and Paul Leroux
A Novel Modular Radiation Hardening Approach Applied to a Synchronous Buck Converter
Reprinted from: *Electronics* **2019**, *8*, 513, doi:10.3390/electronics8050513 . . . . **155**

Leonardo Maria Reyneri, Alejandro Serrano-Cases, Yolanda Morilla, Sergio Cuenca-Asensi, Antonio Martínez-Álvarez
A Compact Model to Evaluate the Effects of High Level C++ Code Hardening in Radiation Environments
Reprinted from: *Electronics* **2019**, *8*, 653, doi:10.3390/electronics8060653 . . . . **166**

Honorio Martin, Pedro Martin-Holgado, Yolanda Morilla, Luis Entrena and Enrique San Millan
Total Ionizing Dose Effects on a Delay-Based Physical Unclonable Function Implemented in FPGAs
Reprinted from: *Electronics* **2018**, *7*, 163, doi:10.3390/electronics7090163 . . . . **179**

Luis Alberto Aranda, Pedro Reviriego and Juan Antonio Maestro
Protecting Image Processing Pipelines against Configuration Memory Errors in SRAM-Based FPGAs
Reprinted from: *Electronics* **2018**, *7*, 322, doi:10.3390/electronics7110322 . . . . **190**

About the Special Issue Editor

Paul Leroux is Professor in the Department of Electrical Engineering (ESAT) of KU Leuven (University of Leuven), Belgium. He received M.Sc. and Ph.D. degrees in electronic engineering from KU Leuven in 1999 and 2004, respectively. From November 2011 to July 2016, he headed the Electrical Engineering Technology Cluster. Since August 2016 he had been the Campus Chair of KU Leuven, Geel Campus. His current research activities within the ESAT-ADVISE research group focus on radiation-hardened analog, mixed-signal, and RF IC design for communication and instrumentation in space, high-energy physics, and nuclear energy applications. His group is part of the CERN CMS collaboration, where Prof. Leroux is the KU Leuven team leader. Prof. Leroux has (co-)authored over 200 papers in international journals and conference proceedings. In 2010, he received the SCK-CEN Prof. Roger Van Geen Award from the FWO/FNRS for his highly innovative work on IC design for harsh radiation environments.

Editorial

Radiation Tolerant Electronics

Paul Leroux

KU Leuven, Dept. Electrical Engineering (ESAT) - ADVISE, 2440 Geel, Belgium; paul.leroux@kuleuven.be

Received: 20 June 2019; Accepted: 25 June 2019; Published: 27 June 2019

1. Introduction

Research on radiation tolerant electronics has increased rapidly over the last few years, resulting in many interesting approaches to model radiation effects and design radiation hardened integrated circuits and embedded systems. This research is strongly driven by the growing need for radiation hardened electronics for space applications, high-energy physics experiments such as those on the large hadron collider at CERN, and many terrestrial nuclear applications including nuclear energy and safety management. With the progressive scaling of integrated circuit technologies and the growing complexity of electronic systems, their ionizing radiation susceptibility has raised many exciting challenges, which are expected to drive research in the coming decade. Even though total ionizing dose effects in bulk CMOS are well known, little is still known on the radiation performance of advanced (FD-)SOI and FinFET technologies. Regarding single-event effects, the continued scaling has drastically increased the number of multiple-transistor or multiple-cell upsets, which requires not only new solutions to reduce the error rate in digital and mixed-signal ASICs, but also for FPGAs. The radiation hardness assurance of complex systems with multiple components in mixed technologies also necessitates new testing paradigms and verification methodologies to limit the time and cost for evaluation.

2. The Present Issue

This Special Issue features fifteen articles highlighting recent breakthroughs in modeling radiation effects for the design of radiation hardened integrated circuits, radiation hardening in embedded systems, and radiation hardening assurance. The contents of these papers are introduced here.

Two papers discuss the effect of radiation on advanced semiconductor devices such as dedicated power MOSFETs and double-polysilicon self-aligned bipolar transistors. In [1], the effects of cell structure adjustment on the performance of a power MOSFET were examined by first analyzing the design parameters. Next, a SEE- and TID-hardened power MOSFET was designed and fabricated. Results of the investigation confirmed the achievement of excellent radiation hardness and decent specific on-resistance for the device. Article [2] discussed the mechanism of degradation on the Irradiated Double-Polysilicon Self-Aligned Bipolar Transistor with a dose rate of 50 rad (Si)/s and 0.05 rad (Si)/s. The comparison of the high and low dose rate showed that the increase of the base current caused by low dose rate irradiation was larger than that caused by high dose rate irradiation, resulting in greater current gain degradation than that caused by the high dose rate, highlighting that the ELDRS effect may occur.

Several papers discuss the total-dose and/or single-event radiation effects on custom designed analog, mixed signal, and RF integrated circuits. A radiation-hardened instrumentation amplifier for sensor readout integrated circuits was presented in [3] to target nuclear fusion applications. The circuit boasts TID effect monitoring and adaptive reference control functions. The radiation tolerance was verified through SPICE simulations with radiation-aware transistor models. In [4], the authors presented the proton induced SEE characterization of a highly integrated RF transceiver in 65 nm CMOS. The exposed proton energies were split into two test campaigns to induce high energy protons (up to 184 MeV) and low energy protons down to 4 MeV. The results showed a very low sensitivity to proton irradiation, independent of the proton energy. The total ionizing dose radiation assessment of a 15.6 ps single-shot Time-to-Digital Converter was presented in [5]. Two samples were irradiated and were able to reach a dose of 2.2 MGy before failing to meet specification due to an increased non-linearity error, originating from the increased mismatch in the charge-pump of the sampling circuit. A comprehensive evaluation of two subthreshold voltage reference circuits with respect to their resilience to SEE, TID, and TID/DD was performed in [6]. The evaluation was supported by measured results with γ-rays, x-rays, protons, and heavy ions. The high total doses applied in this range of experiments provide a complete evaluation of subthreshold circuits in the whole range of space applications, radiation physics instruments, and medical applications. The authors in [7] discussed a time-variant on Single-Event Transients (SETs) in integrated CMOS ring oscillators. The Impulse Sensitive Function (ISF) of the oscillator was used to analyze the impact of the relative moment when a particle hit the circuit. The analysis was based on simulations and verified experimentally with a Two-Photon Absorption (TPA) laser setup. Article [8] presented a comprehensive study of the effects of SETs and SEUs on a frequency synthesizer for the IEEE 802.15.4 standard. The blocks that work at low frequencies were not affected by ion impacts. However, high frequency circuits such as the VCO were more vulnerable. The VCO's radiation tolerance was improved by using RC-filtering and a capacitive divider was introduced to improve the degraded phase noise.

Two articles focus on the radiation hardening of digital circuits. A new approach to implement fine-grain circuit hardening was developed and validated in [9]. This offers a dedicated VHDL package as a new tool for mitigating soft errors on digital circuits, with minimal code modifications as the designer only has to select which signals or ports should be hardened and then change their datatype accordingly. Article [10] presented a novel method for the physical implementation of Triple Modular Redundant high-speed digital circuits. The method uses a distributed constraining approach for TMR branches to avoid long interconnects between voters. The method was tested with increasingly complex digital modules and showed results that improved as the design size increased.

Three papers of this Special Issue target embedded radiation hardening in FPGA or microcontroller systems. In [11] single-event radiation hardening techniques for SRAM-based FPGAs in 65 nm CMOS technology were discussed. Both layout hardening techniques and configuration hardening techniques including ECC and TMR were employed for this FPGA. The heavy-ion results indicated a satisfactory radiation tolerance, especially for the DICE CRAMs. A novel four module radiation hardening approach for FPGA was presented in [12]. This was implemented on a zynq-7000 development board (Zybo) and it was shown that the proposed method could be used for a radiation tolerant synchronous buck converter design for applications requiring a relatively longer mission time, compared to the TMR and FMR techniques. In [13], a compact model was presented to evaluate the effects of high-level C++ code radiation hardening. The use of appropriate C++ classes facilitated the use of TMR. Additionally, the availability of an easy-to-use performance estimation model could be used for quick and effective radiation tolerance optimization of microcontroller systems.

Finally, two articles presented a link between the research fields of cryptography and image processing, respectively. In [14] the authors presented the total ionizing dose effects on a delay-based physical unclonable function implemented in FPGAs for authentication and key generation in space systems. Article [15] discussed a novel method to protect series and parallel line-buffer-based image processing pipelines against configuration memory errors in SRAM-Based FPGAs. The proposed technique presented lower FPGA resource usage, and fewer false positive detections than the other techniques; moreover, the image processing system did not have to be stopped and rebooted upon errors due to the partial reconfiguration.

3. Future

The wide range of articles in this Special Issue exemplifies the broadness of the field of radiation hardened micro-electronics. The dream to enable high performance computing, signal processing, and communication in the harshest and most diverse radiation environments presents the community with many research challenges. It inevitably brings researchers together from several disciplines ranging from nuclear and solid-state physics over advanced modeling approaches and creative circuit design techniques to the application of progressive algorithms and deep learning to optimize system performance for the most diverse applications under the harshest of conditions.

Acknowledgments: I would like to thank all of the researchers who submitted articles to this Special Issue for their excellent contributions. I am also grateful to all of the reviewers who helped in the evaluation of the manuscripts and made very valuable suggestions to improve the quality of the contributions. I would like to acknowledge the editorial board of MDPI Electronics, who invited me to guest edit this Special Issue. I am also grateful to the Electronics Editorial office staff who worked thoroughly to maintain the rigorous peer-review schedule and timely publication.

References

1. Wang, T.; Wan, X.; Jin, H.; Li, H.; Sun, Y.; Liang, R.; Xu, J.; Zheng, L. Optimization of the Cell Structure for Radiation-Hardened Power MOSFETs. *Electronics* **2019**, *8*, 598. [CrossRef]
2. Liu, M.; Lu, W.; Yu, X.; Wang, X.; Li, X.; Yao, S.; Guo, Q. Mechanism of Degradation Rate on the Irradiated Double-Polysilicon Self-Aligned Bipolar Transistor. *Electronics* **2019**, *8*, 657. [CrossRef]
3. Jeong, K.; Ro, D.; Lee, G.; Kang, M.; Lee, H.-M. A Radiation-Hardened Instrumentation Amplifier for Sensor Readout Integrated Circuits in Nuclear Fusion Applications. *Electronics* **2018**, *7*, 429. [CrossRef]
4. Budroweit, J.; Jaksch, M.P.; Sznajder, M. Proton Induced Single Event Effect Characterization on a Highly Integrated RF-Transceiver. *Electronics* **2019**, *8*, 519. [CrossRef]
5. Van Bockel, B.; Prinzie, J.; Leroux, P. Radiation Assessment of a 15.6ps Single-Shot Time-to-Digital Converter in Terms of TID. *Electronics* **2019**, *8*, 558. [CrossRef]
6. Andreou, C.M.; González-Castaño, D.M.; Gerardin, S.; Bagatin, M.; Gómez Rodriguez, F.; Paccagnella, A.; Prokofiev, A.V.; Javanainen, A.; Virtanen, A.; Liberali, V.; et al. Low-Power, Subthreshold Reference Circuits for the Space Environment: Evaluated with γ-rays, X-rays, Protons and Heavy Ions. *Electronics* **2019**, *8*, 562. [CrossRef]
7. Prinzie, J.; Smedt, V.D. Single Event Transients in CMOS Ring Oscillators. *Electronics* **2019**, *8*, 618. [CrossRef]
8. Díez-Acereda, V.L.; Khemchandani, S.; del Pino, J.; Mateos-Angulo, S. RHBD Techniques to Mitigate SEU and SET in CMOS Frequency Synthesizers. *Electronics* **2019**, *8*, 690.
9. Muñoz-Quijada, M.; Sanchez-Barea, S.; Vela-Calderon, D.; Guzman-Miranda, H. Fine-Grain Circuit Hardening Through VHDL Datatype Substitution. *Electronics* **2019**, *8*, 24. [CrossRef]
10. Prinzie, J.; Appels, K.; Kulis, S. Optimal Physical Implementation of Radiation Tolerant High-Speed Digital Integrated Circuits in Deep-Submicron Technologies. *Electronics* **2019**, *8*, 432. [CrossRef]
11. Cai, C.; Fan, X.; Liu, J.; Li, D.; Liu, T.; Ke, L.; Zhao, P.; He, Z. Heavy-Ion Induced Single Event Upsets in Advanced 65 nm Radiation Hardened FPGAs. *Electronics* **2019**, *8*, 323. [CrossRef]
12. Banteywalu, S.; Khan, B.; De Smedt, V.; Leroux, P. A Novel Modular Radiation Hardening Approach Applied to a Synchronous Buck Converter. *Electronics* **2019**, *8*, 513. [CrossRef]
13. Reyneri, L.M.M.; Serrano-Cases, A.; Morilla, Y.; Cuenca-Asensi, S.; Martínez-Álvarez, A. A Compact Model to Evaluate the Effects of High Level C++ Code Hardening in Radiation Environments. *Electronics* **2019**, *8*, 653. [CrossRef]
14. Martin, H.; Martin-Holgado, P.; Morilla, Y.; Entrena, L.; San-Millan, E. Total Ionizing Dose Effects on a Delay-Based Physical Unclonable Function Implemented in FPGAs. *Electronics* **2018**, *7*, 163. [CrossRef]
15. Aranda, L.A.; Reviriego, P.; Maestro, J.A. Protecting Image Processing Pipelines against Configuration Memory Errors in SRAM-Based FPGAs. *Electronics* **2018**, *7*, 322. [CrossRef]

© 2019 by the author. Licensee MDPI, Basel, Switzerland. This article is an open access article distributed under the terms and conditions of the Creative Commons Attribution (CC BY) license (http://creativecommons.org/licenses/by/4.0/).

Article

Optimization of the Cell Structure for Radiation-Hardened Power MOSFETs

Teng Wang [1], Xin Wan [2,3,*], Hu Jin [2], Hao Li [2], Yabin Sun [4], Renrong Liang [5], Jun Xu [3,5] and Lirong Zheng [1]

1. School of Information Science and Technology, Fudan University, Shanghai 200433, China; tengwang13@fudan.edu.cn (T.W.); lrzheng@fudan.edu.cn (L.Z.)
2. Aurorachip Co. Ltd., Zhejiang 314000, China; h.jin@aurorachip.com (H.J.); h.li@aurorachip.com (H.L.)
3. Center for High Reliability Power Semiconductor, Yangtze Delta Region Institute of Tsinghua University, Zhejiang 314000, China; junxu@tsinghua.edu.cn
4. School of Information Science Technology, East China Normal University, Shanghai 200241, China; ybsun@ee.ecnu.edu.cn
5. Institute of Microelectronics, Tsinghua University, Beijing 100084, China; liangrr@tsinghua.edu.cn
* Correspondence: x.wan@aurorachip.com; Tel.: +86-151-2000-1156

Received: 16 March 2019; Accepted: 25 May 2019; Published: 28 May 2019

Abstract: Power MOSFETs specially designed for space power systems are expected to simultaneously meet the requirements of electrical performance and radiation hardness. Radiation-hardened (rad-hard) power MOSFET design can be achieved via cell structure optimization. This paper conducts an investigation of the cell geometrical parameters with major impacts on radiation hardness, and a rad-hard power MOSFET is designed and fabricated. The experimental results validate the devices' total ionizing dose (TID) and single event effects (SEE) hardness to suitably satisfy most space power system requirements while maintaining acceptable electrical performance.

Keywords: radiation-hardened; single event gate rupture (SEGR); SEB; power MOSFETs

1. Introduction

Power MOSFETs are widely applied in space power systems [1]. However, they are vulnerable to particle from galactic cosmic rays, solar flares, and radiation belts, which may cause total ionizing dose effects, single event gate rupture (SEGR) effects and single event burnout (SEB) effects [2,3]. There has been a substantial research on such radiation effects [4–7], whereas radiation hardening on power MOSFETs, the more necessary resolve, has only been discussed in a few articles [8–12] whose content mostly focused on a single hardening issue, such as SEB, SEGR, and TID. Apparently, these radiation effects, along with electrical performance, are essential considerations during the design and fabrication stage of a power MOSFET; moreover, many trade-offs should be decided when balancing between several electrical parameters and radiation survivability. This paper entails a description of the design and fabrication of TID-, SEB-, and SEGR-hardened power MOSFETs, on the basis of a careful optimization of the devices' cell structure and doping profile. Experimental verifications conducted show excellent radiation hardness and acceptable electrical performance of such devices for space power systems.

2. Design Considerations

2.1. Cell Structure

A power MOSFET chip is composed of several regions, including cell region, termination structure, gate bus, and gate pad. Of these, the cell region determines many electrical parameters and typically accounts for the majority of the chip area. However, it is also the most vulnerable region to irradiation. Normally, SEGR, SEB, and TID effects should be simultaneously mitigated in the cell region, whereas in other regions, only one of these effects is considered.

The cell structures and geometrical parameters of a power MOSFET are detailed in classic textbooks on power semiconductors [13,14]. Such geometrical parameters, together with the doping profiles, determine most of the device's electrical parameters, such as on-resistance (R_{on}), threshold voltage (V_{th}), and breakdown voltage (BV_{ds}). Nonetheless, the present study does not consider detailed discussions regarding the effect of these parameters on the performance of the device. However, the electrical performance must be reasonably reserved when radiation-hardened power MOSFETs are designed.

2.2. Oxide Thickness

Gate oxide thickness is affected by three major factors, namely threshold voltage, SEGR effects and TID effects, and secondary factors as device capacitance and electro-static discharge (ESD) robustness. TID effects are mitigated by keeping the gate oxide as thin as possible [15]. Conversely, a thin gate oxide exhibits a reduced ability to withstand the SEGR effects [16]. Most power MOSFETs are designed within a pre-irradiation threshold voltage (V_{th}) of 2–4 V. Certain radiation hardness requires V_{th} to remain within such specifications after receiving a specified dose, followed by high-temperature annealing. On this basis, the chosen V_{th} is greatly influenced by the shifting behaviors. The shifts could be negative or positive, depending on the dominant type of radiation-induced charge [15]. For negative-shifting-dominated cases, a higher V_{th} can save additional room for V_{th} shifting and is thus preferred. By contrast, for positive-shifting-dominated cases, a lower V_{th} is preferred for the same reason. Once the gate oxide t_{OX} is given, V_{th} can be adjusted by changing the doping density in the channel region.

Likewise, SEGR effects are mitigated by keeping the t_{OX} large enough to avoid dielectric breakdown. During a heavy ion strike, the dielectric strength is temporarily reduced. Models with more physical insight were proposed by Javanainen et al. [17], although a simple empirical expression with little physical justification is adopted in this work, as follows [16]:

$$E_{CRIT} = \frac{V_{GS}}{t_{OX}} = \frac{E_{BD}}{\left(1 + \frac{Z}{44}\right)}, \quad (1)$$

where E_{CRIT} is the critical electric field of gate oxide that must withstand heavy-ion injection; E_{BD} is the intrinsic dielectric breakdown strength of gate oxide, which is 10^7 V/cm for most thermal oxides; and Z is the atomic number of the injected heavy ions.

In rad-hard power MOSFETs' datasheets, SEE resistance ability is illustrated as a safe operating area under certain heavy-ion injection (SEE SOA) [18,19]. In principle, SEE SOA is expressed as a series of gate and drain voltage bias conditions. The negative gate bias is directly applied to the gate to contribute all its value to the gate dielectrics, whereas only a portion of the drain bias is coupled to the gate dielectrics after heavy-ion injection [20]. Therefore, the minimum gate oxide bounded by SEGR effects can be calculated as follows:

$$t_{OX,min} = \frac{(\alpha V_{DS} - V_{GS})\left(1 + \frac{Z}{44}\right)}{E_{BD}}, \quad (2)$$

where α is the coupled ratio of drain voltage related to the device design, as discussed later. Note that the bias conditions considered here are the worst bias conditions for SEGR production and are, hence, used for SEGR testing. The shift in the threshold voltage due to TID effects is a major problem for all metal-oxide-semiconductor (MOS) devices. For power MOSFETs, the relatively thick gate oxide makes this issue more severe. The V_{th} shift has been attributed to two kinds of radiation-induced charges, namely oxide charges and interface traps [15]. Therefore, the V_{th} shift (ΔV_{th}) is the sum of the oxide-charge-induced negative shift, named ΔV_{ot}, and the interface-trap-induced positive shift, named ΔV_{it}. Both ΔV_{ot} and ΔV_{it} are strongly related to t_{OX}. The relationship can be expressed as follows [21]:

$$\Delta V_{ot,it} = \frac{1}{C_{OX}} \times \frac{-1}{t_{OX}} \int_0^{t_{OX}} \rho_{ot,it}(x) x dx, \quad (3)$$

where $\rho_{ot,it}$ is the charge distribution of radiation-induced oxide-trapped or interface-trapped charge. Reduction of t_{OX} entails a two-fold effect. First, reducing t_{OX} can reduce the V_{th} shift for a given charge density, which is attributed to a larger C_{OX} resulting from a thinner t_{OX}. Second, it can reduce charge generation for a given dose, as shown in Equation (3). The integration term can be simplified by introducing a uniform charge generation for the oxide charge, resulting in the expression [22]:

$$\Delta V_{ot} = -\frac{\Delta Q_{ot}}{C_{OX}} = -\frac{qg_0 Dt_{OX} Y_h \sigma_h}{C_{OX}} = -\frac{qg_0 D Y_h \sigma_h}{\varepsilon_{OX}} t_{OX}^2, \quad (4)$$

where q is the electric charge (expressed in Coulomb), g_0 is the electron–hole pair generation rate in SiO_2 (in pairs/cm^3/rad(SiO_2)), D is the total dose level in units of rad(SiO_2), Y_h is charge yield of holes, σ_h is trapping cross section for holes captured by hole traps in oxide, and ε_{OX} is the dielectric constant of SiO_2. Note that Y and σ are affected by the electric field presented during irradiation, and the trapped charges can also be annealed with elevated temperature.

Interface traps generation is much more complicated. However, protons are considered to play a key role in the formation of interface traps. Moreover, the process of proton generation in the oxide is intimately related to the transport of holes. By introducing the parameter Y_p, which is the product of $N_{D'H}$ (concentration of hydrogen-containing defects) and $\sigma_{D'H}$ (cross section for proton release from these defects) [23], ΔV_{it} can be expressed in a similar manner as ΔV_{ot}, as follows:

$$\Delta V_{it} = \frac{\Delta Q_{it}}{C_{OX}} = \frac{qg_0 Dt_{OX} Y_h Y_p \sigma_p}{C_{OX}} = \frac{qg_0 D Y_h Y_p \sigma_p}{\varepsilon_{OX}} t_{OX}^2, \quad (5)$$

where σ_p is the cross section of protons captured by the traps at interface. Note that for one to get a relatively simple solution, a uniform distribution of $N_{D'H}$ (and, hence, the Y_p) in terms of space has been assumed, which may not be true for all cases. Moreover, $N_{D'H}$ is space-and-time-dependent and σ_p is field-dependent. Therefore, a simple method for quantitatively calculating the radiation-induced interface traps for all cases seems impractical, if not impossible. However, as an analytical model, Equation (5) does reflect the relationship of the interface trap generation with the hole transport, as widely accepted by society. Combining Equations (4) and (5) allows the maximum t_{OX} bounded by the TID effects to be expressed as follows:

$$t_{OX,max} = \sqrt{\frac{\Delta V_{th,max} \times \varepsilon_{OX}}{qg_0 D Y_h |Y_p \sigma_p - \sigma_h|}}, \quad (6)$$

where $\Delta V_{th,max}$ is the maximum allowed threshold shift. Note that even for a given dose, $|Y_p \sigma_p - \sigma_h|$ varies with dose rate, bias condition, and temperature and is strongly related to the fabrication process. Given this limitation and the uncertainties, $|Y_p \sigma_p - \sigma_h|$ remains a useful parameter to be extracted from the experimental perspective and can thus be used as a starting point in the device's design. Manipulating Equations (2) and (6) yields the lower and upper bounds of t_{OX}. Once t_{OX} is chosen, channel doping density can be fixed with equations governing V_{th}.

2.3. JFET Region Width

Parameter α has been introduced in Section 2.2 to account for the coupling of drain voltage to the gate dielectric. Based on Equation (2), the lower bound of t_{OX} can be reduced with reduced α, which means that a larger range of t_{OX} is available at the design stage. Moreover, α has been demonstrated to correlate with JFET region width (L_{JFET}) and thus can be reduced, with a reduced L_{JFET} [10], as illustrated in Figure 1.

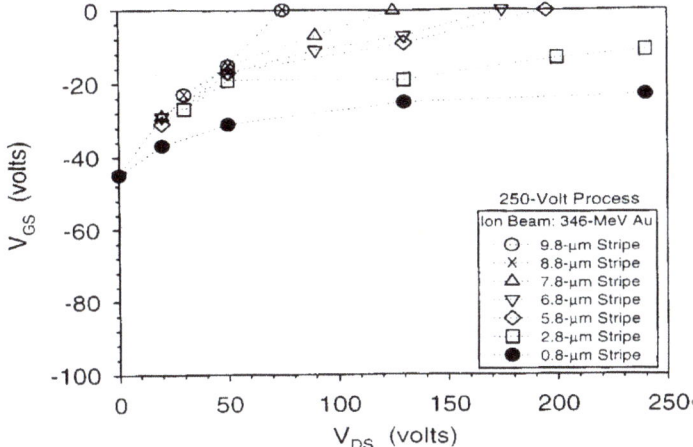

Figure 1. Single event gate rupture (SEGR) failure threshold responses for various L_{JFET}, after Reference [10].

Although it greatly improves SEGR hardness, a small L_{JFET} creates a negative impact on R_{on}. The specific resistance contributed by the JFET region ($R_{JFET,SP}$) can be expressed as follows:

$$R_{JFET,SP} = \rho_{JFET} H_{JP} \frac{L_{CELL}}{L_{JFET}}, \qquad (7)$$

where ρ_{JFET} is the resistivity of the JFET region, which is inversely proportional to JFET region doping; H_{JP} is body junction depth; L_{CELL} is the cell pitch; and L_{JFET} is the JFET region width. According to Equation (7), $R_{JFET,SP}$ is inversely proportional to L_{JFET}; thus, decreasing L_{JFET} will greatly increase $R_{JFET,SP}$, leading to worse resistance. Fortunately, the JFET region's resistance is only a portion of the total resistance. Therefore, the increasing on-resistance can be tolerated, as long as L_{JFET} is not extremely small. Nevertheless, the chosen L_{JFET} remains a critical element. A previous study [10] asserted that for 250 V power MOSFETs, L_{JFET} should be less than 5.8 µm to achieve a full V_{DS} range under zero gate bias. The JFET region should be carefully designed for SEGR-hardened devices.

2.4. P Body and P+ Well Doping

Several models have been proposed to describe the SEB process [24–26]. For instance, the parasitic BJT has been postulated to play a key role in SEB production. With the P-body region of the power MOSFET acting as the base region of the parasitic BJT, the body's doping profile becomes essential for hardening the device against SEB effects. In general, larger P-body depth (H_{JP}) and higher doping concentrations (N_{BODY}), as well as a reduced length between N+ source edge and P+ well edge (L_{BODY}), are desirable for an SEB-hardened cell design. However, as expressed in Equation (7), a deeper P-body has negative effects on $R_{JFET,SP}$, whereas a high N_{BODY} or a short L_{BODY} may affect the channel doping concentration.

3. Results

TID- and SEE-hardened power MOSFETs were designed on account of the trade-offs mentioned above. The key geometrical parameters and doping concentrations essential for the design are summarized in Table 1. Buffer layer technology was employed to improve SEB hardness [11]. The values of other parameters were chosen as common non-rad-hard power MOSFET designs. The whole chip area was 12 mm², whereas the active area (cell region) was approximately 8.5 mm². Stripe cell topology [10] was considered.

Table 1. Key geometrical parameters and doping concentrations for device design.

Symbol in Figure 1	Value	Unit
t_{ox}	80	nm
H_{JP}	3	μm
L_{CELL}	10	μm
L_{BODY}	~2	μm
L_{JFET}	~3	μm
N_{BODY}	~5 × 10^{16}	cm^{-3}

The designed power MOSFETs were fabricated by Tianjin Zhonghuan Semiconductor Co., Ltd., with 6-inch wafers. Processes with high thermal budget, such as the P-body driven process, were adjusted prior to gate oxidation to improve TID hardness. Diced devices were packaged in TO-220. Ninety devices were randomly selected for testing under a Keysight B1506 power semiconductor analyzer. Figure 2 illustrates the distributions of the testing results, with median BV_{ds} around 120 V and median R_{on} around 44 mΩ. For this cell design, for a 120 V maximum blocking ability, the specific resistance was 3.74 mΩ-cm². All the V_{th} values fell in the range of 2.36–2.62 V, of which more than 80% were roughly 2.40–2.50 V (not depicted in the figure). The ESD endurance exceeded 2000 V in human body model (HBM) mode, and the maximum avalanche energy was 662.5 mJ.

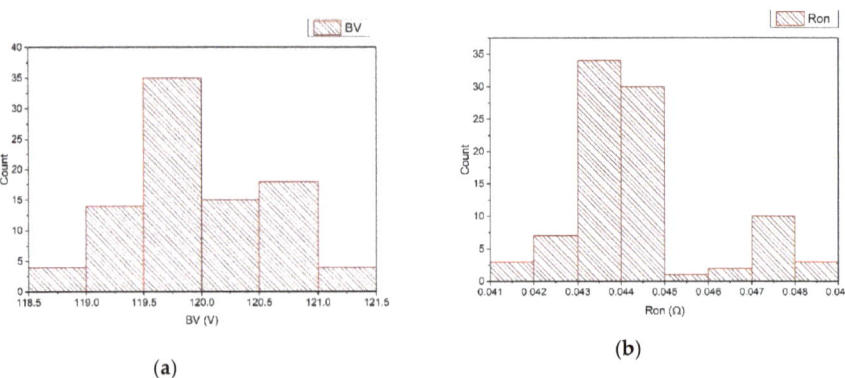

Figure 2. Test results for 90 randomly selected devices: (a) Breakdown voltage results; (b) On-resistance results.

Figure 3 illustrates the typical output and transfer curves, as measured with the Keysight B1506. The B1506 testing system has two modes, namely high-current and low-current modes. On the one hand, the high-current mode is able to test current up to 20 A; the plateau is caused by this limitation. However, this mode is not suitable for testing low current because of the leakage issue. On the other hand, the low-current mode is able to test current under a picoampere, although the maximum current in this mode is 1 A. The transfer curves in Figure 3 combined the results for both testing modes.

The fabricated devices were irradiated with Co-60 at the Shanghai Institute of Applied Physics, Chinese Academy of Sciences. The devices were placed on especially designed PCB boards, allowing separately the gate and drain node biases. The PCB boards were separated from the radiation source by approximately 30 cm, thus yielding a calculated dose rate of 100 rad(Si)/s. Additionally, the PCB boards were made to be as small as possible to minimize the dose rate inhomogeneous. Subsequently, the devices were irradiated under room temperature and then annealed at 100 °C for 168 h under the same bias condition after irradiation. For the gate bias condition (GB), the gate was biased at 12 V, with the drain and the source connected to ground. For the drain biased condition (DB), the drain was biased at 80 V, with the gate and the source connected to ground. Three devices were tested under each bias condition. Results of the TID experiment are displayed in Figure 4.

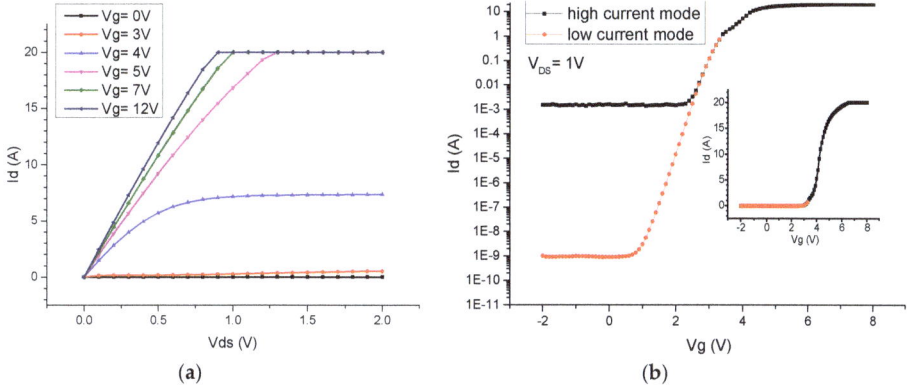

Figure 3. (a) Output curves and (b) transfer curves for the fabricated devices.

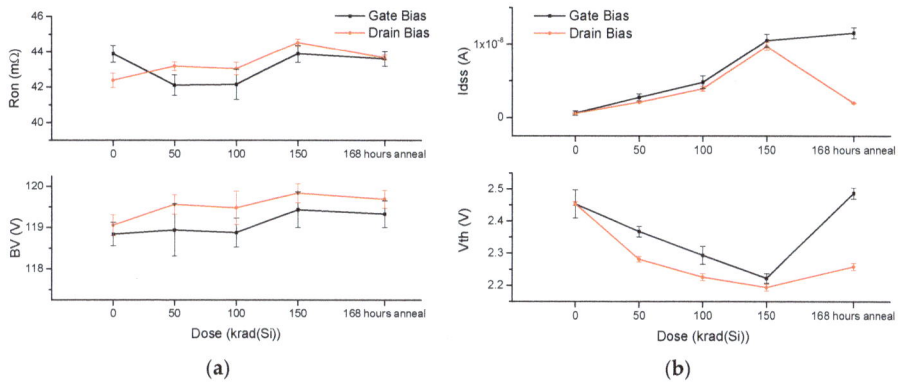

Figure 4. Parameters change with total ionizing dose (TID) dose and annealing time: (a) R_{on} and BV_{dss}; (b) I_{dss} and V_{th}.

Based on the figure, the on-resistance and breakdown voltage indicated negligible change after 150 krad(Si) TID irradiation and 168 h annealing, for both bias conditions. I_{dss} also increased with the dose for GB and DB, whereas I_{dss} increased after annealing under GB and consequently decreased under DB. For the threshold voltage, V_{th} decreased with the dose for both bias conditions, but with different annealing behavior. More specifically, V_{th} rebounded to a value slightly higher than its initial value under GB, whereas the rebound was much less under DB. Nonetheless, for each bias condition, at 150 krad(Si) dose, the annealing-induced V_{th} shift was less than 0.25 V. The shift behaviors of V_{th} during irradiation and annealing are described further under the discussion section. The terminations used in these devices included a traditional floating ring and filed plate structure, with optimized parameters [9]. The small BV_{ds} and I_{dss} change (Figure 4) indicate that the design of the termination was also radiation-hardened.

SEE experiments were conducted at Institute of Modern Physics, Chinese Academy of Sciences. The chips were packaged in TO-39, with the cap removed. 794 MeV Xe ions with a surface linear energy transfer (LET) of approximately 66 MeV·cm^2/mg were used. During the experiment, V_{GS} was set to 0 V, and V_{DS} was increased in steps of 10 V. The flux was roughly within 5000–10,000 ions/cm²s; the pass criteria was both gate and drain leakage current stay within the specification value after 2×10^6 ions/cm² irradiation [19]. Neither SEB nor SEGR was observed under V_{DS} = 100 V with a V_{GS} = 0 V bias condition.

4. Discussion

In space applications, the dose rate is much lower than the high-dose rate (HDR) experiment typically performed in laboratory. Such disparity may cause a significant difference of $|Y_p\sigma_p - \sigma_h|$ used in Equation (6) in the two cases. However, low-dose rate (LDR) experiments are relatively time-consuming and expensive. Therefore, the present study adopted an accelerated aging test to estimate the worst-case degradation of MOS devices [27,28], as it has been proven applicable to power MOSFETs [29]. Initially, the devices were irradiated with HDR for a relatively short time. Since the interface traps took a longer time to form, hole trapping in oxide defects dominated in this stage, thereby yielding $|Y_p\sigma_p - \sigma_h| \approx \sigma_h$ and a negative ΔV_{th}. In the annealing stage, the build-up of interface traps dominated while the trapped oxide charges decreased with time, yielding a recovery or even a rebound of ΔV_{th}. Therefore, the HDR+ high-temperature annealing procedure eliminated charge compensation in the LDR environment and produced worse (conserved) results. To further investigate the details of the behavior of radiation-induced charges, a mid-gap method was used to separate these two charges [30,31], where V_T is the threshold voltage extracted by using the maximum transconductance method. Here, note that V_T was different from V_{th} in Figure 4b, which was basically the gate voltage as the drain current reached 1 mA. Therefore, it was convenient for the engineer to monitor V_{th}. On the other hand, V_T has a physical meaning and is more accurate for parameter calculation. The mobility was extracted as follows:

$$\sqrt{I_D(sat)} = \sqrt{\frac{W\mu_n C_{OX}}{2L}}(V_{GS} - V_T), \tag{8}$$

where $I_{D(sat)}$ is drain current in the saturation region, W is the total channel width, L is the channel length, and C_{OX} is the gate oxide capacitance. Since Figure 4 depicts that the sample-to-sample variations were acceptable, a single device was randomly selected to perform extraction for each bias condition. Table 2 presents the extracted parameters of the device pre-irradiation, at 150 krad(Si) irradiation, and after annealing.

Table 2. Extracted parameters for device pre-irradiation.

	Unit	Virgin		150 krad(Si)		Anneal	
		Gate Bias	Drain Bias	Gate Bias	Drain Bias	Gate Bias	Drain Bias
V_T	V	3.75	3.82	3.61	3.68	3.88	3.66
V_{ot}	V	0.96	0.99	0.45	0.30	0.68	0.69
μ_n	cm²/V·s	319.34	339.00	273.30	252.45	219.05	279.15
ΔV_T	V	0.00	0.00	−0.14	−0.14	0.13	−0.15
ΔV_{ot}	V	0.00	0.00	−0.51	−0.69	−0.28	−0.30
ΔV_{it}	V	0.00	0.00	0.37	0.55	0.41	0.14
ΔN_{ot}	cm⁻²	0.00	0.00	1.37×10^{11}	1.86×10^{11}	0.76×10^{11}	0.80×10^{11}
ΔN_{it}	cm⁻²	0.00	0.00	1.00×10^{11}	1.48×10^{11}	1.11×10^{11}	0.39×10^{11}
$\Delta \mu_n$	cm²/V·s	0.00	0.00	−46.05	−86.85	−100.29	−59.85

The TID-induced oxide-charge density was 1.86×10^{11} cm⁻² for the drain bias condition, whereby such oxide charges should lead to a −0.69 V V_T shift. However, the negative shift was partially compensated by an interface-trap-induced positive shift, resulting in a net shift of −0.14 V. For the GB, both ΔN_{ot} and ΔN_{it} were 30% less than those for the DB. During the annealing process, almost half the generated oxide charges were reduced for both bias conditions. Nevertheless, the annealing behaviors of N_{it} for both conditions were different; N_{it} increased by approximately 10% for the GB and reduced by roughly 75% for the DB. Such similarity between N_{it} and I_{dss} during the annealing stage indicates that the increasing trend for I_{dss} might be related to the generation of interface traps. Moreover, as expected, the N_{it} generation and annealing was qualitatively consistent with the extracted mobility value [32]. The data in Table 2 can be used to calculate $Y_p\sigma_p$ and σ_h, as a starting point in the device design. However, these parameters are highly process-dependent and are, therefore, only valid for this specific process flow.

The parameter selection was further evaluated through fabrication of devices with t_{OX} = 100 nm, which were later subjected to TID experiments. For the other geometric parameters, the process flow and TID experiment setups were kept the same as those for the 80 nm samples. However, note that the oxidation time for the 100 nm samples was longer; thus, worse TID hardness could be expected because of the larger thermal budget and thicker t_{OX}, as illustrated in Figure 5. Here, the bias condition was the same as the gate bias condition described in Section 3. Much larger negative shifts and significant twists in the figure indicate both oxide charges and interface traps being much more in the 100 nm oxide thickness. Figure 6 illustrates the V_T shifts under the two bias conditions after 100 krad irradiation and annealing, where the shifts were higher with thicker t_{OX}, thus reflecting better SEGR hardness. A comparison of SEGR hardness of these devices is a future direction relative to the present study.

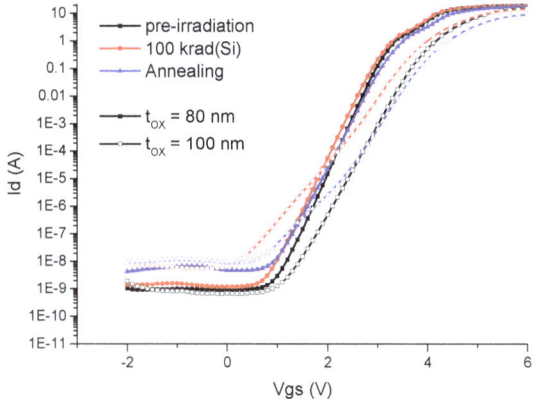

Figure 5. Subthreshold characteristics of power devices with 80 nm and 100 nm gate oxides pre- and post-irradiation and post-annealing.

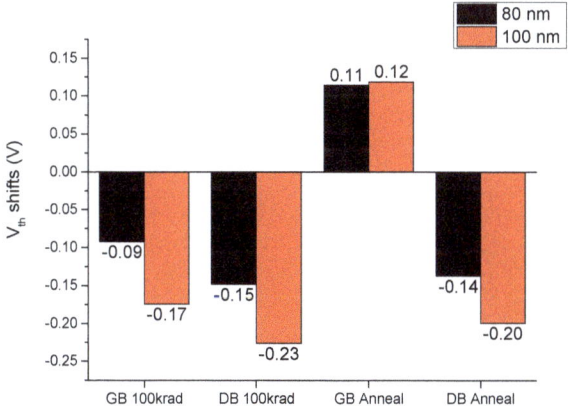

Figure 6. Threshold voltage shifts for 80 nm and 100 nm gate oxide devices under gate bias and drain bias (RB) conditions, after receiving 100-krad(Si) TID and annealing.

5. Conclusions

A rad-hard power MOSFET is appropriately designed through consideration of several radiation effects as TID, SEB, and SEGR, as well as a balance among electrical parameters as R_{on}, V_{th} and BV_{ds}. In this work, the effects of cell structure adjustment on the performance of a power MOSFET were examined, by first analyzing the design parameters. Next, a SEE- and TID-hardened power MOSFET was designed and fabricated by implementing the accompanying design rules. Results of the investigation confirmed the achievement of excellent radiation hardness and decent specific on-resistance for the device. Technically, the V_{th} shifts were less than 0.25 V for 150 krad(Si) irradiation and 168 h annealing. No SEE was observed under $V_{GS} = 0$ V and the $V_{DS} = 100$ V bias condition with LET = 66 MeV·cm^2/mg under Xe ion irradiation. Further investigation on the TID experimental results indicated the estimated charge density induced by radiation and annealing. Moreover, devices with thicker gate oxide were fabricated as the counterpart for the parameter selection evaluation. Experiments with these devices showcased their great potential for application in space power systems.

As a general rule, radiation environments are different for various mission orbits. Notably, a spacecraft in van-Allen belts would suffer more from a TID effect, whereas a deep space mission would require high SEE-hardness devices. Additionally, devices in low-Earth orbits requires lower radiation hardness while they are expected to exhibit better electrical parameters. Hence, various devices with different electrical parameters and radiation hardness are required for different missions. The results in the present study provide an insight for the power semiconductor designer to balance the parameters involved and to design power MOSFET devices based on the application requirements.

Author Contributions: Conceptualization, methodology, writing—original draft preparation, T.W. and X.W.; Chip design, H.J.; Chip fabrication, H.L.; TID and SEE experiment, T.W.; Data analysis, Y.S.; Writing—review and editing, R.L. and J.X.; Supervision, J.X. and L.Z.; Resources and funding acquisition, L.Z.

Funding: This research received no external funding.

Acknowledgments: The authors would like to thank Wanli Wang and Xiaofang Li from Tianjin Zhonghuan Semiconductor Co., Ltd. for their help on device fabrication.

Conflicts of Interest: The authors declare no conflicts of interest. The funders had no role in the design of the study; in the collection, analyses, or interpretation of data; in the writing of the manuscript, or in the decision to publish the results.

Nomenclature

Symbol	Description	Equation
E_{CRIT}	Critical electric field of gate oxide that must withstand heavy-ion injection	(1)
V_{GS}	Applied gate voltage	(1)
t_{OX}	Gate oxide thickness	(1)
E_{BD}	Intrinsic dielectric breakdown strength	(1)
Z	The atomic number of the injected heavy ions	(1)
$t_{OX,min}$	The minimum gate oxide bounded by single event gate rupture (SEGR) effects	(2)
α	The coupled ratio of drain voltage	(2)
V_{DS}	Applied drain voltage	(2)
$\Delta V_{ot,it}$	Threshold shifts induced by oxide-charge or interface traps	(3)
C_{OX}	Gate capacitance	(3)
$\rho_{ot,it}$	The charge distribution of radiation-induced oxide-trapped or interface-trapped charge	(3)
ΔQ_{ot}	Radiation-induced charges in oxide	(4)
q	Electric charge	(4)
g_0	Electron–hole pair generation rate in SiO$_2$	(4)
D	The total dose level	(4)
Y_h	Charge yield of holes	(4)
σ_h	Trapping cross section for holes captured by hole traps in oxide	(4)
ε_{OX}	The dielectric constant of SiO$_2$	(4)

Symbol	Description	Equation
ΔQ_{it}	Radiation-induced interface trap charges	(5)
Y_p	Product of concentration of hydrogen-containing defects and cross section for proton release from these defects	(5)
σ_p	The cross section of protons captured by the traps at interface	(5)
$t_{OX,max}$	The maximum gate oxide bounded by total ionizing dose (TID) effects	(6)
$\Delta V_{th,max}$	The maximum allowed threshold shift	(6)
$R_{JFET,SP}$	Specific resistance contributed by the JFET region	(7)
ρ_{JFET}	The resistivity of the JFET region	(7)
H_{JP}	Body junction depth	(7)
L_{CELL}	Cell pitch	(7)
L_{JFET}	JFET region width	(7)

References

1. Adell, P.C.; Scheick, L.Z. Radiation Effects in Power Systems: A Review. *IEEE Trans. Sci.* **2013**, *60*, 1929–1952. [CrossRef]
2. Barth, J.; Gee, G.; Adolphsen, J.W. First observation of proton induced power MOSFET burnout in space: The CRUX experiment on APEX. *IEEE Trans. Sci.* **1996**, *43*, 2921–2926.
3. George, J.S.; Clymer, D.A.; Turflinger, T.L.; Mason, L.W.; Stone, S.; Koga, R.; Beach, E.; Huntington, K.; Lauenstein, J.-M.; Titus, J.; et al. Response Variability in Commercial MOSFET SEE Qualification. *IEEE Trans. Sci.* **2017**, *64*, 317–324. [CrossRef]
4. Kuboyama, S.; Mizuta, E.; Nakada, Y.; Shindou, H. Physical analysis of damage sites introduced by SEGR in silicon vertical power MOSFETs and implications for post-irradiation gate-stress test. *IEEE Trans. Nucl. Sci.* **2019**, in press. [CrossRef]
5. Singh, G.; Galloway, K.F.; Russell, T.J. Radiation-Induced Interface Traps in Power Mosfets. *IEEE Trans. Sci.* **1986**, *33*, 1454–1459. [CrossRef]
6. Picard, C.; Brisset, C.; Hoffmann, A.; Charles, J.-P.; Joffre, F.; Adams, L.; Siedle, A.H. Use of electrical stress and isochronal annealing on power MOSFETs in order to characterize the effects of 60 Co irradiation. *Microelectron. Reliab.* **2000**, *40*, 1647–1652. [CrossRef]
7. Titus, J.L. An Updated Perspective of Single Event Gate Rupture and Single Event Burnout in Power MOSFETs. *IEEE Trans. Sci.* **2013**, *60*, 1912–1928. [CrossRef]
8. Roper, G.B.; Lowis, R. Development of a radiation hard n-channel power MOSFET. *IEEE Trans. Nucl. Sci.* **1983**, *30*, 4110–4115. [CrossRef]
9. Davis, K.; Schrimpf, R.; Cellier, F.; Galloway, K.; Burton, D.; Wheatley, C. The effects of ionizing radiation on power-MOSFET termination structures. *IEEE Trans. Sci.* **1989**, *36*, 2104–2109. [CrossRef]
10. Savage, M.; Burton, D.; Wheatley, C.; Titus, J.; Gillberg, J. An improved stripe-cell SEGR hardened power MOSFET technology. *IEEE Trans. Sci.* **2001**, *48*, 1872–1878. [CrossRef]
11. Liu, S.; Titus, J.L.; Boden, M. Effect of Buffer Layer on Single-Event Burnout of Power DMOSFETs. *IEEE Trans. Sci.* **2007**, *54*, 2554–2560. [CrossRef]
12. Wan, X.; Zhou, W.S.; Ren, S.; Liu, D.G.; Xu, J.; Bo, H.L.; Zhang, E.X.; Schrimpf, R.D.; Fleetwood, D.M.; Ma, T. SEB Hardened Power MOSFETs With High-K Dielectrics. *IEEE Trans. Sci.* **2015**, *62*, 2830–2836. [CrossRef]
13. Baliga, B.J. *Advanced Power MOSFET Concepts*; Springer: New York, NY, USA, 2010; pp. 23–61.
14. Grant, D.A.; Gowar, J. *Power MOSFETs: Theory and Applications*, 1st ed.; Wiley-Interscience: New York, NY, USA, 1989.
15. Schrimpf, R.; Wahle, P.; Andrews, R.; Cooper, D.; Galloway, K. Dose-rate effects on the total-dose threshold-voltage shift of power MOSFETs. *IEEE Trans. Sci.* **1988**, *35*, 1536–1540. [CrossRef]
16. Titus, J.; Wheatley, C.; Van Tyne, K.; Krieg, J.; Burton, D.; Campbell, A. Effect of ion energy upon dielectric breakdown of the capacitor response in vertical power MOSFETs. *IEEE Trans. Sci.* **1998**, *45*, 2492–2499. [CrossRef]
17. Javanainen, A.; Ferlet-Cavrois, V.; Jaatinen, J.; Kettunen, H.; Muschitiello, M.; Pintacuda, F.; Rossi, M.; Schwank, J.R.; Shaneyfelt, M.R.; Virtanen, A. Semi-Empirical Model for SEGR Prediction. *IEEE Trans. Sci.* **2013**, *60*, 2660–2665. [CrossRef]

18. STMicroelectronics. STRH8N10 Datasheets. Available online: https://www.st.com/resource/en/datasheet/strh8n10.pdf (accessed on 3 May 2019).
19. Iakovlev, S.A.; Anashin, V.S.; Chubunov, P.A.; Koziukov, A.E.; Bu-Khasan, K.B.; Maksimenko, T.A.; Chlenov, A.M. MOSFETs SEB & SEGR qualification results with SOA estimation. In Proceedings of the 17th European Conference on Radiation and Its Effects on Components and Systems (RADECS), Geneva, Switzerland, 2–6 October 2017.
20. Wheatley, C.; Titus, J.; Burton, D. Single-event gate rupture in vertical power MOSFETs; an original empirical expression. *IEEE Trans. Sci.* **1994**, *41*, 2152–2159. [CrossRef]
21. Schwank, J.R.; Shaneyfelt, M.R.; Fleetwood, D.M.; Felix, J.A.; Dodd, P.E.; Paillet, P.; Ferlet-Cavrois, V. Radiation effects in MOS oxides. *IEEE Trans. Nucl. Sci.* **2008**, *55*, 1833–1853. [CrossRef]
22. Fleetwood, D.; Meisenheimer, T.; Scofield, J. 1/f noise and radiation effects in MOS devices. *IEEE Trans. Electron Devices* **1994**, *41*, 1953–1964. [CrossRef]
23. Rashkeev, S.; Cirba, C.; Fleetwood, D.; Schrimpf, R.; Witczak, S.; Michez, A.; Pantelides, S. Physical model for enhanced interface-trap formation at low dose rates. *IEEE Trans. Sci.* **2002**, *49*, 2650–2655. [CrossRef]
24. Hohl, J.H.; Galloway, K.F. Analytical model for single event burnout of power MOSFETs. *IEEE Trans. Electron Devices* **1994**, *41*, 1953–1964. [CrossRef]
25. Wrobel, T.; Beutler, D. Solutions to heavy ion induced avalanche burnout in power devices. *IEEE Trans. Sci.* **1992**, *39*, 1636–1641. [CrossRef]
26. Johnson, G.H.; Palau, J.M.; Dachs, C.; Galloway, K.F.; Schrimpf, R.D. A review of the techniques used for modeling single-event effects in power MOSFET's. *IEEE Trans. Nucl. Sci.* **1996**, *43*, 546–560. [CrossRef]
27. Department of Defense. *MIL-STD-883E, Test Method 1019.4. Ionizing Radiation (Total Dose) Test Procedure*; Defense Supply Center Columbus: Columbus, OH, USA, 1996.
28. Department of Defense. *MIL-STD-750E, Test Method 1019.5 Steady-State Total Dose Irradiation Procedure*; Defense Supply Center Columbus: Columbus, OH, USA, 2006.
29. Khosropour, P.; Galloway, K.F.; Zupac, D.; Schrimpf, R.D.; Calvel, P. Application of test method 1019.4 to non-hardened power MOSFETs. *IEEE Trans. Nucl. Sci.* **1994**, *41*, 555–560. [CrossRef]
30. Winokur, P.S.; Schwank, J.R.; McWhorter, P.J.; Dressendorfer, P.V.; Turpin, D.C. Correlating the Radiation Response of MOS Capacitors and Transistors. *IEEE Trans. Sci.* **1984**, *31*, 1453–1460. [CrossRef]
31. McWhorter, P.J.; Winokur, P.S. Simple technique for separating the effects of interface traps and trapped-oxide charge in metal-oxide-semiconductor transistors. *Appl. Phys. Lett.* **1986**, *48*, 133–135. [CrossRef]
32. Zupac, D.; Galloway, K.; Khosropour, P.; Anderson, S.; Schrimpf, R.; Calvel, P. Separation of effects of oxide-trapped charge and interface-trapped charge on mobility in irradiated power MOSFETs. *IEEE Trans. Sci.* **1993**, *40*, 1307–1315. [CrossRef]

© 2019 by the authors. Licensee MDPI, Basel, Switzerland. This article is an open access article distributed under the terms and conditions of the Creative Commons Attribution (CC BY) license (http://creativecommons.org/licenses/by/4.0/).

Article

Mechanism of Degradation Rate on the Irradiated Double-Polysilicon Self-Aligned Bipolar Transistor

Mohan Liu [1,2,*], Wu Lu [1,*], Xin Yu [1,2], Xin Wang [1], Xiaolong Li [1], Shuai Yao [1,2] and Qi Guo [1]

1. Key Laboratory of Functional Materials and Devices for Special Environments, Xinjiang Technical Institute of Physics and Chemistry, Chinese Academy of Sciences, Urumqi 830011, China; yuxin@ms.xjb.ac.cn(X.Y.); wangxin210@ms.xjb.ac.cn(X.W.); lixl@ms.xjb.ac.cn (X.L.); yaoshuai15@mails.ucas.edu.cn(S.Y.); guoqi@ms.xjb.ac.cn(Q.G.)
2. University of Chinese Academy of Sciences, Beijing 100049, China
* Correspondence: liumh@ms.xjb.ac.cn (M.L.); luwu@ms.xjb.ac.cn (W.L.); Tel.: +86-15894697376 (M.L.); +86-13659906670 (W.L.)

Received: 8 April 2019; Accepted: 27 May 2019; Published: 11 June 2019

Abstract: The latent enhanced low dose rate sensitivity (ELDRS) effect is observed in the double-polysilicon self-aligned (DPSA) technology PNP bipolar junction transistor (BJT) irradiated with a high and low dose rate gamma ray, which is discussed from the perspective of the three-stage degradation rate of the excess base current. The great degradation rate as a result of the high dose irradiation of the first stage is dominantly ascribed to the positive oxide trap charges accumulated during a short irradiation, and then due to the competition between the recombination of electrons and capture of the hole by the traps. It declined sharply into a degradation rate saturated region of the second stage. However, for the low dose rate, the small increase in the degradation rate in the first stage is caused by the holes escaping from the initial recombination and being transported to the interface to form the interface states. Then, the competition between the steadily increasing interfacial trap charge and the continuously annealed shallow level oxide trap charge leads to the stable increase of degradation under low dose irradiation. Finally, in stage three, the increases of the degradation rates for high and low dose irradiation result from the different amounts of the hydrogen molecules generated by the hole reactive with depassiviated Si suspended bonds, which can interact with the deep level defects and release protons, causing an increase of interfacial trap charges with prolonged irradiation.

Keywords: saturation effect; gain degradation; total ionizing dose; gamma ray; bipolar transistor

1. Introduction

As a state-of-the-art high speed bipolar complementary process, double-polysilicon self-aligned (DPSA) technology has been widely used in high-speed analog integrated circuits. Compared with the traditional bipolar junction transistors (BJTs), DPSA BJTs have a smaller linewidth due to the isolation of local oxidation of silicon (LOCOS) combined with deep trench isolation (DTI). The use of polysilicon emitters can increase the current gain so that the device can achieve vertical scale down without reducing the punch-through voltage of the emitter–collector junction and the loss of the current gain [1–3]. The self-aligning structure and DTI can realize the lateral scale down of the device, greatly reduce the area of the device, the circuit, and the corresponding parasitic capacitance, significantly reduce the power consumption delay product of the circuit, and improve the integration level of the bipolar circuit [4,5]. Therefore, the high performance and speed of the analog integrated circuit made by this technology have wide application prospects in space RF(radio frequency)/microwave communication and other extreme environments [6].

The previous total dose irradiation test results in the literature [6,7] have shown that the direct current gain is the most sensitive parameter of the bipolar device under total ionizing dose radiation, and most of the devices and circuits fabricated by bipolar technology suffer from the so-called enhanced low dose rate sensitivity (ELDRS) effect. For decades, most of the total ionizing dose research was focused on the radiation response and the damage mechanism of the traditional bipolar technology; only a few reports were related to the DPSA technology. In 1999, Flamen et.al presented research about the radiation tolerance on quasi-self aligned (QSA) single polysilicon emitter bipolar technology. The experimental results have shown that this bipolar technology is superior to traditional bipolar technology in structure and related function, and that is has a good tolerance of the radiation without adding special manufacturing steps [8]. Graves investigated the radiation and hot-carrier stress response on polysilicon emitter NPN BJTs fabricated in a bipolar-complementary-metal-oxide-semiconductor (BiCMOS) process [9]. More recently, Zhang et al. presented the radiation response on the DPSA NPN BJTs with Si and silicon-on-insulator (SOI) substrate at high and low injection levels [10–14]. However, all of the above research was focused on the preliminary total ionizing dose response on the NPN BJTs. Studies on the dose rate response and the degradation rate of the electrical parameters are still not enough to apply it to real space applications. Thus, in this paper, the dose rate response on the DPSA BJT has been investigated under high and low dose rate gamma ray irradiation and a preliminary analysis of the radiation effect and the damage mechanism is made from the perspective of the degradation rate of the excess base current.

Based on the understanding of the comprehensive research on the radiation damage effect of bipolar transistors, the current gain degradation saturation phenomenon and generation mechanism of the devices at the total dose of 100 krad (Si) were investigated by using a bipolar transistor which is very resistant to total ionizing dose radiation. Section 2 of the paper introduces the devices and the methods used in the experiments. Detailed experimental results are presented in Section 3. Section 4 explains the experimental results and discusses their effects on practical future space applications. Finally, Section 5 concludes the article.

2. Experimental Devices and Methods

DPSA PNP BJTs studied in this paper adopted a standard bipolar process with the trench isolation technology and the cross-section of the device being depicted in Figure 1. The devices presented here were irradiated with a ^{60}Co gamma ray at the Xinjiang Technical Institute of Physics & Chemistry of Chinese, Academy of Sciences. The devices were exposed at the high and low dose rate of 50 rad (Si)/s and 0.05 rad (Si)/s with a reverse emitter-base bias voltage, which is usually recognized as the worst operation condition [6]. The electrical parameters were measured by the KEITHLEY 4200-SCS semiconductor parameter analyzer by removing the devices under test from the irradiation chamber within 20 minutes before and after the dose accumulated to about 100 krad (Si) at room temperature.

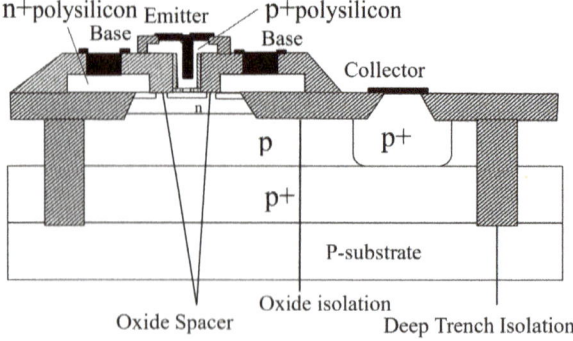

Figure 1. The cross-section of the DPSA BJT.

3. Experimental Results

Figure 2 shows the changes in the Gummel curve of the experimental samples with the cumulative total dose before and after irradiation at dose rates of 50 rad (Si)/s and 0.05 rad (Si)/s to 100 krad (Si). As the dose accumulated, the collector current did not change significantly, while the base current increased slightly when biased at the emitter-base voltage lower than $V_{EB} = 0.7$ V. However, the base current of the irradiated transistor is mainly composed of the initial current and the surface compound current. The oxide trap positive charge and the interface trap charge generated by the irradiation increases the surface recombination rate [7], so the increase of the surface recombination current leads to an increase of the base current, which then induced a declination of the current gain which is defined as the ratio of the collector current and the base current ($\beta = I_C/I_B$).

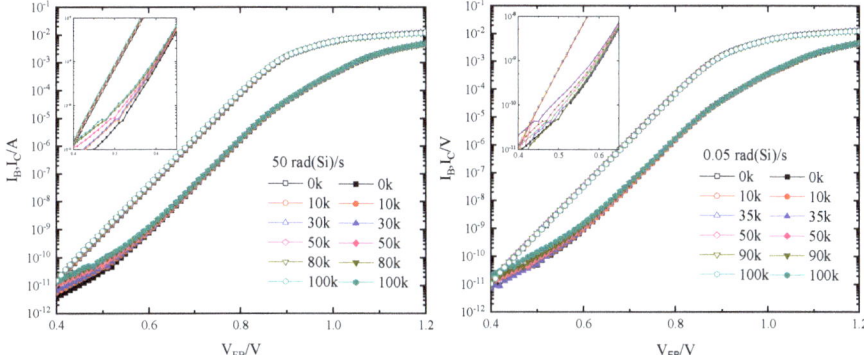

Figure 2. Gummel characteristics of a DPSA BJT before and after irradiation at the dose rate of 50 rad (Si)/s and 0.05 rad (Si)/s. V_{EB} = emitter-base voltage; I_C = collector current; I_B = base current.

For comparison of the degradation level of the base current and the current gain under different dose rates, the excess base current ΔI_B ($\Delta I_B = I_B - I_{B0}$) and normalized current gain β/β_0, are introduced in the characterization analysis as shown in Figures 3 and 4, where I_{B0}, β_0, I_B and β are the base current and current gain before and after irradiation corresponding to $V_{EB} = 0.6$ V, respectively. It can be seen in Figure 3 that when $V_{EB} = 0.6$ V, ΔI_B is increasing continuously with the accumulating total dose. When the irradiation dose increases up to 100 krad (Si), the base current I_B increases by a value of 465 pA ($\Delta I_B = 465$ pA) for the high dose irradiation, while the increment of ΔI_B is about 636 pA for the low dose rate, which is greater than the degradation under the high dose rate. Therefore, this PNP BJT fabricated with the DPSA technology may suffer from the ELDRS effect in the real and lower dose rate space irradiation environment, although the total degradation of the base current is slight. What is notable in Figure 3 is that the decrease of the base current under the high dose rate is above that irradiated with the low dose rate before the total dose reached 50 krad (Si). Moreover, for the first 50 krad (Si) irradiation, the increasing rate of ΔI_B at the high dose rate is more rapid than irradiating at a low dose rate. It also shows the opposite increasing trend above the 50 krad (Si). For the final 10 krad (Si), the increasing speed of ΔI_B soars up under the low dose rate and creates a relatively big gap between the high and low dose rate. These phenomena prove that the amounts of the interface trap charges in the Si/SiO$_2$ interface dominate the attenuation of the base current under the long irradiation with a low dose rate condition [7].

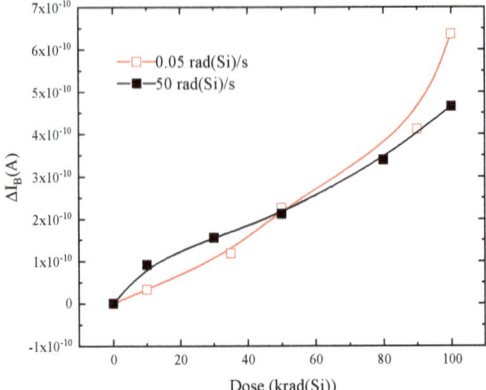

Figure 3. The excess base current of DPSA BJTs irradiated at a dose rate of 50 rad (Si)/s and 0.05 rad (Si)/s for PNP.

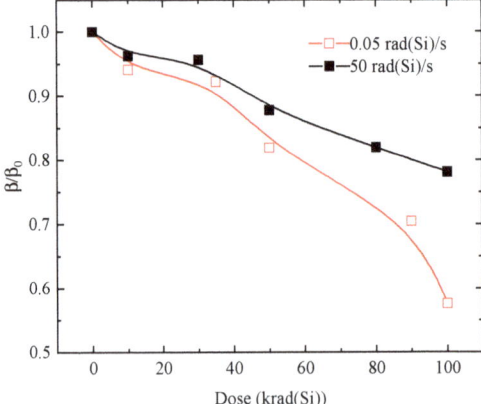

Figure 4. The normalized current gain of a DPSA BJT irradiated at the dose rate of 50 rad (Si)/s and 0.05 rad (Si)/s for PNP.

As the base current increases after irradiation, while the collector current remains basically unchanged, the degradation of the normalized current gain is finally shown in Figure 4. It can be seen clearly from the Figure that with the increase of the accumulated total dose, the current gains of the PNP BJTs all decrease rapidly. However, the declination rates are different between the high and low dose rate and consistent with the ΔI_B depicted in Figure 3. The gain after high and low dose rate irradiation is reduced to 57.6% and 78% of the initial value, respectively. The results also show that the damage of the current gain under low dose rate irradiation is greater than that under a high dose rate, and will hence result in a potential ELDRS effect.

4. Discussion

Figure 5 plots the degradation curve of the excess base current per unit krad (Si) irradiation dose of the DPSA BJTs to explain the radiation response processes mentioned above. Obviously, the degradation rate of the base current can be divided into three stages all through the irradiation process. For the beginning of the first stage, the degradation of the base current up to 10 pA per unit krad (Si) when the dose is accumulated to the first 10 krad (Si) at the high dose rate, and then with the total dose increases to 50 krad (Si), the degradation rate of the base current gradually descends and stabilizes to about 4.25 pA per krad (Si) degradation. On the contrary, under the low dose rate irradiation, the degradation rate of the excess base current was less than 3.5 pA at the initial 35 krad (Si) irradiation. After the dose reached 35 krad (Si) under the low dose rate, it begins to increase until 50 krad (Si), and then stayed as a constant of 4.5 pA/krad (Si) and a little bit above the degradation rate of 4.25 pA under the high dose rate irradiation, which is considered as the second stage in Figure 5.

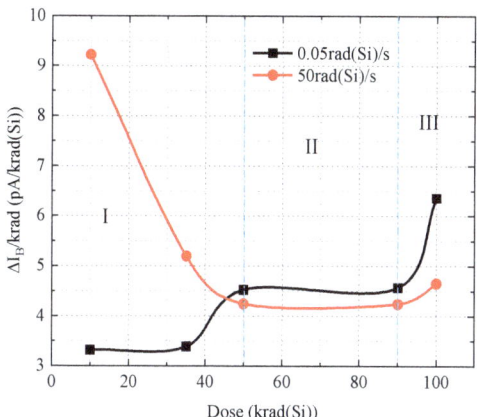

Figure 5. The degradation rate of the excess base current of a DPSA BJT under different dose rates.

The yield of the electron-hole pair under the high dose rate is much higher than that of low dose rate irradiation. A large number of oxide trap charges (V_O^+, V_OH^+) are generated by the combination of the holes (h^+) which escaped from the initial recombination with the electrons resulting in oxygen vacancies (V_O) and hydrogenated oxygen vacancies (V_OH, V_OH^+) that are introduced in the manufacturing process by the following reactions [15,16]:

$$V_O + h^+ \rightarrow V_O^+ \tag{1}$$

$$V_OH + h^+ \rightarrow V_OH^+ \tag{2}$$

At the same time, long-term irradiation at a low dose rate contributes to the annealing of the oxide trap charge, and as a result, the initial value of the oxide trap charge is much higher than that of low dose rate irradiation, making the degradation rate of low dose rate irradiation much lower than that of high dose rate irradiation before 35 krad (Si). This explains the huge difference in degradation rate between high and low dose rates in the initial stage of irradiation. With the increase of irradiation time, the proton release reactions of the large amount of holes which escaped recombination under the low dose rate with the hydrogenated oxygen vacancies in the process of transport towards the Si/SiO$_2$ interface [17] are dominated as shown in Equation (3) below. The released proton starts to arrive to the Si/SiO$_2$ interface and reacts with the silicon dangling bond. The formation of the interface states begin at the accumulated total dose up to 35 krad (Si). Then, the degradation rate of the first stage increases under low dose rate irradiation. At the same time, the probability of recombination at high dose rates is increasing due to the very high concentrations of the electron-hole. Therefore, the proton release processes are being depressed and the interface trap charges are being reduced, which describe the declination of the degradation rate in Figure 5 under the high dose rate.

$$V_OH + h^+ \rightarrow V_O + H^+ \tag{3}$$

The space electric field induced by long-term irradiation at high dose rates is large enough to prevent the subsequent holes or hydrogen from being transported to the nearby interface to form deeper oxide trap charges and interfacial trap charges, while the shallow trap charges can be annealed under long-term room temperature irradiation [18,19]. Moreover, the dimerization of hydrogen increases gradually when the released protons react with the hydrogenated oxygen vacancies, with more and more accumulating ionizing total dose as shown in Equation (4). As discussed above, the degradation rate decreases and tends to be stable at the high dose rate. However, the competition between the stable increased interface trap charges and the annealing of low level oxide trap charges have predominated the degradation under the low dose rate. From Figure 5, it can be seen that the constant degradation rate of stage two is between the dose of 50 krad (Si) and 90 krad (Si), which may differ from process to process.

$$V_OH + H^+ \rightarrow V_O + H_2 \tag{4}$$

When the total dose is accumulated to 90 krad (Si), the degradation rate begins to increase significantly since the holes have enough time to transfer to the Si/SiO$_2$ interface and form a large number of interface trapped charges [19]. However, the increases in degradation rates are different between the high and low dose rate as the amount of holes that reached the Si/SiO$_2$ interface are not identical for the various dose rates [20,21]. There are more holes escaping from the initial electron-hole recombination being transported to the interface, where they release protons and then depassivate the Si dangling bonds [19,20,22] which are hydrogen passivated in the fabrication process, creating more interface trap charges under the low dose rate irradiation than at high dose rates. Furthermore, hydrogen molecules generated by the hydrogen dimer of Equation (4) will crack at the shallow oxide traps as shown in Equation (5) and behave as a source of protons, increasing the interface trap charge [23] and leading to an increase of degradation rate in the third stage under high dose rates.

$$V_O^+ + H_2 \rightarrow V_OH + H^+ \tag{5}$$

5. Conclusions

In conclusion, this paper presents the gamma-ray radiation effect of DPSA PNP BJTs with a dose rate of 50 rad (Si)/s and 0.05 rad (Si)/s and discusses the mechanisms of different dose rate responses at different stages from the perspective of the degradation rate of the base current. The experimental results showed that the base current of the transistor did not change significantly under the irradiation conditions, showing a relatively good tolerance of radiation. The comparison of high and low dose rate showed that the increase of base current caused by low dose rate irradiation was larger than that caused by high dose rate irradiation, resulting in greater current gain degradation than that caused by high dose rate, highlighting that the ELDRS effect may occur. The three-stage degradation rate illustrated for the lower dose level of high dose rate irradiation demonstrate that the oxide trap charges are responsible for the rapid decrease. The competition of hole recombination by the electrons or captured by the traps resulted in a decrease of the degradation rate. As the competition reaches equilibrium, the degradation rate tends to be stable. The accumulated hydrogen molecular induced interface state is accountable for the increase of the degradation rate in the third stage for all irradiation dose rates.

Author Contributions: Conceptualization, W.L. and X.Y.; Data curation, M.L.; Formal analysis, M.L. and W.L.; Funding acquisition, W.L. and Q.G.; Investigation, M.L., X.L. and S.Y.; Methodology, W.L., X.Y., X.W., X.L. and S.Y.; Project administration, W.L. and X.Y.; Writing—original draft, M.L.

Funding: The work is supported by the National Natural Science Foundation of China (Grant No. U1532261).

Acknowledgments: The authors would like to acknowledge the great support offered by the staff of the Irradiation Center of the Xinjiang Technical Institute of Physics and Chemistry, Chinese Academy of Sciences.

Conflicts of Interest: The authors declare no conflict of interest.

References

1. Miura-Mattausch, M.; Rüstig, J.; Kircher, R. Dependence of current gain β on spacer geometry and emitter size in polysilicon self-aligned bipolar transistors. *Solid State Electron.* **1990**, *33*, 325–331. [CrossRef]

2. Glenn, J.; Neudeck, G. A Double Self-Aligned Silicon Bipolar Transistor Utilizing Selectively-Grown Single Crystal Extrinsic Contacts. *Dep. Electr. Comput. Eng. Tech. Rep.* **1992**.
3. Post, I.R.C.; Ashburn, P.; Wolstenholme, G.R. Polysilicon emitters for bipolar transistors: a review and re-evaluation of theory and experiment. *IEEE Trans. Electron. Devices* **1992**, *39*, 1717–1731. [CrossRef]
4. Uchino, T.; Shiba, T.; Kikuchi, T.; Tamaki, Y.; Watanabe, A.; Kiyota, Y. Very-high-speed silicon bipolar transistors with in-situ doped polysilicon emitter and rapid vapor-phase doping base. *IEEE Trans. Electron. Devices* **1995**, *42*, 406–412. [CrossRef]
5. Nakamura, T.; Nishizawa, H. Recent progress in bipolar transistor technology. *IEEE Trans. Electron. Devices* **1995**, *42*, 390–398. [CrossRef]
6. Pease, R.L.; Schrimpf, R.D.; Fleetwood, D.M. ELDRS in Bipolar Linear Circuits: A Review. *IEEE Trans. Nucl. Sci.* **2009**, *56*, 1894–1908. [CrossRef]
7. Fleetwood, D.M. Total Ionizing Dose Effects in MOS and Low-Dose-Rate-Sensitive Linear-Bipolar Devices. *IEEE Trans. Nucl. Sci.* **2013**, *60*, 1706–1730. [CrossRef]
8. Flament, O.; Synold, S.; de Pontcharra, J.; Niel, S. Radiation tolerance of NPN bipolar technology with 30 GHz Ft. *IEEE Trans. Nucl. Sci.* **2000**, *47*, 654–658. [CrossRef]
9. Graves, R.J.; Schmidt, D.M.; Kosier, S.L.; Wei, A.; Schrimpf, R.D.; Galloway, K.F. Visualization of ionizing-radiation and hot-carrier stress response of polysilicon emitter BJTs. In Proceedings of the 1994 IEEE International Electron Devices Meeting, San Francisco, CA, USA, 11–14 December 1994; pp. 233–236.
10. Wu, X.; Zhang, P.; Tang, Z.; Tan, K.; Lu, W.; Jia, J. Reliability of DPSA bipolar junction transistors in radiation environment. In Proceedings of the International NanoElectronics Conference, INEC, Chengdu, China, 9–11 May 2016; pp. 1–2.
11. Yi, Q.N.; Zhang, P.J.; Wu, X.; Chen, W.S.; Yang, Y.H.; Zhu, K.F.; Zhong, Y. Characteristics of ELDRS at high and low-level injection in double polysilicon self-aligned NPN bipolar transistors. In Proceedings of the 2016 13th IEEE International Conference on Solid-State and Integrated Circuit Technology (ICSICT), Hangzhou, China, 25–28 October 2016; pp. 1056–1058.
12. Zhang, P.J.; Wu, X.; Jia, J.C.; Yi, Q.N.; Chen, W.S.; Yang, Y.H.; Zhu, K.F.; Lu, W.; Zhong, Y. A comparison of the reliability of 60Coγ ray irradiation on bulk-Si substrate and SOI substrate DPSA bipolar transistors. In Proceedings of the 2016 13th IEEE International Conference on Solid-State and Integrated Circuit Technology, ICSICT 2016, Hangzhou, China, 25–28 October 2016; pp. 1182–1184.
13. Zhang, P.; Wu, X.; Yi, Q.; Chen, W.; Yang, Y.; Zhu, K.; Tan, K.; Zhong, Y. A comparison of the effects of cobalt-60 γ ray irradiation on DPSA bipolar transistors at high and low injection levels. *Microelectron. Reliab.* **2017**, *71*, 86–90. [CrossRef]
14. Liu, M.; Lu, W. Dependence of emitter size on dose rate effect of double polysilicon self-aligned bipolar transistors. *He Jishu/Nuclear Tech.* **2018**, *41*, 1–5.
15. Lenahan, P.M.; Dressendorfer, P.V. Hole traps and trivalent silicon centers in metal/oxide/silicon devices. *J. Appl. Phys.* **1984**, *55*, 3495–3499. [CrossRef]
16. Freitag, R.K.; Brown, D.B.; Dozier, C.M. Experimental evidence of two species of radiation induced trapped positive charge. *IEEE Trans. Nucl. Sci.* **1993**, *40*, 1316–1322. [CrossRef]
17. Graves, R.J.; Cirba, C.R.; Schrimpf, R.D.; Milanowski, R.J.; Michez, A.; Fleetwood, D.M.; Witczak, S.C.; Saigne, F. Modeling low-dose-rate effects in irradiated bipolar-base oxides. *IEEE Trans. Nucl. Sci.* **1998**, *45*, 2352–2360. [CrossRef]
18. Boch, J.; Saigne, F.; Touboul, A.D.; Ducret, S.; Carlotti, J.F.; Bernard, M.; Schrimpf, R.D.; Wrobel, F.; Sarrabayrouse, G. Dose rate effects in bipolar oxides: Competition between trap filling and recombination. *Appl. Phys. Lett.* **2006**, *88*, 19–22. [CrossRef]
19. Fleetwood, D.M.; Schrimpf, R.D.; Pantelides, S.T.; Pease, R.L.; Dunham, G.W. Electron Capture, Hydrogen Release, and Enhanced Gain Degradation in Linear Bipolar Devices. *IEEE Trans. Nucl. Sci.* **2008**, *55*, 2986–2991. [CrossRef]
20. Rashkeev, S.N.; Fleetwood, D.M.; Schrimpf, R.D.; Pantelides, S.T. Effects of hydrogen motion on interface trap formation and annealing. *IEEE Trans. Nucl. Sci.* **2004**, *51*, 3158–3165. [CrossRef]
21. Freitag, R.K.; Brown, D.B. Study of low-dose-rate radiation effects on commercial linear bipolar ICs. *IEEE Trans. Nucl. Sci.* **1998**, *45*, 2649–2658. [CrossRef]
22. Boch, J.; Saign, F.; Schrimpf, R.D.; Vaill, J.-R.; Dusseau, L.; Lorfvre, E. Physical Model for the Low-Dose-Rate Effect in Bipolar Devices. *IEEE Trans. Nucl. Sci.* **2006**, *53*, 3655–3660. [CrossRef]

23. Li, X.-L.; Lu, W.; Wang, X.; Yu, X.; Guo, Q.; Sun, J.; Liu, M.-H.; Yao, S.; Wei, X.-Y.; He, C.-F. Estimation of enhanced low dose rate sensitivity mechanisms using temperature switching irradiation on gate-controlled lateral PNP transistor. *Chin. Phys. B* **2018**, *27*, 036102. [CrossRef]

© 2019 by the authors. Licensee MDPI, Basel, Switzerland. This article is an open access article distributed under the terms and conditions of the Creative Commons Attribution (CC BY) license (http://creativecommons.org/licenses/by/4.0/).

Article

A Radiation-Hardened Instrumentation Amplifier for Sensor Readout Integrated Circuits in Nuclear Fusion Applications

Kyungsoo Jeong [1], Duckhoon Ro [1], Gwanho Lee [2], Myounggon Kang [2,*] and Hyung-Min Lee [1,*]

[1] School of Electrical Engineering, Korea University, Seoul 02841, Korea; jksoo2002@korea.ac.kr (K.J.); roduckhoon@korea.ac.kr (D.R.)
[2] Department of Electronics Engineering, Korea National University of Transportation, Chungju 27469, Korea; ghlee@ut.ac.kr
* Correspondence: mgkang@ut.ac.kr (M.K.); hyungmin@korea.ac.kr (H.-M.L.); Tel.: +82-43-841-5164 (M.K.); +82-2-3290-3219 (H.-M.L.)

Received: 22 November 2018; Accepted: 9 December 2018; Published: 12 December 2018

Abstract: A nuclear fusion reactor requires a radiation-hardened sensor readout integrated circuit (IC), whose operation should be tolerant against harsh radiation effects up to MGy or higher. This paper proposes radiation-hardening circuit design techniques for an instrumentation amplifier (IA), which is one of the most sensitive circuits in the sensor readout IC. The paper studied design considerations for choosing the IA topology for radiation environments and proposes a radiation-hardened IA structure with total-ionizing-dose (TID) effect monitoring and adaptive reference control functions. The radiation-hardened performance of the proposed IA was verified through model-based circuit simulations by using compact transistor models that reflected the TID effects into complementary metal–oxide–semiconductor (CMOS) parameters. The proposed IA was designed with the 65 nm standard CMOS process and provides adjustable voltage gain between 3 and 15, bandwidth up to 400 kHz, and power consumption of 34.6 µW, while maintaining a stable performance over TID effects up to 1 MGy.

Keywords: radiation-hardened; instrumentation amplifier; sensor readout IC; total ionizing dose; nuclear fusion

1. Introduction

Radiation effects on electronic components are critical issues in various fields, such as space, medical imaging, and nuclear applications. Among them, nuclear fusion has been considered a safe and effective solution to generate massive energy, while requiring accurate sensing systems to precisely control environmental parameters in the nuclear fusion reactor, such as temperature, pressure, electromagnetic field, etc. [1–3]. Thus, a sensor readout system, which amplifies sensor signals and provides digitized codes to the back-end control system, plays an important role to guarantee reliability and safety of the nuclear fusion system.

The sensor readout integrated circuit (IC) typically consists of four circuit blocks as shown in Figure 1: instrumentation amplifier (IA), filter, analog-to-digital converter (ADC), and multiplexer (MUX). The IA amplifies the small sensor signals, and the filter passes the signals in the frequency band of interest. Then, the ADC converts the analog signals to digital codes, which are serialized through the MUX and provided to the back-end control system. The IA is one of the most critical circuits that needs to amplify the sensor signal accurately at the first stage of the sensor readout IC. However, the IA typically consists of variation-sensitive analog circuits, and its performance easily suffers from parameter variations under radiation effects.

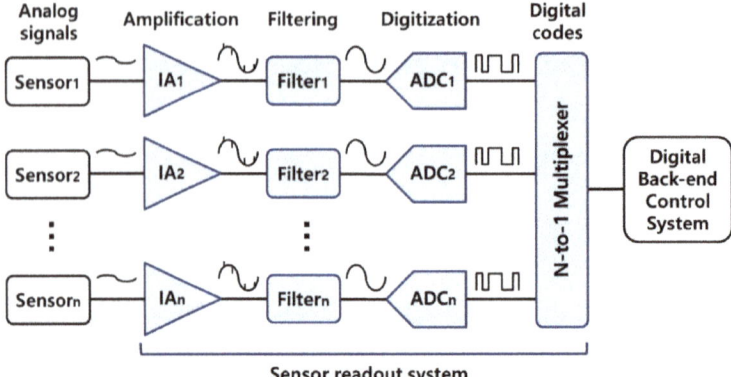

Figure 1. Block diagram of the sensor readout integrated circuit (IC) system. IA: instrumentation amplifier; ADC: analog-to-digital converter.

To reduce the radiation effects on the electronic components, three radiation-hardening methods have been widely considered: radiation hardening by process (RHBP), radiation hardening by shielding (RHBS), and radiation hardening by design (RHBD) [4–6]. While RHBP and RHBS, which improve the radiation tolerance by enhancing the process parameters and using shielded packages, respectively, have been effective ways for space and medical imaging applications, the nuclear fusion reactor suffers from more harsh radiation environments with high integral dose of MGy or higher [2,3]. Therefore, RHBD, which utilizes the optimized circuit structure against radiation effects, should also be considered for the sensor readout IC, especially, sensitive analog circuits, in nuclear fusion systems [6–12].

Silicon-based transistors in ICs, such as CMOS and bipolar junction transistor (BJT), can be affected by electrons, protons, and neutrons in radiation environments, which change the transistor parameters and degrade the circuit performance. These radiation effects on transistors can be categorized into three effects, i.e., total ionizing dose (TID), single event effect (SEE), and displacement damage (DD), as summarized in Table 1 [11–15]. The analog circuits with CMOS transistors mainly suffer from TID effects, which change the transistor parameters over time and are less vulnerable to SEE and DD effects. Thus, the proposed IA focuses on improving the radiation tolerance against the TID effects.

Table 1. Radiation effects to silicon-based transistors in ICs.

Radiation Effects	Cause	Effects to Analog Circuits
Total ionizing dose (TID)	- High-energy particles get through devices and produce electron–hole pair - The holes are trapped in gate oxide	- Changes threshold voltages - Increases leakage currents - Changes transconductance - Increases 1/f noise
Single event effect (SEE)	- High-energy particles impact a device in a moment and change the voltage in a device	- Changes voltages in capacitors - Upsets data in memory or flip-flop
Displacement damage (DD)	- Silicon ions are deviated from crystal lattice by high-energy particles.	- Critical in BJTs and Diodes - Increases leakage current - Less effects to CMOS

In addition, it is important to estimate the IA performance against TID effects during the design stage. To accurately reproduce radiation effects on CMOS transistors, we utilized the compact transistor models, whose parameters were degraded by TID and applied those compact models to SPICE circuit simulations. This compact model-based simulation methodology enables the precise estimation of the IA performance before conducting experiments in actual radiation environments.

The rest of this paper focuses on detailed techniques for the radiation-hardened IA design and performance verification through the compact model-based circuit simulation. Section 2 explains design considerations for choosing the IA topology for radiation environments. Section 3 proposes circuit techniques to improve the radiation tolerance in IAs. Section 4 describes how to use the compact transistor models for SPICE circuit simulations with radiation effects. Section 5 shows model-based simulation results, followed by concluding remarks in Section 6.

2. Radiation-Hardened IA Design

2.1. IA Topology Comparison

There is a variety of topologies of instrumentation amplifiers (IA) for sensor readout front-end ICs, such as capacitive-feedback IA, current-feedback IA, and three-op-amp IA, depending on users' requirements [16]. To design a radiation-hardened IA, it is essential to compare the performances in radiation environments and choose the optimum IA topology that is robust against TID and SEE effects.

The capacitive-feedback IA uses a couple of capacitors in the feedback loop, and the voltage gain is determined by the ratio between the capacitors. However, the voltage values across the capacitors can vary because of unwanted charge injection by SEE, which results in inaccurate output voltages. The current-feedback IA utilizes transconductance amplifiers at input and feedback paths to define its voltage gain with the ratio of transconductance (G_m). The current-feedback IA has the advantages of high common-mode rejection ratio (CMRR) and large input range, but the gain accuracy suffers from TID effects, which change the transconductance values.

Compared to those IAs, the three-op-amp IA enables relatively stable voltage gain against TID effects, since the gain is determined by the ratio between feedback resistors. While the three-op-amp IA has the advantages of high-input impedance and good linearity over wide input–output ranges, it is also less affected by SEE because the DC bias current flowing through the feedback loop keeps the voltage values across resistors from instantaneous charge injection by SEE. Therefore, the three-op-amp IA can be used as the radiation-hardened IA topology, which is less affected by both TID and SEE, compared to other IAs. Table 2 compares the performance of various IA topologies against TID and SEE.

Table 2. Performance comparison of IA topologies against total ionizing dose (TID) and single event effect (SEE). G_m: transconductance.

IA Topologies	Capacitive-Feedback IA	Current-Feedback IA	Three-op-amp IA
TID tolerance	O (Gain ∝ capacitor ratio)	X (Gain ∝ CMOS G_m ratio)	O (Gain ∝ resistor ratio)
SEE tolerance	X (Capacitor voltage changes)	O (DC bias on feedback)	O (DC bias on feedback)

2.2. Radiation-Hardened IA Structure

For the radiation-hardened IA, the op-amp circuits in the three-op-amp IA topology should also operate properly against radiation effects. For accurate readout of sensor signals, the two-stage op-amp with p-type metal–oxide–semiconductor (PMOS) input stages has been widely used thanks to its low noise, high gain, and wide output range [16]. However, the op-amp performance, such as voltage gain, bandwidth, and power consumption, can be degraded due to TID effects as follows: (1) threshold voltage (V_{th}) variation due to TID effects leads the transistors to operate in improper triode regions instead of saturation regions, especially, a tail current transistor in the input stage, and (2) bias currents flowing through the op-amp vary by TID effects, affecting the amplifier performance.

To overcome these limitations, we propose a radiation-hardened IA structure, which adopts the three-op-amp topology and fully-differential structure, while employing TID effect monitoring, V_{th}-insensitive current generator, and adaptive reference control. Figure 2 shows the conceptual block diagram of the proposed radiation-hardened IA. The TID effect monitoring circuit, which is reliably biased by the V_{th}-insensitive current generator, senses the V_{th} variation due to TID effects. Then, the adaptive reference control circuit automatically adjusts the sensor reference voltage, V_{REF}, keeping the tail current transistors in op-amps, A_1 and A_2, to operate in saturation regions regardless of V_{th} variation in the transistors. The voltage gain, which is defined as $A_V = [(R_1 + R_{sel} + R_1)/R_{sel}] \times [R_3/R_2]$, can be adjusted by digitally tuning the R_{sel} value, and the op-amp, A_3, provides fully differential output voltages, V_{OUTP} and V_{OUTN}, to the following ADC for accurate signal digitization. The V_{th}-insensitive current generator also supplies the bias currents not only to the op-amps, A_1–A_3, but also to the ADC for robust DC biasing against TID effects.

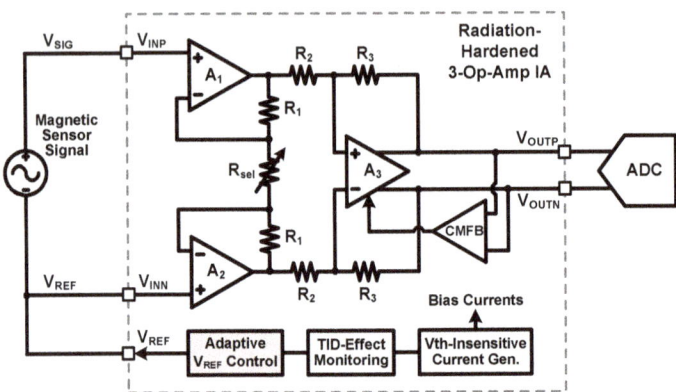

Figure 2. Conceptual block diagram of the proposed radiation-hardened IA.

3. Circuit Details for Radiation-Hardened IA

3.1. TID Effect Monitoring

Figure 3 shows the conceptual and schematic diagrams of the TID effect monitoring circuit, which can monitor the V_{th} variation of the CMOS transistor depending on the integral amount of TID effects. In Figure 3a, the TID effect monitoring consists of the PMOS monitoring transistor, M_M, and the current source, I_{REF}. Then, the monitoring voltage, V_M, can be expressed as follows:

$$V_M = V_{DD} - V_{SG,MM} = V_{DD} - V_{ov,MM} - V_{th,MM} \tag{1}$$

where V_{DD} is the supply voltage, and $V_{ov,MM}$ and $V_{th,MM}$ are the overdrive and threshold voltages of M_M, respectively. If I_{REF} has little variation against TID, then $V_{ov,MM}$ can be relatively constant, and $V_{th,MM}$ variation can be observed by monitoring V_M, which changes as TID increases.

To generate a constant I_{REF} against V_{th} variation by TID, we adopted a beta multiplier structure to implement the TID effect monitoring circuit, as shown in Figure 3b. The n-type metal–oxide–semiconductor (NMOS) transistors, M_6 and M_7, have different size ratio of 1:K, and the amplifier, which consists of M_1–M_4, ensures that drain and gate voltages of M_6 and M_7 are the same. M_M and M_5 have the same size ratio, flowing the same bias current of I_{REF} to M_6 and M_7, respectively. Then, I_{REF}, which flows through M_M, can be defined relatively constant, regardless of V_{th} variation as follows:

$$V_{GS6} = V_{GS7} + I_{REF}R_2 = \sqrt{\frac{2I_{REF}}{\beta_N}} + V_{th,M6} = \sqrt{\frac{2I_{REF}}{K\beta_N}} + V_{th,M7} + I_{REF}R_2 \tag{2}$$

$$I_{REF} = \frac{1}{\beta_N}\left(\frac{2}{R_2^2}\right)\left(1 - \frac{1}{\sqrt{K}}\right)^2 \tag{3}$$

where β_N is $\mu_n C_{ox}(W/L)$, which are the NMOS transistor parameters, and assuming NMOS threshold voltages, $V_{th,M6}$ and $V_{th,M7}$, are affected by TID in the same way. Therefore, V_{th} variation of the monitoring transistor, M_M, which depends on TID over time, can be monitored by observing V_M. The bias current, I_{BIAS}, which supplies the op-amps in the IA, can be generated through M_{10}.

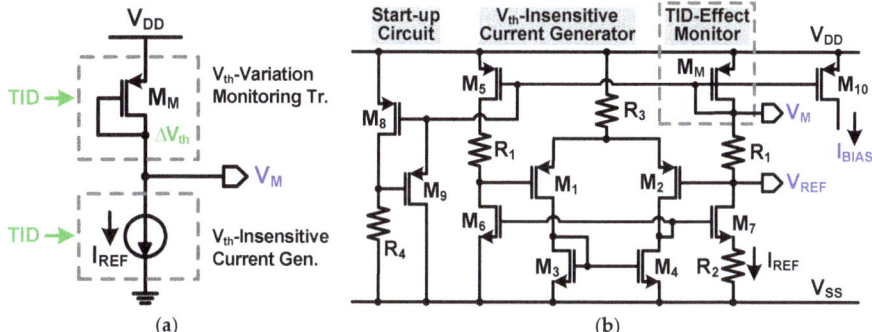

Figure 3. (a) Conceptual diagram and (b) schematic diagram of the TID effect monitoring circuit with the V_{th}-insensitive current generator.

3.2. Adaptive Reference Control

The maximum input voltage level of the op-amps with PMOS input transistors, such as A_1 and A_2 in Figure 2, is limited as $V_{DD} - V_{SG,in} - V_{ov,tail}$, where $V_{SG,in}$ is the source-gate voltage of the PMOS input transistor, and $V_{ov,tail}$ is the overdrive voltage of the tail current source transistor. However, the TID effects can change V_{th} of the transistors (typically increase V_{th} of PMOS transistors), decreasing the maximum input levels, leading the tail current transistor to operate in the triode region and finally degrading the op-amp performance.

To circumvent this situation, the adaptive reference control was utilized to automatically adjust the sensor reference voltage, V_{REF}, which is the common-mode input level of A_1 and A_2, as shown in Figure 4. The adaptive reference control utilizes the TID effect monitoring and the additional resistor R_1 to generate V_{REF} as $V_{DD} - V_{SG,MM} - I_{REF}R_1$. For example, if V_{th} of PMOS transistors increases due to TID, which decreases the maximum input level of the op-amps, V_{REF} (i.e., op-amp input levels) also adaptively decrease to ensure that op-amp input stages are operating properly in saturation regions. The detailed circuit to generate V_{REF} is shown in Figure 3b, and the buffer amplifier, A_4, drives the sensor reference node with V_{REF}.

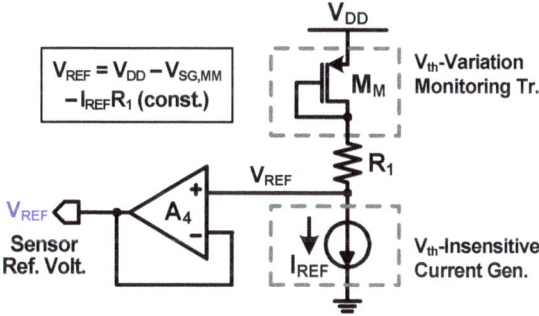

Figure 4. Conceptual diagram of the adaptive reference control.

4. Compact Transistor Modeling with Radiation Effects

In order to observe the circuit performance with the TID effects, the Berkeley short-channel IGFET Model (BSIM) 4 SPICE model was used in this work. The BSIM4 model is widely used as a standard compact model in the industry and has been developed for silicon-based MOS transistors [17,18]. Figure 5a shows a 65 nm device structure using a 3D technology computer-aided design (TCAD) simulation with the Silvaco Victory Device software. We evaluated the electrical characteristics considering various channel widths (W) and channel lengths (L) of the device structure. Figure 5b shows V_{GS} versus I_D characteristics with W = 1 µm and L = 65 nm. The TCAD simulation (circle symbols) showed excellent agreement with the circuit simulation (lines). This indicates that our TCAD simulation exactly reflected the devices used in the 65 nm CMOS process.

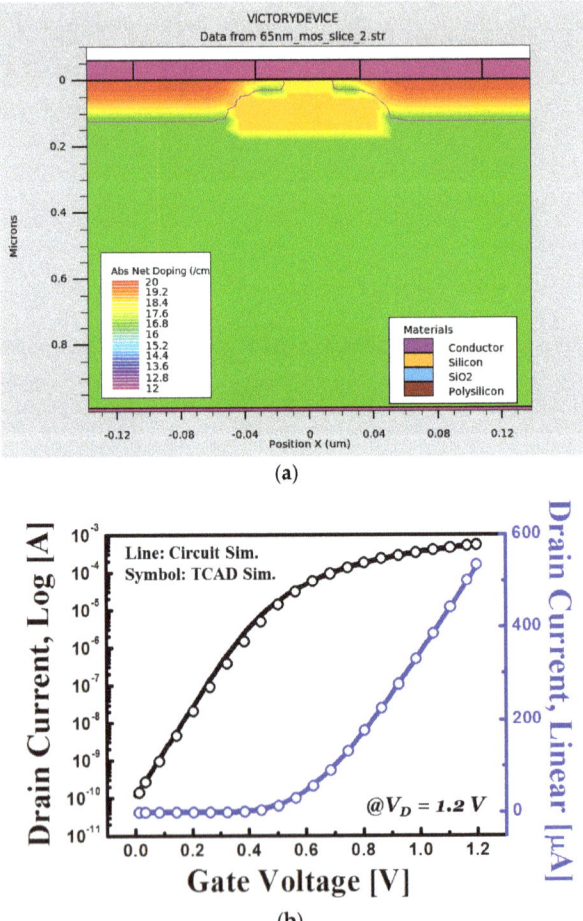

Figure 5. The 65 nm CMOS (**a**) structure and (**b**) comparison of V_{GS} vs. I_D between technology computer-aided design (TCAD) and circuit simulation.

Figure 6 shows the electrical characteristics of the device with TID effects for each Gy level. Figure 6a shows V_{GS} versus I_D, and Figure 6b shows V_{DS} versus I_D. To obtain compact models for each TID quantity, several levels of TID effects were applied to CMOS transistors through the TCAD simulation with the Silvaco Victory Device software, which generated corresponding I-V curves. Then, BSIM parameters, such as $VTH0$ (long channel threshold voltage), VFB (flatband voltage), $VSAT$ (saturation velocity), CIT (interface trap capacitance), etc., which affect the threshold voltage, subthreshold swing, and leakage current, were extracted from those I-V curves and utilized to develop the compact models for each TID. The V_{th} shift phenomenon and the off-current increase, which were caused by the TID effects, were confirmed. The I-V curve in Figure 6 was used for compact modeling in the BSIM4 parameter extraction process for each cumulative dose. The BSIM4 parameters were extracted by using Silvaco Utmost IV software. The flow chart of BSIM4 model parameter extraction was detailed in a previous work [18]. In order to extract the BSIM4 parameters, the V_{GS} versus I_D curve in linear and log scales and the V_{DS} versus I_D curve in linear scale were simultaneously considered. The BSIM4 parameters were extracted by matching the linear and saturation regions of the I-V curve by adjusting parameters such as V_{th0}, V_{sat}. [19,20].

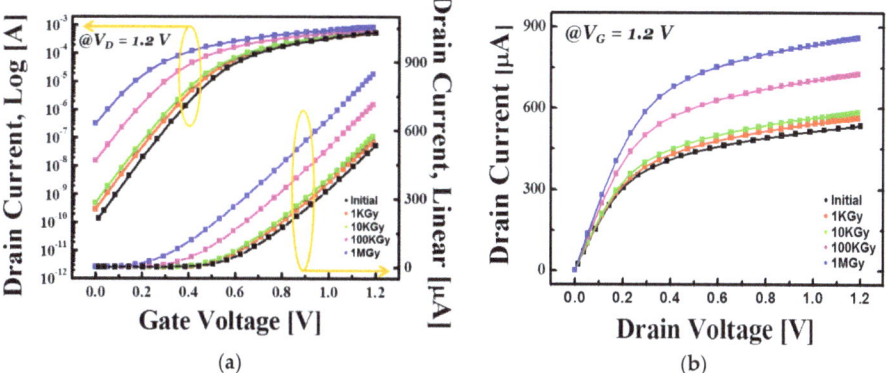

Figure 6. CMOS transistor characteristics depending on TID effects: (**a**) V_{GS} vs. I_D and (**b**) V_{DS} vs. I_D.

5. Simulation Results with Compact Transistor Models

The radiation-hardened IA in Figure 2 was designed in a 65 nm standard CMOS process with a supply voltage, V_{DD}, of 1.2 V and verified through the SPICE simulation. To emulate the TID effects on circuit simulation, we also utilized the compact transistor models described in Section 4. Each compact model included the TID effects of 1 kGy, 10 kGy, 100 kGy, and 1 MGy.

Figure 7 shows the reference current (I_{REF}), monitoring voltage (V_M), and sensor reference voltage (V_{REF}) of Figure 3 against the TID effects. While I_{REF} was relatively constant at higher TID, V_M showed the V_{th} variation of the TID-monitoring PMOS transistor (M_M in Figure 3). Then, V_{REF}, which also decreased at higher TID, could adaptively control the sensor reference level, ensuring TID-tolerant IA operation. Figure 8 shows the voltage gain of the radiation-hardened IA against the TID effects. In Figure 8a, the voltage gain of the radiation-hardened IA was set to 5 and showed a little variation as TID increased. However, the conventional IA, which also had the three-op-amp structure but its sensor reference level (V_{REF}) was fixed to half V_{DD}, had a significant drop of the voltage gain with TID above 10 kGy, because some transistors in the op-amps could operate in triode regions, and their bias currents significantly changed. On the contrary, the radiation-hardened IA could provide an adjustable voltage gain between 3 and 15 over high TID effects, as shown in Figure 8b.

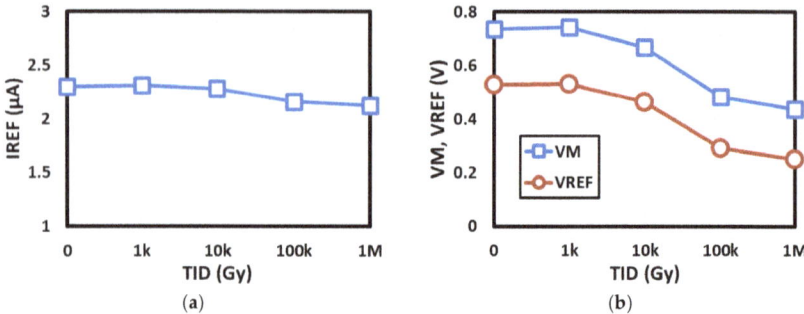

Figure 7. Model-simulation results showing (**a**) reference current (I_{REF}) vs. TID and (**b**) monitoring voltage (V_M) and sensor reference voltage (V_{REF}) vs. TID.

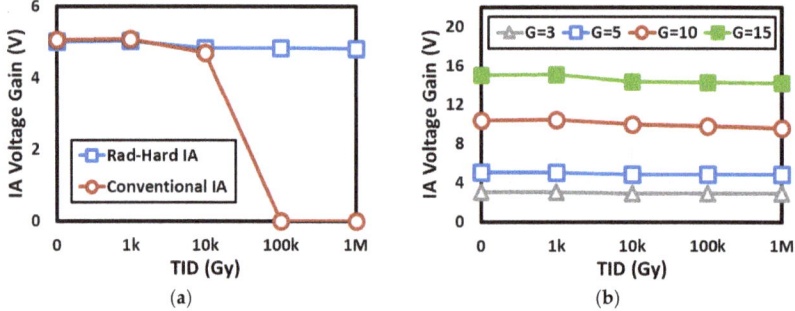

Figure 8. Model simulation results showing (**a**) voltage gain comparison between proposed and conventional IAs by TID and (**b**) adjustable voltage gain (3, 5, 10, and 15) of the proposed IA by TID.

The proposed IA aims for magnetic sensor signals in nuclear fusion reactors, in which the amplitude can be up to 100 mV, so that the IA adopts the adjustable voltage gain range between 3 and 15. Table 3 summarizes the overall performance of the radiation-hardened IA when the voltage gain was set to 5 and TID was 0 and 1 MGy. While the proposed IA maintained similar levels of voltage gains at the high TID of 1 MGy, the power consumption of the IA increased mainly as a consequence of V_{th} variations and leakage currents of the transistors. The bandwidth of the IA decreased to 80 kHz at TID of 1 MGy, but the proposed IA could still operate properly with sensor signals, whose frequencies were typically of the kHz order or lower. Also, the proposed IA provides fully differential output voltages, i.e., V_{OUTP} and V_{OUTN}, as in Figure 2, which enables a high power supply rejection ratio (PSRR). When intended offsets of 5 mV were applied to the amplifiers in experimental practical cases, the proposed IA achieved the PSRR of 81 dB, which could be maintained to 77.7 dB at TID up to 1 MGy.

Table 3. Overall performance of the radiation-hardened IA.

Specification	TID = 0 Gy	TID = 1 MGy
Process	65-nm standard CMOS	
Supply voltage (V)	1.2	
Voltage gain (V/V) *	5.008	4.812
Bandwidth (kHz) **	240	80
Power consumption (µW) **	34.6	98.3
Input referred noise (µV/√Hz) **	0.94	1.12
Power supply rejection ratio (dB) ***	81	77.7

* Adjustable between 3 and 15, ** model-simulated when the voltage gain was set to 5. *** Intended offsets of 5 mV were applied to the amplifiers in experimental practical cases.

While there have been few previous studies about radiation-hardened IAs, the radiation-hardening performance can be roughly compared with that of other analog circuits in sensor front-end systems, such as ADCs and voltage references. The radiation-hardened delta-sigma ADC in reference [11] showed 2.8% degradation (from 109 to 106 dB) in signal-to-noise-distortion ratio (SNDR) at TID up to 1.36 MGy. In radiation-hardened voltage references, the bandgap reference as reported [7] showed ±0.8% variation (±1.5 mV) in reference voltages at TID up to 0.44 MGy, and the bandgap reference reported in another study [12] achieved about ±3% variation (±18 mV) in reference voltages at TID up to 4.5 MGy. Compared to those performances, the proposed radiation-hardened IA achieved 3.9% degradation (from 5.008 to 4.812) in voltage gain at TID up to 1 MGy, showing competitive performance of the circuit design techniques for radiation hardening. Also, it should be noted that the proposed radiation hardening by design (RHBD) can be used along with RHBP and RHBS to further improve the radiation tolerance of the electronic components.

6. Conclusions

A radiation-hardened instrumentation amplifier (IA), which needs to ensure a robust operation against radiation effects such as TID and SEE, is an essential component of sensor readout systems in harsh radiation environments such as nuclear fusion reactors. This paper studied design considerations for choosing the IA topology for radiation environments and proposed the radiation-hardened IA circuit with TID effect monitoring and adaptive reference control functions. The radiation tolerance of the proposed IA was verified through the SPICE circuit simulations by adopting compact transistor models that reflected the TID effects into CMOS parameters.

Author Contributions: Methodology, Validation, Paper Writing, K.J., D.R. and G.L.; Conceptualization, Funding Acquisition, Investigation, Supervision, Writing-Review & Editing, M.K. and H.-M.L.

Funding: This research was supported by National R&D Program through the National Research Foundation of Korea (NRF) funded by the Ministry of Science & ICT (NRF-2017M1A7A1A01016260). This research was also supported by National R&D Program through the National Research Foundation of Korea (NRF) funded by the Ministry of Science & ICT (NRF-2017M1A7A1A01016265).

Acknowledgments: The EDA tool was supported by the IC Design Education Center (IDEC), Korea. The authors would like to thank Suk-Ho Hong and Heung Su Kim of the National Fusion Research Institute (NFRI), Daejeon, Korea, for technical discussion.

Conflicts of Interest: The authors declare no conflict of interest.

References

1. Martin, V.; Bertalot, L.; Drevon, J.M.; Reichle, R.; Simrock, S.; Vayakis, G.; Walsh, M.; Verbeeck, J.; Cao, Y.; Van Uffelen, M. Electronic components exposed to nuclear radiation in ITER diagnostic systems: Current investigations and perspectives. In Proceedings of the EPS Conference on Plasma Diagnostics (ECPD), Frascati, Italy, 14–17 April 2015; pp. 1–7.
2. Leroux, P.; Van Koeckhoven, W.; Verbeeck, J.; Van Uffelen, M.; Esqué, S.; Ranz, R.; Damiani, C.; Hamilton, D. Design of a MGy radiation tolerant resolver-to-digital convertor IC for remotely operated maintenance in harsh environments. *Fusion Eng. Des.* **2014**, *89*, 2314–2319. [CrossRef]
3. Verbeeck, J.; Cao, Y.; Van Uffelen, M.; Casellas, L.M.; Damiani, C.; Morales, E.R.; Santana, R.R.; Meek, R.; Hais, B.; Hamilton, D.; et al. Qualification method for a 1 MGy-tolerant front-end chip designed in 65 nm CMOS for the read-out of remotely operated sensors and actuators during maintenance in ITER. *Fusion Eng. Des.* **2015**, *96–97*, 1002–1005. [CrossRef]
4. Haddad, N.F.; Kelly, A.T.; Lawrence, R.K.; Li, B.; Rodgers, J.C.; Ross, J.F.; Warren, K.M.; Weller, R.A.; Mendenhall, M.H.; Reed, R.A. Incremental enhancement of SEU hardened 90 nm CMOS memory cell. *IEEE Trans. Nucl. Sci.* **2011**, *58*, 975–980. [CrossRef]
5. Clark, L.T.; Mohr, K.C.; Holbert, K.E.; Yao, X.; Knudsen, J.; Shah, H. Optimizing radiation hard by design SRAM cells. *IEEE Trans. Nucl. Sci.* **2007**, *54*, 2028–2036. [CrossRef]

6. Gatti, U.; Calligaro, C.; Pikhay, E.; Roizin, Y. Radiation-hardened techniques for CMOS flash ADC. In Proceedings of the IEEE International Conference on Electronics Circuits and Systems (ICECS), Marseille, France, 7–10 December 2014.
7. Gromov, V.; Annema, A.J.; Kluit, R.; Visschers, J.L. A radiation hard bandgap reference circuit in a standard 0.13 μm CMOS technology. *IEEE Trans. Nucl. Sci.* **2007**, *54*, 2727–2733. [CrossRef]
8. Dang, L.D.T.; Kim, J.S.; Chang, I.J. We-Quatro: Radiaiton-hardened SRAM cell with parametric process variation tolerance. *IEEE Trans. Nucl. Sci.* **2017**, *64*, 2489–2496. [CrossRef]
9. Galib, M.M.H.; Chang, I.J.; Kim, J.S. Supply voltage decision methodology to minimize SRAM standby power under radiation environment. *IEEE Trans. Nucl. Sci.* **2015**, *62*, 1349–1356. [CrossRef]
10. Dang, L.D.T.; Kang, M.; Kim, J.S.; Chang, I.J. Studying the variation effects of radiation hardened Quatro SRAM bit-cell. *IEEE Trans. Nucl. Sci.* **2016**, *63*, 2399–2401. [CrossRef]
11. Verbeeck, J.; Van Uffelen, M.; Steyaert, M.; Leroux, P. 17 bit 4.35 mW 1 kHz delta sigma ADC and 256-to-1 multiplexer for remote handling instrumentation equipment. *Fusion Eng. Des.* **2013**, *88*, 1942–1946. [CrossRef]
12. Cao, Y.; Cock, W.D.; Steyaert, M.; Leroux, P. A 4.5 MGy TID-tolerant CMOS bandgap reference circuit using a dynamic base leakage compensation technique. *IEEE Trans. Nucl. Sci.* **2013**, *60*, 2819–2824. [CrossRef]
13. Virmontois, C.; Goiffon, V.; Magnan, P.; Girard, S.; Inguimbert, C.; Petit, S.; Rolland, G.; Saint-Pé, O. Displacement damage effects due to neutron and proton irradiations on CMOS image sensors manufactured in deep submicron technology. *IEEE Trans. Nucl. Sci.* **2010**, *57*, 3101–3108. [CrossRef]
14. Barnaby, H.J. Total-ionizing-dose effects in modern CMOS technologies. *IEEE Trans. Nucl. Sci.* **2006**, *53*, 3103–3121. [CrossRef]
15. Martin, H.; Martin-Holgado, P.; Morilla, Y.; Entrena, L.; San-Millan, E. Total ionizing dose effects on a delay-based physical unclonable function implemented in FPGAs. *Electronics* **2018**, *7*, 163. [CrossRef]
16. Wu, R.; Huijsing, J.H.; Makinwa, K.A.A. *Precision Instrumentation Amplifiers and Read-out Integrated Circuits*; Springer: New York, NY, USA, 2013.
17. Xi, X.; Dunga, M.; He, J.; Liu, W.; Cao, K.M.; Jin, X.; Ou, J.J.; Chan, M.; Niknejad, A.M.; Hu, C. *BSIM4.3.0 MOSFET Model, User's Manual*; Department of Electrical Engineering and Computer Sciences, University of California: Berkeley, CA, USA, 2003.
18. Jeong, Y.J.; Jeon, J.; Lee, S.; Kang, M.; Jhon, H.; Song, H.J.; Park, C.E.; An, T.K. Development of organic semiconductors based on quinacridone derivatives for organic field-effect transistors: High-voltage logic circuit applications. *IEEE J. Electron Devices Soc.* **2017**, *5*, 209–213. [CrossRef]
19. Kang, M.; Lee, K.; Chae, D.H.; Park, B.-G.; Shin, H. The compact modeling of channel potential in sub-30-nm NAND flash cell string. *IEEE Electron Device Lett.* **2012**, *33*, 321–323. [CrossRef]
20. Kang, M.; Park, I.H.; Chang, I.J.; Lee, K.; Seo, S.; Park, B.-G.; Shin, H. An accurate compact model considering direct-channel interference of adjacent cells in sub-30-nm NAND flash technologies. *IEEE Electron Device Lett.* **2012**, *33*, 1114–1116. [CrossRef]

© 2018 by the authors. Licensee MDPI, Basel, Switzerland. This article is an open access article distributed under the terms and conditions of the Creative Commons Attribution (CC BY) license (http://creativecommons.org/licenses/by/4.0/).

Article

Proton Induced Single Event Effect Characterization on a Highly Integrated RF-Transceiver

Jan Budroweit [1,*], Mattis Paul Jaksch [1] and Maciej Sznajder [2]

1. Avionic Systems, German Aerospace Center (DLR), Institute of Space Systems, 28359 Bremen, Germany; Mattis.Jaksch@dlr.de
2. Mechanics and Thermal Systems, German Aerospace Center (DLR), Institute of Space Systems, 28359 Bremen, Germany; Maciej.Sznajder@dlr.de
* Correspondence: Jan.Budroweit@dlr.de; Tel.: +49-421244201297

Received: 12 April 2019; Accepted: 7 May 2019; Published: 9 May 2019

Abstract: Radio frequency (RF) systems in space applications are usually designed for a single task and its requirements. Flexibility is mostly limited to software-defined adaption of the signal processing in digital signal processors (DSP) or field-programmable gate arrays (FPGA). RF specifications, such as frequency band selection or RF filter bandwidth are thereby restricted to the specific application requirements. New radio frequency integrated circuit (RFIC) devices also allow the software-based reconfiguration of various RF specifications. A transfer of this RFIC technology to space systems would have a massive impact to future radio systems for space applications. The benefit of this RFIC technology allows a selection of different RF radio applications, independent of their RF parameters, to be executed on a single unit and, thus, reduces the size and weight of the whole system. Since most RF application sin space system require a high level of reliability and the RFIC is not designed for the harsh environment in space, a characterization under these special environmental conditions is mandatory. In this paper, we present the single event effect (SEE) characterization of a selected RFIC device under proton irradiation. The RFIC being tested is immune to proton induced single event latch-up and other destructive events and shows a very low response to single failure interrupts. Thus, the device is defined as a good candidate for future, highly integrated radio system in space applications.

Keywords: single event effects; proton irradiation; RFIC; SEE testing; space application

1. Introduction

The German Aerospace Center (DLR), Institute of Space System, is currently working on the development of a highly integrated multi-band software-defined radio (SDR) platform for space application [1,2]. Compared to state-of-the-art SDR systems, the design shall allow the reconfiguration of relevant radio frequency (RF) parameters, such as the RF bandwidth, mixing frequency or the sample rate for analog to digital conversion (and vice versa). Usually, most of the RF parameters depend on the executed application and only the digital signal processing is adjustable by software. For this reason, the RF front-end is typically designed with discrete components to the specific application requirements. Some radio systems allow multiplexing between different RF front-end modules, such as that presented in [3,4], but will increase the overall size and weight of the system. To allow a software-based reconfiguration on RF front-end related parameters, a new radio frequency integrated circuit (RFIC) technology must be used and needs to be investigated, particularly the performance in a radiation environment for the utilization in space applications.

In this paper, we present the single event effect (SEE) characterization of an RFIC device for the multi-band radio platform purpose under proton irradiation. In Section 2, the device under test (DUT) and the general test method is presented. The test requirements and test site are presented in

Section 3 and the test setup and procedures are presented in Section 4. The test results of this SEE characterization are described in Section 5 and are later discussed in Section 6. Finally, the conclusion is made in Section 7.

2. Device Under Test

The selected RFIC as a demonstrator for the implementation into the highly integrated multi-band SDR platform is the AD9361 agile RF Transceiver from Analoge Devices [5]. The integrated circuit (IC) device is a 2 × 2 RF transmitter and receiver (transceiver) including individual RF front-ends, a mixed-signal baseband (BB) unit with an integrated frequency synthesizer and a selectable low voltage differential signaling (LVDS) or complementary metal-oxide-semiconductor (CMOS) digital interface. Any functionality, whether its RF or BB related, can be configured by software over a serial peripheral interface (SPI). A block diagram of the device is presented in Figure 1.

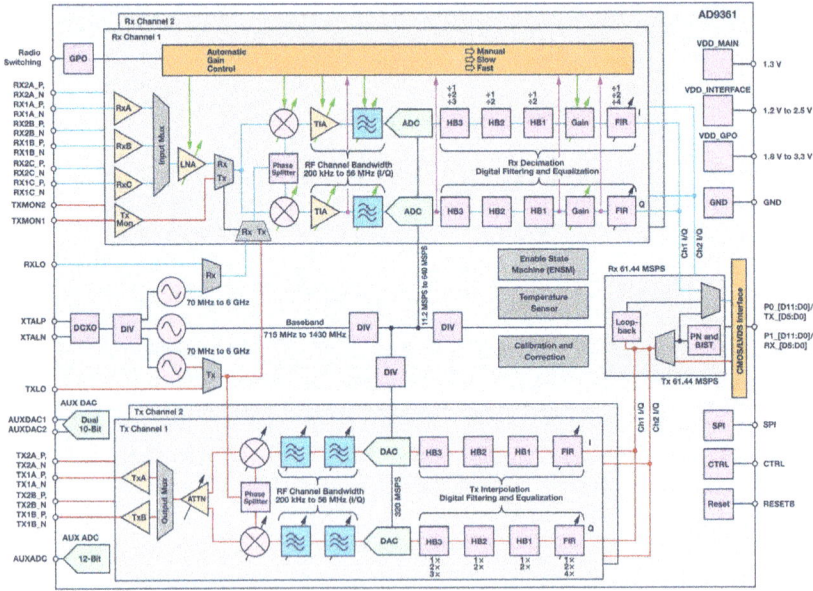

Figure 1. AD9361 block diagram [5].

Besides plenty of configurations, another major advantage of the device is the automatic self-calibration sequences invoking each time the RFIC is reconfigured to a new BB setting or power-cycled/reset. The device can be operated in time division duplex (TDD) mode, where only transmission or reception is possible, or in frequency division duplex (FDD) mode, where transmission and reception are both enabled. The different functions of the DUT are described in more detail in the following sections.

2.1. RF Front-End

The AD9361 includes two independent transmit and receive paths. Each transmitter has two multiplexable outputs (four in total) and each receiver chain consist of three selectable inputs (six in total). Thus, multiple band selection is possible by the design of individual front-ends, consisting of additional application specific filters, mixers, low noise amplifiers (LNA) or power amplifiers (PA). Each transmitter and receiver chain includes series of filters (analog and digital) and amplifiers, where the gain can be selected manually or controlled by an automatic gain control (AGC). The RF front-end calibration includes RF DC-offset calibration, a quadrature calibration for the receive and transmission chain and the calibration of the RF clock synthesizer charge pump and voltage controlled oscillator (VCO).

2.2. Control System

All configurations of the RFIC can be programmed via an SPI interface to its 8-bit register map (0x000 to 0x3F6). In addition to the automatic self-calibration and correction procedures, the AD9361 consists of an enable state machine (ENSM), which allows the user to select between different operation modes (e.g., TDD or FDD). The available ENSM modes are:

- Sleep mode: Clocks and BB phased-locked loop (PLL) disabled
- TX: Transmitter enabled (only on TDD)
- RX: Receiver enabled (only on TDD)
- FDD: Transmitter and receiver enabled
- Alert: Synthesizer enabled only
- Wait: Synthesizer disabled (power saving mode)

2.3. Direct Baseband Conversion

The RFIC supports a direct BB conversion for the receive and transmit chain, without using any intermediate frequencies. The mixing frequency can be selected individually for transmission and reception. The AD9361 uses a quadrature demodulation for the direct down-conversion to the BB and a vice versa for the transmission. Thus, a complex pair (I and Q) is processed individually (e.g., filtered and amplified) in the BB and is also digitized by separated analog digital converters (ADCs). The ADC and digital to analog converters (DACs) are adjustable in their sample rate and controlled by the BB synthesizer.

2.4. Clocking

The AD9361 requires a crystal or oscillator clocking source to generate the clock frequencies inside of the RFIC. The RF clocks are derived via a local VCO that is controlled by an internal low-drop out (LDO) voltage regulator. The BB frequencies are also generated by a PLL synthesizer. The BB PLL synthesizer and the RF synthesizer are calibrated at the same time.

2.5. Power Application

The power supply of the AD9361 is separated into two 1.3 V power domains for the main function of the device and one 1.8 V rail for the interface voltage (e.g., LVDS). A third domain of 3.3 V can be supplied for the device general purpose output (GPO) if they are required to be used. The 1.3 V power domain requires a stability of less than ± 30 mV and an ultra low phase noise conducted by the selected LDO voltage regulator to enhance the RF performance of the device.

2.6. Device Packaging and Chip Technology

The RFIC is encapsulated in a 144-pin chip scale package ball grid array housing. The size of the device is given with $10 \times 10 \times 1.7$ mm^3. The semiconductor die is based on a 65 nm silicon on insulator (SOI) CMOS process. An X-ray picture of the device is presented in Figure 2 and shows that the die is located on a printed circuit board (PCB) stack.

The die is assembled faced-up with pads connected to the redistribution layers by wire-bonds. The die has a size of approximately $500 \times 500 \times 220$ µm^3. The molding compound is based on silica (86.20%), epoxy resin (6.00%), phenol resin (6.00%), metal hydroxide (1.50%) and carbon black (0.30%) [6]. The encapsulation thickness on top of the die is calculated to 280 µm, which is an important value, since the DUT is not going to be encapsulated to expose the die for the proton test.

Figure 2. Side view: X-ray image of the AD9361.

3. Test Requirements and Conditions

In this section, the test requirements, firstly derived from previous DLR missions and which are typically desired for radio applications in space, are presented. In the second part of this section, the test site is presented as well as calibration results for the selected proton energies are given.

3.1. Test Requirements

For this test purpose, we firstly limited our radiation environment to proton irradiation. The reason for this is that many applications for the DUT are suitable for earth observation missions, primarily in low earth orbit (LEO). For LEO, the population of charged particles stably trapped by the Earth magnetic field is high, mainly consisting of protons (100 keV to hundreds of MeV) and electrons (few tens of keV to 10 MeV) is high. There are models available (e.g., NASA AP-8 and AE-8) predicting the proton flux for given attitude profiles and depending on the solar activities [7,8]. Thus, we decided to take several reference missions in LEO for the test requirements in terms of proton energies and the total fluence.

To estimate the proton fluxes at the LEO environment, two altitudes (400 km and 800 km) and three different inclination angles $\alpha = \{0°, 51.64°, 98°\}$ were taken into account. In addition, solar maximum and minimum activities were considered. The AP-8 model was used for estimating the flux of trapped protons [9] and the CREME-96 model for the galactic cosmic rays (GCR) [10]. The analysis results are presented in Figure 3. The proton fluence (see Figure 3b) is a product of the flux and the time. Here, a one-year period was considered. The proton flux magnitude (see Figure 3a) at the altitude of 800 km is approximately two orders of magnitude larger than at 400 km.

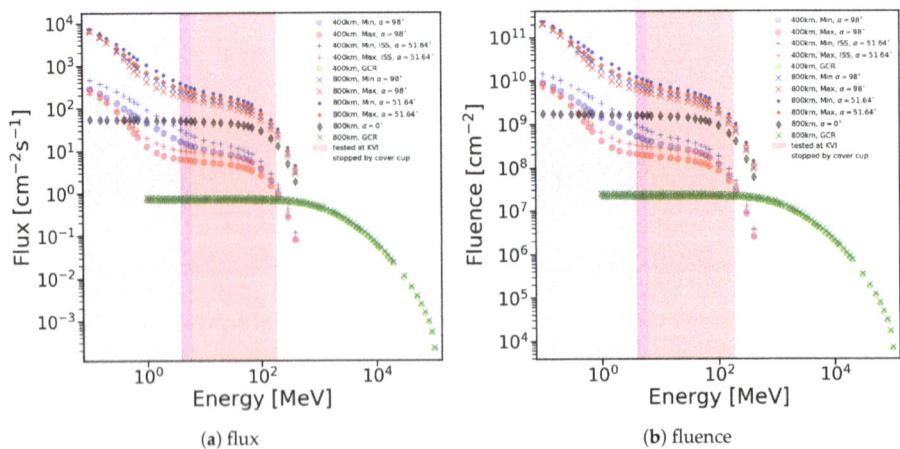

Figure 3. Integral flux and fluence spectra.

It is strictly related to the strength of the magnetic field and topology of the trapped proton belt that plays a dominant role. Proton fluxes for two inclination angles of $\alpha = \{51.64°, 98°\}$ are comparable. In both cases, a satellite would pass the so-called South Atlantic anomaly (SAA) [11] and for the inclination of 98° also the Earth's polar regions. Influence of the SAA and the Earth's polar regions can be easily examined by looking at the proton flux for an inclination angle of 0°. Here, a satellite would rotate the Earth only along the equator and would not pass through the mentioned high radiation regions. Solar activity counted in 11-year cycle results with variation of number of particles sent throughout the interplanetary space, and, therefore, trapped by the Earth's magnetic field. As one can see, the proton flux for minimum Sun activity is actually higher than at its maximum state because, at solar maximum, the increased UV irradiation causes the atmosphere to expand. The thicker is the atmosphere, the more collision events there are with the incident protons at the high altitudes, and, therefore, fewer protons can reach the low altitudes. The GCR flux level for two considered altitudes is comparable and it is up to four orders of magnitude lower than for the trapped solar protons. However, the energy range of the GCRs expands almost up to 10^4 MeV. The shape of the GCR-flux curve is unaffected by a satellite orbit and its inclination because the GCR origin is the Galaxy and beyond, thus the particles are coming from all possible directions. It is then easier to compute equivalent fluence for any number of years. The DUT was exposed to a discrete set of proton energies from 4 MeV up to 184 MeV. The range, in both figures, is marked by the light red area. In space, the DUT is nominally covered with 280 µm thick silica cup, which results in stopping of the protons with energies lower than ≈6.1 MeV [12].

In addition to the reference missions derived test requirements, we worked according the European space component coordination (ESCC) single event effects test method and guidelines, ESCC basic Specification No. 25100 [13]. The ESCC No. 25100 requires a total fluence of 10^{11} protons/cm² on five different energies in a range of 20–200 MeV. Depending on the DUT SEE response, we expanded the proton irradiation to 10^{11} protons/cm² if the numbers of event was not too high for the target fluence of the reference missions.

3.2. Test Site

The selected test site for the proton irradiation was the Kernfysisch Versneller Instituut (KVI), located on the Zernike Campus of the University of Groningen, Netherlands. To test again a wide range of proton energies, we decided to split the test purpose into two test campaigns to avoid high degrading of the primary beam energy.

3.2.1. Beam Energies

In the first configuration, we selected a primary beam energy of 190 MeV. The irradiation field was produced by scattering the primary proton beam using a double scatter foil method (1.44 mm Pb foil and a 0.9 mm W inhomogeneous scatter foil) located 3 m up stream of the irradiation position. The scatter foils together with beam optics and a 100 mm diameter collimator and the KVI-degraders determined the field at the DUT.

A 20 × 20 mm² rectangular collimator was inserted to protect sensitive equipment. Due to energy loss in air, beam intensity monitor and scatter foil, the maximum beam energy at the DUT position was 184 MeV. Lower energies were produced by inserting an amount of degrader material (Aluminum), as given in Table 1. In a second beam configuration, we used a proton beam with a primary energy of 66.5 MeV. Due to energy loss in the scatter foils, air, and beam ionization device, the maximum beam energy at the DUT was 59.5 MeV based on calculations with SRIM 2013 [14]. This energy is denoted as 60 MeV. By inserting a degrader material (Table 2), the beam energy at the DUT could be reduced. At low proton energies, the beam could hardly considered as mono-energetic, as it had have a very large energy spread. The field was produced using a 0.3 mm homogeneous lead scatter foil at 3 m from the DUT position. The field size was limited by a 50 mm diameter field collimator and a 30 × 30 mm² collimator in front of the DUT.

KVI establishes that one can measure the flux using a small plastic scintillation detector of 1 cm diameter placed at the position of the DUT.

Flux calibrations were measured for both the 20 × 20 mm² and the 30 × 30 mm² square collimator. The flux calibration values are listed in Tables 1 and 2. The statistical accuracy of these values was better than 1%. The systematic errors were estimated to be smaller than 10% on the basis of dose measurements, earlier measurements for different collimator sizes and aluminum activation analysis.

Table 1. Specification of the high energies at the DUT position (MeV) and resulting calibration factor (protons per cm² per monitor unit (MU)) for the desired configuration and the amount of degrader material (mm Aluminum) that needs to be used.

Nominal Energy [MeV]	Al Degrader [mm]	Calibration [Protons/cm²/MU]
184	0	218.57
150	31.5	191.07
120	55.5	162.11
100	69.5	134.57
70	86.5	98.79

Table 2. Specification of the low energies at the DUT position (MeV) and resulting calibration factor (protons per cm² per MU) for the desired configuration and the amount of degrader material (mm Aluminum) that needed to be used.

Nominal Energy [MeV]	Al Degrader [mm]	Calibration [Protons/cm²/MU]
60	0	97.55
50	4.0	93.68
40	7.5	89.17
30	10.5	80.10
25	11.7	75.34
20	12.7	63.85
15	13.5	56.46
10	14.2	55.67
7	14.5	53.56
4	14.7	50.94

3.2.2. Field Size and Homogeneity

The field uniformity was measured using a LANEX scintillation screen that was placed at the position of the DUT. The intensity of the scintillation light had a linear correlation to the fluence that was applied.

In general, a homogeneity of minimum 10% was desired for the collimator field size on 20 × 20 mm², or 30 mm × 30 mm², respectively. For high energy degradation (e.g., down to 10 MeV), 10% was not perfectly achieved. Assuming that the die of the DUT had a dimension of 500 × 500 µm (half the collimator size), this issue was determined to be negligible.

4. Test Setup And Procedures

In this section, firstly the test setup at the test site is presented. Secondly, the test procedures are discussed in detail, with the prioritization of different type of SEE and the required actions for functional recovery and to prevent permanent damages of the DUT and the test setup.

4.1. Test Setup

The schematic of the test setup is presented in Figure 4. An Ethernet connection was used from the control room to interface the DUT and the test equipment, which needed to be placed in the radiation area, close to the DUT to reduce cable losses and mismatch effects in the setup. On the other hand, due to scattering effects, the test equipment inside of the radiation area needed to be located as far away as possible from the beam line and the DUT.

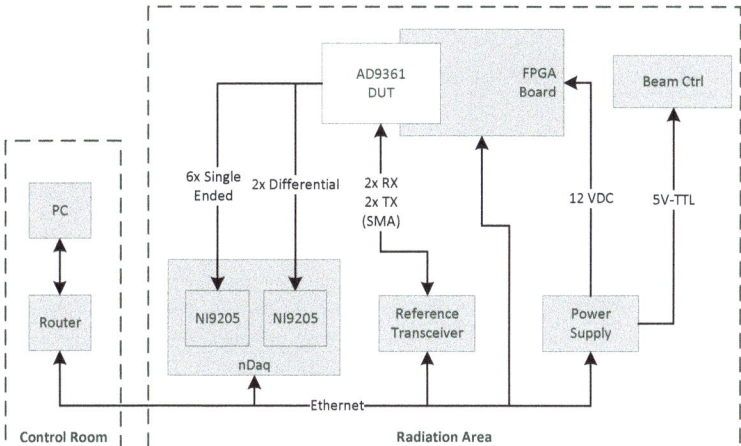

Figure 4. Schematic of the SEE test setup.

Additional shielding with lead and paraffin blocks would improve the test equipment safety. The distance from the DUT and the test equipment was approximately 12 m. The test setup consisted of the DUT, a FPGA board, a power supply and control unit (PCU), a reference transceiver, a voltage monitoring device (nDaq) and a control PC inside of the control room. The DUT itself was mounted on a test PCB, which interfaced with the FPGA board for data processing and power supply purposes. The FPGA in the setup was a system on chip from Xilinx (San José, USA) (Zynq-7000), which combines a FPGA fabric and a dual-core ARM processor [15]. The processor was used for the operating system (OS) and executed the test software. The FPGA board used a FMC (FPGA Mezzanine Card) connector to interface the DUT and provides an Ethernet interface to connect the DUT setup to the control equipment.

A picture of the top (Figure 5, left) and bottom (Figure 5, right) view of the test PCB, including the DUT (yellow box), is given in Figure 5. A major advantage of this PCB was the separated and isolated location of the DUT from other, known to be radiation sensitive devices (e.g., power supply devices). The radiation exposed area is highlighted with red (30×30 mm^2) and orange (20×20 mm^2) frames.

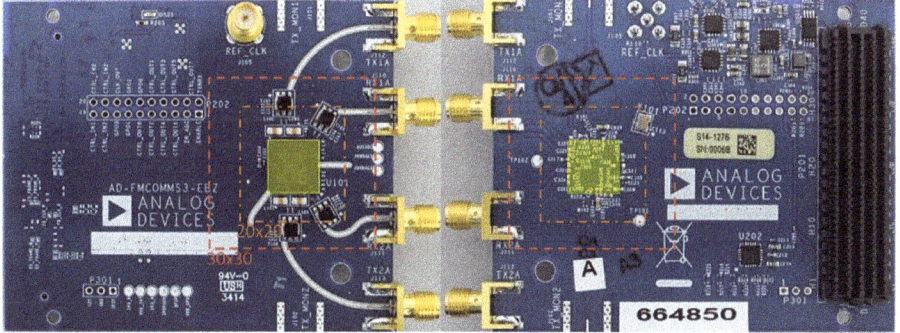

Figure 5. Top (**left**); and bottom (**right**) view of the DUT test board.

4.2. Test Procedures

The automatic test procedure for each test run is illustrated in the flow chart in Figure 6. In the beginning of each test run, the DUT was initialized and configured. A test bench was running, which captured the data from DUT. A register scrubbing and functional verification was performed in terms of soft SEE detection and to enable certain recovery processes. The signal processing and SEE monitoring were executed by a tailored OS, running on the FPGA board. Firstly, the DUT register configuration was scrubbed and then compared with the initial generated register values. Thus, register-based single event upset (SEU) or multiple bit upsets (MBU) could be detected. When a SEU or MBU occurred, the verification of the DUT functionality (by reading out the driver-depending configurations) was performed. If there was also an incongruity detected, the system tried to rewrite the function (by the driver) in a first step. If this reconfiguration failed, a re-initialization was triggered by a dedicated pin of the DUT. In summary, two types of single event failure interrupts (SEFI) were categorized: (a) SEFIs, which are recovered by reconfiguration; and (b) a recovering by re-initialization. A reconfiguration required a simple SPI commanding, while a re-initialization required also a recalibration procedure of the DUT ($\approx 10\times$ longer). The test was observed for different types of SEE. The most critical ones were single event latch-ups (SEL), which might lead to a destructive damage of the device. Thus, a shutdown or reboot was required in the case of an SEL event. The numbers of SEL events were counted by the control program in LabView. SELs were monitored by the voltage drop-off on a shunt resistor placed on each 1.3 V power rail of the DUT. The voltage was captured by an ADC module (NI9205/nDAQ—100,000 samples per second), which was analyzing the levels and performed a hard shutdown of the PCU output when a level was running out of the limit boundary.

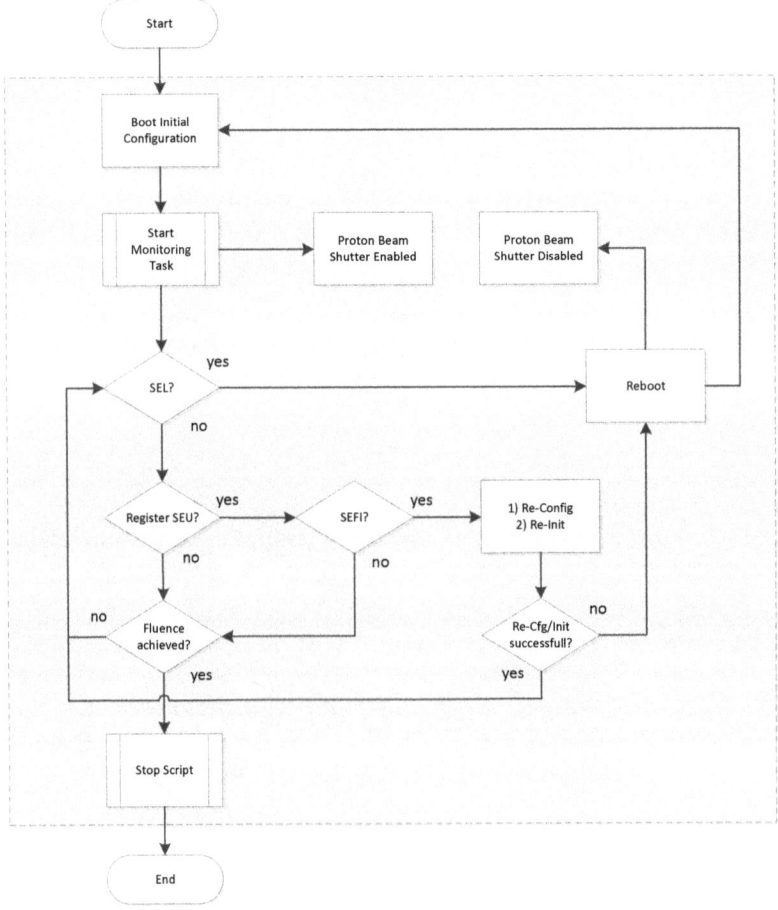

Figure 6. SEE test procedure flow chart.

After one second, the PCU enabled its output again and the system started rebooting. The beam stayed turned off during a SEL detection and was automatically enabled when the setup was functional again. The current limitation for the SEL detection was set to twice the nominal value (0.9 A) in a hold time for one second. SEUs, MBUs and SEFIs had second priority. The general software control architecture and flow is presented in Figure 7. A start script on the control PC enabled the power supply output for the DUT and controlled the beam shut down mechanism.

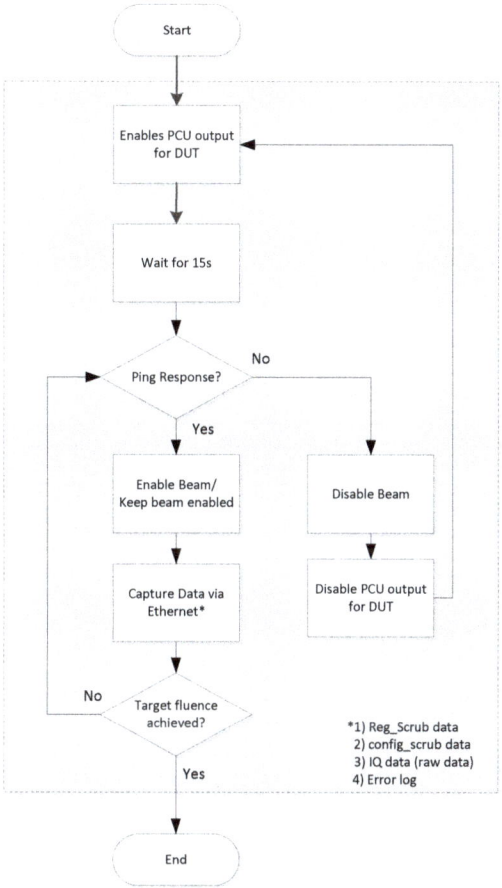

Figure 7. SEE software control flow chart.

A ping command to the DUT via Ethernet verified the connection and operational status. When the DUT was operational (10–15 s after power on), it responded to the ping request of the PC and the beam was enabled. The ping was then triggered every second and, if the response was interrupted, it was assumed that the OS crashed and the DUT was then power cycled. During the reboot process, the proton beam was disabled.

5. Experimental Results

In this section, the radiation test results are presented. Two samples, according to the recommendation of the ESCC test guideline No. 25100 [13], were exposed to proton irradiation at room temperature (19.5 °C) in air. The sample information, e.g., device code, serial numbers (SN) of the DUT board and the DUT itself, are given in Table 3.

Table 3. DUT sample information.

Sample #	DUT	Board SN#	DUT Lot#	DUT SN#	Fabricated in
1	AD9361BBCZ	00095	#1350	2769606.1	Singapore
2	AD9361BBCZ	733353	#1446	3014880.1	Singapore

Since the DUT is only manufactured on a single fabrication site and no changes in the fabrication process have been announced (public service by Analog Devices), it was assumed that the SEE response for both samples would be similar. As described in the test procedure in Section 4.2, different kinds of SEE were investigated with different categories and priority. The cross sections for SELs, SEUs, MBUs and SEFIs are presented in the following sections.

5.1. Single Event Latchup

Neither destructive nor non-destructive events of SELs were observed during the test with a proton energy of maximum 184 MeV.

5.2. Single Event Upset and Multiple Bit Upsets

In Tables 4 and 5, the SEU and MBU cross sections for both DUTs are presented. The cross sections (σ) are given for all tested energies in a range from 4 MeV to 184 MeV.

Table 4. SEU and MBU rate for DUT Sample 1.

Energy [MeV]	LET [MeV·cm^2·mg^{-1}]	Avg. Flux [#·cm^{-2}·s^{-1}]	Fluence [#·cm^{-2}]	SEU [#]	σ_{SEU} [cm^2]	MBU [#]	σ_{MBU} [cm^2]
184	3.83×10^{-3}	9.11×10^{7}	1.00×10^{11}	15	1.50×10^{-10}	1	1.00×10^{-11}
150	4.38×10^{-3}	9.48×10^{7}	1.00×10^{11}	24	2.40×10^{-10}	1	1.00×10^{-11}
120	5.84×10^{-3}	6.85×10^{7}	1.00×10^{11}	9	9.00×10^{-11}	0	0
100	7.61×10^{-3}	6.80×10^{7}	1.00×10^{11}	16	1.60×10^{-10}	0	0
60	8.56×10^{-3}	9.94×10^{8}	1.00×10^{11}	13	1.30×10^{-10}	0	0
50	9.85×10^{-3}	9.56×10^{8}	1.00×10^{11}	20	2.00×10^{-10}	1	1.00×10^{-11}
40	1.17×10^{-2}	9.09×10^{8}	1.00×10^{11}	15	1.50×10^{-10}	0	0
30	1.47×10^{-2}	8.47×10^{8}	1.00×10^{11}	10	1.00×10^{-10}	0	0
20	2.02×10^{-2}	7.81×10^{8}	1.00×10^{11}	9	9.00×10^{-11}	0	0
10	3.46×10^{-2}	6.00×10^{8}	1.00×10^{11}	11	1.10×10^{-10}	1	1.00×10^{-11}
7	4.53×10^{-2}	5.41×10^{8}	1.00×10^{11}	10	1.10×10^{-10}	1	1.00×10^{-11}
4	6.86×10^{-2}	5.50×10^{8}	1.00×10^{11}	17	1.70×10^{-10}	6	6.00×10^{-11}

The target fluence for each energy was set to 10^{11} protons/cm^2, since the SEE response was too low for the selected reference mission parameters in Figure 3.

Depending on the selected proton beam energy, the flux was between 6.8×10^7 and 1×10^9 protons/cm^2/s. The numbers of SEU counted ranged 4–36 bit-flips of the 8-bit configuration registers. The event rate did not monotonically increase with energy in the studied range. Thus, a threshold or saturation energy could not be determined. MBUs were counted if multiple bit-flips of a DUT configuration register were detected. The numbers of events were independent of the numbers of bits-flips (at least two bits). The MBUs counted for both DUTs were in the range from 0 to 10. A detailed evaluation of the cross sections for the SEUs and MBUs is given in Section 6.

The cross section values for both SEU and MBU events as a function of the incident proton energy is presented in Figure 8. The SEU events are presented in Figure 8a and the MBU events in Figure 8b. Sample 1 is in blue while Sample 2 is represented in green.

Standard deviation of the proton energy at the DUT site was provided by the facility and the values depended on the energy of the primary beam, thickness of the degrader, and the distance of the degrader to the DUT. The values are presented in Table 6. The two primary beams of energy 60 MeV and 184 MeV are marked by bold font.

Table 5. SEU and MBU rate for DUT Sample 2.

Energy [MeV]	LET [MeV·cm²·mg⁻¹]	Avg. Flux [#·cm⁻²·s⁻¹]	Fluence [#·cm⁻²]	SEU [#]	σ_{SEU} [cm²]	MBU [#]	σ_{MBU} [cm²]
184	3.83×10^{-3}	1.08×10^{8}	1.00×10^{11}	30	3.00×10^{-10}	10	1.00×10^{-10}
150	4.38×10^{-3}	8.76×10^{7}	1.00×10^{11}	11	1.10×10^{-10}	1	1.00×10^{-11}
120	5.84×10^{-3}	8.73×10^{7}	1.00×10^{11}	18	1.80×10^{-10}	3	3.00×10^{-11}
100	7.61×10^{-3}	5.00×10^{7}	1.00×10^{11}	11	1.10×10^{-10}	0	0
60	8.56×10^{-3}	9.88×10^{8}	1.00×10^{11}	25	2.50×10^{-10}	2	2.00×10^{-11}
50	9.85×10^{-3}	1.01×10^{9}	1.00×10^{11}	9	9.00×10^{-11}	0	0
40	1.17×10^{-2}	9.54×10^{8}	1.00×10^{11}	8	8.00×10^{-11}	0	0
30	1.47×10^{-2}	8.57×10^{8}	1.00×10^{11}	17	1.70×10^{-10}	1	1.00×10^{-11}
20	2.02×10^{-2}	6.87×10^{8}	1.00×10^{11}	36	3.60×10^{-10}	5	5.00×10^{-11}
10	3.46×10^{-2}	5.81×10^{8}	1.00×10^{11}	27	2.70×10^{-10}	3	3.00×10^{-11}
7	4.53×10^{-2}	5.38×10^{8}	1.00×10^{11}	24	2.40×10^{-10}	5	5.00×10^{-11}
4	6.86×10^{-2}	5.19×10^{8}	1.00×10^{11}	4	4.00×10^{-11}	0	0

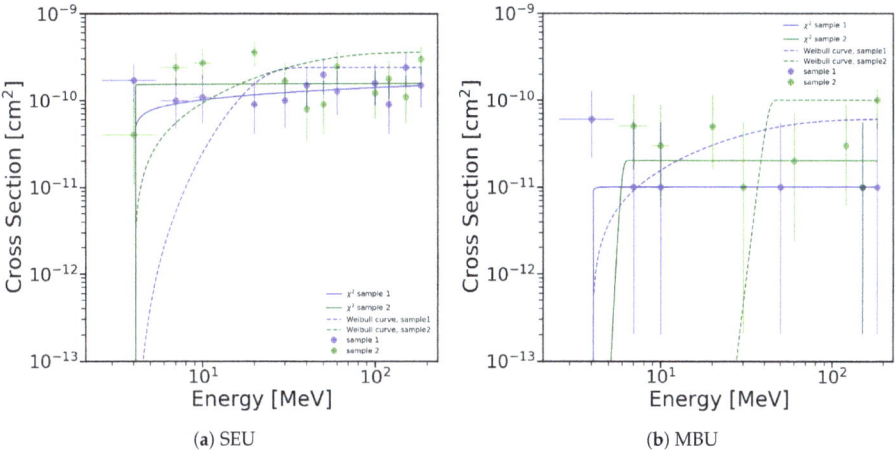

(a) SEU

(b) MBU

Figure 8. SEU and MBU cross section plots together with the Weibull fit function.

Table 6. Proton energy deviation δE at the DUT site.

Energy [MeV]	Degrader (Al) Thickness [mm]	δE [MeV]
184	0	0.58
150	31.5	0.90
120	55.6	1.00
100	69.5	1.18
60	0	0.44
50	4.0	0.57
40	7.5	0.66
30	10.5	0.77
20	12.7	0.90
10	14.2	1.2
7	14.5	1.3
4	14.7	1.4

Uncertainty of the cross section was calculated according to recommendations given in the ESCC standard [13]. It is a product of uncertainty of number of events and proton fluence:

$$\frac{\delta\sigma_{\text{lower, upper}}}{\sigma} = \sqrt{\left(\frac{\delta N_{\text{events lower, upper}}}{N_{\text{events}}}\right)^2 + \left(\frac{\delta F}{F}\right)^2}. \qquad (1)$$

The facility states that the fluence uncertainty is better than 10% and this number was used for calculations. Deviation in the number of events δN_{events} were calculated using the chi-square distribution function for a given confidence level, which, as recommended by the ESCC standard, was taken as 95%. Accordingly, the left-tailed and the right-tailed part of the function were used to determine the lower and the upper number of the events. Then, the lower and upper deviation of number of events was calculated by subtraction of the lower and upper events number and number of events measured during the irradiation test, i.e.,

$$\delta N_{\text{events lower}} = N_{\text{events}}^{\text{measured}} - N_{\text{events lower}}, \quad \delta N_{\text{events upper}} = N_{\text{events upper}} - N_{\text{events}}^{\text{measured}}. \qquad (2)$$

Then, both numbers, $\delta N_{\text{events lower}}$ and $\delta N_{\text{events upper}}$, were used together with Equation (1) to calculate the $\delta\sigma_{\text{lower}}$ and $\delta\sigma_{\text{upper}}$.

The cross section values were then used to fit the so-called Weibull function into the data:

$$\sigma(x) = \sigma_{\text{SAT}}\left[1 - exp\left(-\frac{x - x_0}{W}\right)^S\right], \qquad (3)$$

where σ_{SAT} is a saturation cross section, x states for the proton energy, x_0 is the proton energy threshold, W is the so-called width parameter given in a units of MeV, and S is the unit-less exponent parameter.

The proton energy threshold for all of the curve fit calculations was taken as the lowest proton energy which returns at least one event effect. The remaining three Weibull function parameters were used to fit the curve to the data. A χ^2 test was used for fitting procedure [16]:

$$\chi^2 = \sum_{i=1}^{N} \frac{(Model - Measurement_i)^2}{Measurement_i}. \qquad (4)$$

Here, *Model* represents the $\sigma(x)$ described by the Weibull function, *Measurement$_i$* represents the ith cross section value for examined event type, and N represents number of cross sections taken into account. The smaller is the χ^2 value, the better does the *Model* represent the experimental data. Table 7 contains all four parameters together with the χ^2 values:

Table 7. Weibull function parameters for SEU and MBU event types based on the χ^2 method.

Sample No.	Type of Events	σ_{SAT} [cm^2]	x_0 [MeV]	W	S	χ^2
1	SEU	1.70×10^{-10}	4.0	8.91	0.24	2.81×10^{-10}
2	SEU	2.40×10^{-10}	4.0	0.21	0.01	4.94×10^{-10}
1	MBU	1.00×10^{-11}	4.0	0.39	0.01	0.60×10^{-10}
2	MBU	2.00×10^{-11}	7.0	9.91	2.01	1.59×10^{-10}

An alternative method to fit the Weibull curve, Equation (3), was used for the dataset. OMERE software (in version 5.2.4) has an algorithm that can determine the W and S parameters of the curve [17]. They are presented by dashed lines in Figure 8. Table 8 contains the fit parameters.

Table 8. Weibull function parameters for SEU and MBU event types based on the OMERE software algorithm.

Sample No.	Type of Events	σ_{SAT} [cm^2]	x_0 [MeV]	W	S
1	SEU	2.40×10^{-10}	4.0	15.78	2.295
2	SEU	3.60×10^{-10}	4.0	24.33	0.86
1	MBU	6.00×10^{-11}	4.0	24.33	0.86
2	MBU	1.00×10^{-10}	7.0	34.85	13.12

Both methods gave slightly different results. The χ^2 method fit the curve more according to the data while the OMERE software algorithm seemed to omit cross sections for the lowest tested energies.

5.3. Single Event Failure Interrupt

Compared to the SEU and MBU events, SEFIs are more important, since the device loses its initial functionality and a recovery process needs to be performed. For SEFIs, two categories were defined, as already described in Section 4.2: (1) reconfiguration (SEFI$_{CFG}$); and (2) re-initialization (SEFI$_{INIT}$). The cross sections and SEFI event counts are presented in Table 9 for DUT Sample 1 and Table 10 for DUT Sample 2.

Table 9. SEFI rate for DUT Sample 1.

Energy [MeV]	LET [MeV·cm^2·mg^{-1}]	Avg. Flux [#·cm^{-2}·s^{-1}]	Fluence [#·cm^{-2}]	SEFI$_{CFG}$ [#]	σ_{CFG} [cm^2]	SEFI$_{INIT}$ [#]	σ_{INIT} [cm^2]
184	3.83×10^{-3}	9.11×10^7	1.00×10^{11}	0	0	0	0
150	4.38×10^{-3}	9.48×10^7	1.00×10^{11}	1	1.00×10^{-11}	0	0
120	5.84×10^{-3}	6.85×10^7	1.00×10^{11}	0	0	1	1.00×10^{-11}
100	7.61×10^{-3}	6.80×10^7	1.00×10^{11}	1	1.00×10^{-11}	0	0
60	8.56×10^{-3}	9.94×10^8	1.00×10^{11}	1	1.00×10^{-11}	0	0
50	9.85×10^{-3}	9.56×10^8	1.00×10^{11}	2	2.00×10^{-11}	0	0
40	1.17×10^{-2}	9.09×10^8	1.00×10^{11}	2	2.00×10^{-11}	1	1.00×10^{-11}
30	1.47×10^{-2}	8.47×10^8	1.00×10^{11}	0	0	0	0
20	2.02×10^{-2}	7.81×10^8	1.00×10^{11}	0	0	0	0
10	3.46×10^{-2}	6.00×10^8	1.00×10^{11}	2	2.00×10^{-11}	0	0
7	4.53×10^{-2}	5.41×10^8	1.00×10^{11}	1	1.00×10^{-11}	0	0
4	6.86×10^{-2}	5.50×10^8	1.00×10^{11}	1	1.00×10^{-11}	0	0

SEFIs emerge quite rarely compared to SEUs. We observed that SEFIs were only caused by a SEU or MBU event and never occurred randomly. In most cases, the DUTs were reconfigurable and no re-initialization was required. The SEFI rates for reconfiguration ranged 0–6 events.

Table 10. SEFI rate for DUT Sample 2.

Energy [MeV]	LET [MeV·cm^2·mg^{-1}]	Avg. Flux [#·cm^{-2}·s^{-1}]	Fluence [#·cm^{-2}]	SEFI$_{CFG}$ [#]	σ_{CFG} [cm^2]	SEFI$_{INIT}$ [#]	σ_{INIT} [cm^2]
184	3.83×10^{-3}	1.08×10^8	1.00×10^{11}	3	3.00×10^{-11}	0	0
150	4.38×10^{-3}	8.76×10^7	1.00×10^{11}	2	2.00×10^{-11}	0	0
120	5.84×10^{-3}	8.73×10^7	1.00×10^{11}	3	3.00×10^{-11}	1	1.00×10^{-11}
100	7.61×10^{-3}	5.00×10^8	1.00×10^{11}	2	2.00×10^{-11}	0	0
60	8.56×10^{-3}	9.88×10^8	1.00×10^{11}	0	0	0	0
50	9.85×10^{-3}	1.01×10^9	1.00×10^{11}	3	3.00×10^{-11}	0	0
40	1.17×10^{-2}	9.54×10^8	1.00×10^{11}	3	3.00×10^{-11}	0	0
30	1.47×10^{-2}	8.57×10^8	1.00×10^{11}	1	1.00×10^{-11}	0	0
20	2.02×10^{-2}	6.87×10^8	1.00×10^{11}	6	6.00×10^{-11}	0	0
10	3.46×10^{-2}	5.81×10^8	1.00×10^{11}	1	1.00×10^{-11}	0	0
7	4.53×10^{-2}	5.38×10^8	1.00×10^{11}	0	0	0	0
4	6.86×10^{-2}	5.19×10^8	1.00×10^{11}	0	0	0	0

Only three re-initializations were required over the full energy range and on both DUTs. A more detailed discussion about the SEFI rate and their interpretation is presented in Section 6.

The cross section of the SEFI$_{CFG}$ as a function of proton energy for both samples is shown in Figure 9. The SEFI$_{INIT}$ cross sections are not plotted, since Sample 1 indicates two and Sample 2 one event throughout the whole energy range. The Weibull function was fitted to the data using the same χ^2-test procedure as presented for the SEU and MBU events.

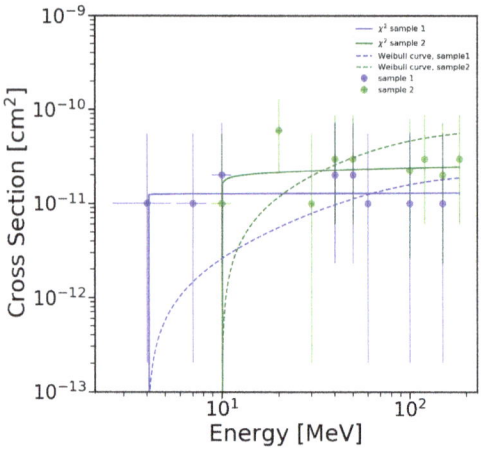

Figure 9. SEFI cross section together with the Weibull fit function.

Table 11 contains the Weibull function parameters and χ^2 values. In addition, as for the SEU and MBU event type, the OMERE software was used to fit the curve into the dataset. Table 12 contains the curve fit parameters. The curves for both of the samples are plotted using the dashed lines.

Table 11. Weibull function parameters for SEFI$_{CFG}$ event type based on the χ^2 method.

Sample No.	Type of Events	σ_{SAT} [cm^2]	x_0 [MeV]	W	S	χ^2
1	SEFI$_{CFG}$	1.99×10^{-11}	4.0	7.01	0.01	0.21×10^{-10}
2	SEFI$_{CFG}$	3.00×10^{-11}	10.0	1.41	0.11	0.56×10^{-10}

As for the SEU and the MBU event types, the fit of the OMERE function made by the OMERE software algorithm seemed to omit the cross sections recorded for the low energy protons.

Table 12. Weibull function parameters for SEFI$_{CFG}$ event type fit by the OMERE software algorithm.

Sample No.	Type of Events	σ_{SAT} [cm^2]	x_0 [MeV]	W	S
1	SEFI$_{CFG}$	2.00×10^{-11}	4.0	58.40	0.87
2	SEFI$_{CFG}$	6.00×10^{-11}	10.0	58.40	0.87

6. Analysis of The Results

In this section, we discuss the SEE test results presented in Section 5. Additionally, the SEE cross sections are correlated to dedicated space missions, which are usually intended to be the reference missions for the test requirements. The corresponding SEE event rates are presented.

6.1. See Test Result Interpretation

Different kinds of SEE events are discussed and interpreted. We observed several SEUs, MBUs and rare SEFIs during the test campaign. Below, error logs for different kinds of events are presented.

Listing 1: Error log (SEU/MBU) cutout for Sample 1, run 10

```
- SNIP -
Time: 1514764812      Register: 250    New Value: e2    Old Value: ea    Reg Counter: 1    SEU Counter: 1
Time: 1514764821      Register: 481    New Value: 19    Old Value: 8     Reg Counter: 2    SEU Counter: 3
- SNIP -
Time: 1514764838      Register: 583    New Value: 82    Old Value: 2     Reg Counter: 3    SEU Counter: 5
Time: 1514764838      Register: 566    New Value: 18    Old Value: 58    Reg Counter: 4    SEU Counter: 6
- SNIP -
```

In Listing 1, a cutout of the console output for a SEU and a MBU event is presented. As mentioned above, a MBU was counted if at least two bits inside of the 8-bit configuration registers were flipped. In the above error log (Listing 1), cutout (Sample 1, Run 10) at *Time: 1514764812*, firstly, one SEU occurred and afterwards (*Time: 1514764821*) we observed that on register 481_{dec} the value changed from 8_{hex} $(0000\ 1000)_b$ to 19_{hex} $(0001\ 1001)_b$. Thus, two bit-flips were detected and clarified as one MBU. In Listing 2, an error log cutout (Sample 1, Run 12) for a reconfiguration SEFI is presented.

Listing 2: Error Log (SEFI - Re-Config)

```
- SNIP -
Time: 1514764831         Register: 562    New Value: 4    Old Value: 0    Reg Counter: 2    SEU Counter: 2
Error in Register Function
out\_altvoltage0 RX LO frequency Old Line: 2450000000
New Line: 8589934590
*** Re-Configuration successful ***
- SNIP -
```

Firstly, a SEU in register 562_{dec} (part of the RX synthesizer registers) occurred, which caused a change in the receiver local oscillator frequency (RX LO frequency). This SEFI could be recovered by a simple reconfiguration, commanding the initial RX LO frequency to the device. In some minor cases, we observed that even a reconfiguration was not successful to recover the DUT initial functionality. A snip of the error log, including a re-initialization required SEFI, is presented in Listing 3.

Listing 3: SEFI - Re-Init

```
- SNIP -
Time: 1514764849      Register: 347    New Value: 20    Old Value: 0    Reg Counter: 9     SEU Counter: 9
Time: 1514764849      Register: 410    New Value: 0     Old Value: 1    Reg Counter: 10    SEU Counter: 10
Error in Register Function
out\_altvoltage1 TX LO fastlock save Old Line: 0 87,247,119,52,23,39,23,23,70,255,159,3,127,29,31,63
New Line: 0 87,247,119,52,23,39,23,23,70,255,159,3,127,29,27,63
Error in Register Function
out\_altvoltage1 TX LO fastlock save Old Line: 0 87,247,119,52,23,39,23,23,70,255,159,3,127,29,31,63
New Line: 0 87,247,119,52,23,39,23,23,70,255,159,3,127,29,27,63
*** Re-Configuration failed, performing Re-Initialization ***
*** Re-Initialization successful ***
- SNIP -
```

Firstly, two SEUs were detected in register 347_{dec} and 410_{dec}. Register 347_{dec} is an open register for the receive signal strength indicator (RSSI) measurement and is not declared to be responsible for the following SEFI. Register 410_{dec} is part of the Rx BB DC Offset register configuration and its SEU changed the function of the TX LO fastlock saving. The TX LO fastlock saving stores the parameters for the TX synthesizer. After detection of the SEFI, a reconfiguration was performed. At this specific event, a re-initialization was required after the reconfiguration attempt failed. The initial setting were restored successfully and the DUT was functional again.

6.2. Event Rate Calculation

The worst case event rate is defined as a product of maximum calculated cross section of examined event type and proton flux which corresponds to minimum proton energy for at least one event. It is the co-called threshold energy E_{th}. The flux, however, was taken from radiation analysis of reference missions. A calculation was made to estimate maximum number of possible events per year. Schematically, the rate can be represented as:

$$Rate = flux(E_{th}) \cdot \sigma_{max}. \tag{5}$$

For SEU events, the highest cross section was recorded for the second sample at 3.6×10^{-10} cm^2. The threshold energy was 4.0 MeV. Table 13 contains a short reference mission description, the corresponding flux of trapped protons and GCR for the mentioned E_{th} and the SEU rate given in number of failures per device per year. For the trapped protons, the highest rate of 4.4 can be expected for a reference mission scenario at 800 km altitude, minimum Sun activity, and orbit inclination angle of 51.64°. The lowest rate of ≈0.1 can be expected at 400 km altitude, maximum Sun activity and inclinations angle of $\alpha = \{98°, 51.56°\}$. Therefore, at ISS orbit and maximum Sun activity, one could expect ≈1 SEU event after 10 years of mission period. The GCR input to the SEU events is expected to be negligibly small, only one SEU event caused by the GCRs would happen after 100 years of DUT operation in the LEO environment.

Table 13. SEU event rates for considered reference missions.

Reference Mission	Proton Flux [cm^{-2}s^{-1}]	SEU Rate [Failure/Device/Year]
400 km, Min, $\alpha = 98°$	17	0.19
400 km, Max, $\alpha = 98°$	7	0.08
400 km, Min, ISS, $\alpha = 51,64°$	31	0.35
400 km, Max, ISS, $\alpha = 51,64°$	10	0.11
800 km, Min, $\alpha = 98°$	243	2.76
800 km, Max, $\alpha = 98°$	190	2.16
800 km, Min, $\alpha = 51,64°$	388	4.40
800 km, Max, $\alpha = 51,64°$	301	3.42
800 km, $\alpha = 0°$	50	0.57
GCR	1	0.01

The highest cross section for the MBU andr the SEFI$_{CFG}$ events was 6.0×10^{-11} cm^2 and it was recorded for the first and the second sample, respectively. It was six times smaller than the SEU events. Therefore, the corresponding number of event rates was lowered by the same factor. The energy threshold for the MBU events was 4.0 MeV. Table 14 contains, as for the SEU events, reference mission description, flux of trapped protons and GCR for the E_{th}, and the corresponding event rate. For the trapped protons, the highest event rate of 0.73 per year can be expected for a reference mission of 800 km altitude, minimum Sun activity, and inclination angle of 51.64°. Only one MBU and one SEFI$_{CFG}$ event per 100 years can be expected at 800 km altitude, maximum Sun activity and inclination angle of 98°. For the ISS orbit, the DUT would indicate 2–6 MBU and SEFI$_{CFG}$ events per 100 years.

Table 14. MBU and SEFI$_{CFG}$ event rates for considered reference missions.

Reference Mission	Proton Flux [cm^{-2}s^{-1}]	MBU and SEFI$_{CFG}$ Rate [Failure/Device/Year]
400 km, Min, $\alpha = 98°$	17	0.03
400 km, Max, $\alpha = 98°$	7	0.01
400 km, Min, ISS, $\alpha = 51,64°$	31	0.06
400 km, Max, ISS, $\alpha = 51,64°$	10	0.02
800 km, Min, $\alpha = 98°$	243	0.46
800 km, Max, $\alpha = 98°$	190	0.36
800 km, Min, $\alpha = 51,64°$	388	0.73
800 km, Max, $\alpha = 51,64°$	301	0.57
800 km, $\alpha = 0°$	50	0.09
GCR	1	0.002

Only three SEFI$_{INIT}$ events were recorded for two test samples and within the whole considered energy range. The single event corresponded to a cross section of 1.0×10^{-11} cm^2. Energy threshold for the SEFI$_{INIT}$ events was 20.0 MeV. Since E_{th} was much larger than for the other event types, the corresponding flux of trapped protons and GCR was also much lower (see Table 15). For all of the considered reference missions, a time period of at least 50 years would result with recognizable number of events. For the GCR, a time period of 3000 years would be needed to generate one SEFI$_{INIT}$ event.

Table 15. SEFI$_{INIT}$ event rates for considered reference missions.

Reference Mission	Proton Flux [cm^{-2}s^{-1}]	SEFI$_{INIT}$ Rate [Failure/Device/Year]
400 km, Min, $\alpha = 98°$	7	0.002
400 km, Max, $\alpha = 98°$	4	0.001
400 km, Min, ISS, $\alpha = 51.64°$	10	0.003
400 km, Max, ISS, $\alpha = 51.64°$	6	0.002
800 km, Min, $\alpha = 98°$	104	0.032
800 km, Max, $\alpha = 98°$	86	0.027
800 km, Min, $\alpha = 51.64°$	149	0.047
800 km, Max, $\alpha = 51.64°$	123	0.039
800 km, $\alpha = 0°$	40	0.013
GCR	1	0.0003

6.3. Further Detected Abnormalities

During the test, we observed some abnormalities, which are described in this section. These abnormalities include SEU-based current condition changes of the DUT and SEUs in masked registers, which are changing their value continuously without any radiation-based event.

6.3.1. Influence of a SEU to the DUT Current Conditions

Even though we did not observed SELs or other destructive events, there were some abnormalities in the supply voltage domain that should be discussed. As expected, we observed some changes in the current condition of the DUT when a SEE occurred. We observed that conducted current could change with a single SEU, without triggering a SEFI of the DUT. An example for such an event, a SEU-based current change, is presented in Figure 10a for Sample 1 on the sixth test run with 100 MeV. The current on power rail A dropped marginally, whereas the current on power rail B increased by 100 mA. This was not declared as critical, but somehow interesting, since no obvious malfunctions were observed.

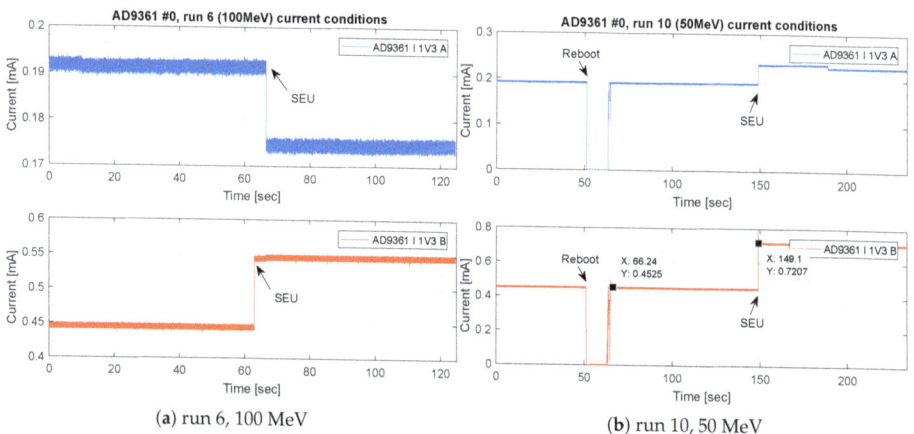

(a) run 6, 100 MeV   (b) run 10, 50 MeV

Figure 10. Current conditions on the DUT 1.3V power rails (a,b), after SEU events (Sample 1, Runs 6 and 10).

As shown in Figure 10b, we recorded the current condition of both power rails on a SEU event for Run 10 at 50 MeV. The initial current for power rail B was 450 mA. The SEU event, which also did not force a SEFI, led to an increased current of 720 mA. This current boost was also observed for power rail A. This phenomenon was not classified as a SEL, since we assumed that an SEL would have a longer rise time instead of the observed jump on the current value. Even though the device was specified to a maximum current of 1 A, we would recommend performing a reconfiguration or even a reboot, to avoid thermal stress of the DUT, particularly in vacuum. Figure 10b also presents a power cycle, performed on the FPGA board. We observed that, even with the use of collimator and additional lead brick in front of the FPGA board, the OS running on the FPGA board crashed and needed to be power cycled. The numbers of power cycles of the FPGA board were observed to increase as the proton beam energy became more degraded. We assumed that, by the degrading and the collimator, generated particles such as neutrons were hitting the FPGA through the shielding and forces a system crash. Since we were controlling the beam activity with the OS functionality on the FPGA, a correct total achieved fluence on the DUT could be ensured.

6.3.2. SEU in Masked/Non-Scrubbed Configuration Registers

During test preparation, we recognized that several configuration registers were changing their values independently of a radiation effect. These registers are so-called *masked-registers*. As an example, the register values/status over time for the RX phase and gain correction is presented in Figure 11.

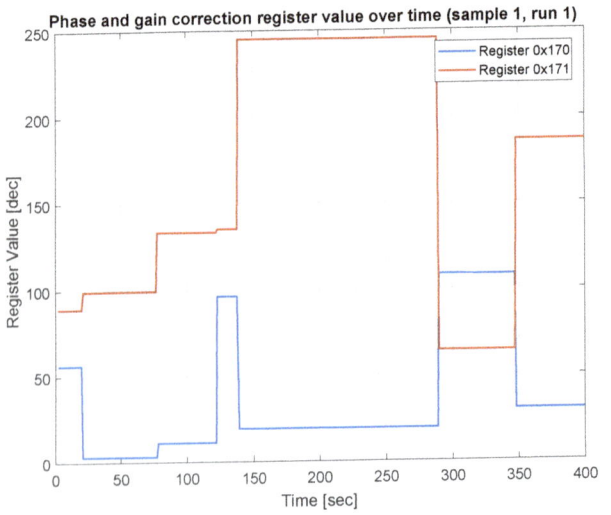

Figure 11. Masked configuration register (0x170 and 0x171) for gain and phase correction.

Since there is no periodical behavior visible, and no short range of the register values to define an upper and lower threshold was found, it was almost impossible to determine an SEU inside of these registers.

Another example for continuous alternation of the values is the configuration register responsible for an internal temperature sensor. In Figure 12a, the nominal behavior for the temperature registers of the DUT is shown (Sample 1, Run 1). In Run 2 for Sample 1, we observed a drop from 41_{dec} to 34_{dec}, as presented in Figure 12b. Such a temperature drop was unlikely for the DUT and was thus declared as a SEU in the masked-register 0xE.

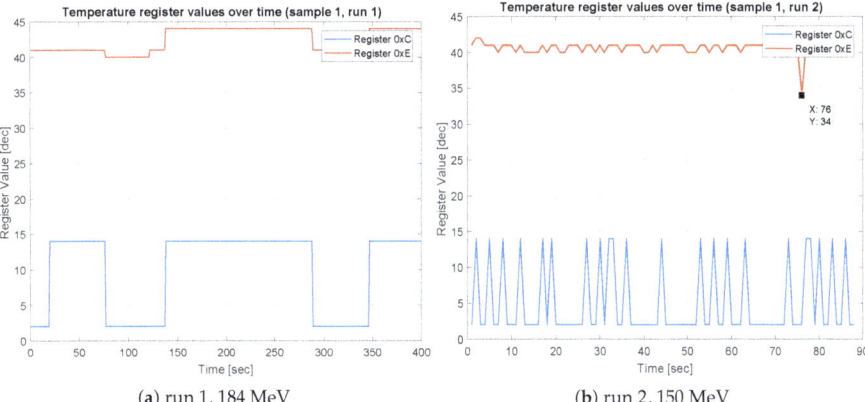

Figure 12. Temperature register values nominal (**a**); and with bit-flip event (**b**) (Sample 1, Runs 1 and 2).

7. Conclusions

In this paper, we present the proton induced SEE characterization of a highly integrated RF transceiver (AD9361). The DUT was fabricated on a 65 nm CMOS process and was therefore categorized to be sensitive of proton irradiation. Due to the DUT complexity, a special test approach/method was required to classify different kinds of events. The exposed proton energies were split into two test campaigns, to induce high energy protons (up to 184 MeV) and low energy protons down to 4 MeV. The results show a very low response to proton irradiation, independent of the proton energy. The total fluence of 1.00×10^{11} #·cm^{-2}, however, was not enough to achieve a number of failures for desirable error statistic ($\geq$100 failures). Two DUTs was tested and the results show a similar behavior. The worst case event rate calculations show that, depending on different type of reference missions/orbits, a SEU or MBU is expected to occur once in 10 years and a SEFI around every 100 years in LEO. Further activity will involve the SEE characterization of the DUT on heavy ion irradiation. Additionally, we also focus on the evaluation of the transceivers integrated ADCs and DACs, which also might be affected by radiation (SEUs, single event effects transients and SEFIs) and has not been taken into account during this test campaign.

Author Contributions: Main idea, J.B.; conceptualization, J.B. and M.P.J.; test requirements, J.B. and M.S.; test method and procedures, J.B. and M.P.J.; testing and results, J.B., M.S. and M.P.J.; event rate calculation, M.S.; and writing, review and editing, J.B., M.P.J. and M.S.

Funding: This research received no external funding.

Conflicts of Interest: The authors declare no conflict of interest.

References

1. Budroweit, J. Design of a Highly Integrated and Reliable SDR Platform for Multiple RF Applications on Spacecrafts. In Proceedings of the GLOBECOM 2017—2017 IEEE Global Communications Conference, Singapore, 4–8 December 2017; pp. 1–6. [CrossRef]
2. Budroweit, J.; Koelpin, A. Design challenges of a highly integrated SDR platform for multiband spacecraft applications in radiation enviroments. In Proceedings of the 2018 IEEE Topical Workshop on Internet of Space (TWIOS), Anaheim, CA, USA, 14–17 January 2018; pp. 9–12. [CrossRef]
3. Crowne, M.J.; Haskins, C.B.; Wallis, R.E.; Royster, D.W. Demonstrating TRL-6 on the JHU/APL Frontier Radio for the Radiation Belt Storm Probe mission. In Proceedings of the 2011 Aerospace Conference, Big Sky, MT, USA, 5–12 March 2011; pp. 1–8. [CrossRef]

4. O'Neill, M.B.; Millard, W.P.; Bubnash, B.M.; Mitch, R.H.; Boye, J.A. Frontier Radio Lite: A Single-Board Software-Defined Radio for Demanding Small Satellite Missions. In Proceedings of the 30th Annual AIAA/USU Conference on Small Satellites, SSC16-VII-2, Logan, UT, USA, 8–13 August 2016.
5. Pu, D.; Cozma, A.; Hill, T. Four Quick Steps to Production: Using Model-Based Design for Software-Defined Radio. *Analog. Dialogue* **2015**, *49*, 1–5.
6. AD9361 Material Declaration. Available online: https://www.analog.com/media/en/package-pcb-resources/material-declaration/cspbga/cspbga_10x10(bc-144-7).pdf (accessed on 25 April 2019).
7. Vettel, J.I. *The AE-8 Trapped Electron Model Environment*; NSSDC/WDC-A-R&S, 91-24; NASA: Greenbelt, MD, USA, 1991.
8. Vettel, J.I. *The NASA/National Space Science Data Center Trapped Radiation Environment Model Program (1964–1991)*; NSSDC/WDC-A-R&S 91-29; NASA: Greenbelt, MD, USA, 1991.
9. Jordan, C.E. NASA Radiation Belt Models AP-8 and AE-8. *Sci. Rep.* **1986**, *1*, 1–20.
10. Tylka, A.J.; Adams, J.H.; Boberg, P.R.; Brownstein, B.; Dietrich, W.F.; Flueckiger, E.O.; Petersen, E.L.; Shea, M.A.; Smart, D.F.; Smith, E.C. CREME96: A Revision of the Cosmic Ray Effects on Micro-Electronics Code. *IEEE Trans. Nucl. Sci.* **1997**, *44*, 2150–2160. [CrossRef]
11. Holmes-Siedle, A.; Adams, L. Radiation environments. In *Handbook of Radiation Effects*; Oxford University Press: Oxford, UK, 2007; pp. 18–19.
12. Ziegler, J.F.; Ziegler, M.D.; Biersack, J.P.; SRIM—The stopping and range of ions in matter (2010). *Nucl. Instrum. Methods Phys. Res. Sect. B Beam Interact. Mater. Atoms.* **2010**, *268*, 1818–1823. [CrossRef]
13. ESA. *Single Event Effects Test Method and Guidelines, ESCC Basic Specification No. 25100*; Issue 2; ESA: Paris, France, 2014.
14. The Stopping and Range of Ions in Matter (SRIM). Available online: http://www.srim.org/index.htm#SRIMMENU (accessed on 15 March 2019).
15. Zynq-7000 SoC Product Advantages. Available online: https://www.xilinx.com/products/silicon-devices/soc/zynq-7000.html (accessed on 25 April 2019).
16. Brandt, S. The chi2-Test for Goodness-of-Fit. In *Data Analysis. Statistical and Computational Methods for Scientists and Engineers*; Springer International Publishing: London, UK, 2014; pp. 199–200.
17. The OMERE Software. Available online: http://www.trad.fr/en/space/omere-software/ (accessed on 25 April 2019).

© 2019 by the authors. Licensee MDPI, Basel, Switzerland. This article is an open access article distributed under the terms and conditions of the Creative Commons Attribution (CC BY) license (http://creativecommons.org/licenses/by/4.0/).

Article

Radiation Assessment of a 15.6 ps Single-Shot Time-to-Digital Converter in Terms of TID

Bjorn Van Bockel *, Jeffrey Prinzie and Paul Leroux

Department of Electrical Engineering (ESAT), KU Leuven, 3000 Leuven, Belgium; jeffrey.prinzie@kuleuven.be (J.P.); paul.leroux@kuleuven.be (P.L.)
* Correspondence: bjorn.vanbockel@kuleuven.be

Received: 25 April 2019; Accepted: 14 May 2019; Published: 18 May 2019

Abstract: This article presents a radiation tolerant single-shot time-to-digital converter (TDC) with a resolution of 15.6 ps, fabricated in a 65 nm complementary metal oxide semiconductor (CMOS) technology. The TDC is based on a multipath pseudo differential ring oscillator with reduced phase delay, without the need for calibration or interpolation. The ring oscillator is placed inside a Phase Locked Loop (PLL) to compensate for Process, Voltage and Temperature (PVT) variations- and variations due to ionizing radiation. Measurements to evaluate the performance of the TDC in terms of the total ionizing dose (TID) were done. Two different samples were irradiated up to a dose of 2.2 MGy SiO_2 while still maintaining a resolution of 15.6 ps. The TDC has a differential non-linearity (DNL) and integral non-linearity (INL) of 0.22 LSB rms and 0.34 LSB rms respectively.

Keywords: CMOS; TDC; radiation effects; total ionizing dose (TID); single-shot; PLL; ring oscillator

1. Introduction

Complementary metal oxide semiconductor (CMOS) technology scaling comes with a rapid decrease of supply voltages. Circuits which use voltage domain signal processing become less suitable in these scaled technology nodes, because of the inherent decrease in dynamic voltage range. Therefore, processing signals in the time domain becomes more interesting since their performance enhances due to reduced time-delays in the circuits [1,2]. In this article, the design, simulation and measurement of a radiation hardened single shot time-to-digital converter (TDC) is discussed.

TDCs can be compared to analog-to-digital converters (ADCs) as they digitize analog time differences instead of analog voltage differences. Several applications require precise time measurements. For example time-of-flight (TOF) measurements or particle tracking in high energy physics, where the precision of distance measurements is related to the resolution of the TDC. Also inside other circuits like frequency synthesizers [3], clock generators, clock data recovery circuits (CDRs), time-domain ADCs [4] and jitter measurement circuits [5]. In these applications, the TDC is a critical component to the overall performance of the circuit. This requirement leads to the need for high-performance TDCs with a small quantization delay, low noise, large sampling speed and high linearity. The main challenge in the design of a TDC is to overcome the minimum gate-delay of the technology, this is needed to increase the resolution of the TDC. Commonly used methods to obtain sub-gate-delay resolution are, the Vernier architecture [6], (passive or active) interpolation [7] and the parallel TDC [8]. The problem with these techniques is the matching of the delay-cells and the lack of self-calibration, which is used in delay-locked-loop (DLL) based delay lines [9,10]. Other commonly used techniques are based on oversampling and noise shaping. For example, $\Delta\Sigma$ TDCs [11,12], which require high-performance analog circuitries, gated ring oscilator (GRO) based TDCs [13] and switched ring oscillator (SRO) based TDCs [14]. Where the SRO architecture can achieve a larger oversampling rate (OSR) because the sampling frequency can be higher than the reference frequency, which is not the case for the GRO architecture.

All previously mentioned architectures can be divided into two main types. The first type is a single-shot TDC, which includes the one described in this article. The second type is a multi-shot or oversampled TDC where multiple correlated time-interval measurements contribute to the output. In this article, the focus is on the design of a single-shot TDC because the event that need to be measured will only occur once. The targeted applications are: nuclear energy, high energy physics and space [15]. This article is structured as follows. In Section 2, the architecture of the proposed ring oscillator based TDC is described. In Section 3 the circuit implementation details are presented. In Section 4, the analyses and measurement results of the fabricated prototype before and after irradiation is described. Finally, conclusions are drawn in Section 5.

2. Proposed TDC Architecture

The design presented in this paper is based on a second-order phase locked loop (PLL) with a ring oscillator as voltage controlled oscillator (VCO). Figure 1 shows the VCO which consists of 64 delay cells, based on pseudo differential delay cells [16]. The PLL is locked to a reference clock of 125 MHz with a multiplication factor of 8. The VCO thus runs at a stable frequency of 1 GHz, which leads to a period of 1 ns for the fine detection range. Therefore, once the PLL is locked the average static delay of the delay cells is 15.6 ps. The PLL feedback loop is used to ensure that the TDC is robust against process, voltage and temperature (PVT) variations, and additionally to variations due to ionizing radiation. Ionizing radiation, more specificaly the total ionizing dose (TID), influences the devices by changing the threshold voltage (V_T) and degrading the mobility, due to trapped charges in the devices [17]. This change in V_T and mobility will decrease the current consumption of the circuit which leads to decreased performance [18]. In the case of the VCO, the oscillation frequency would decrease. Nevertheless, the PLL will compensate the degradation of the free running oscillation frequency of the VCO with increasing dose to keep the divided output frequency equal to the 125 MHz reference.

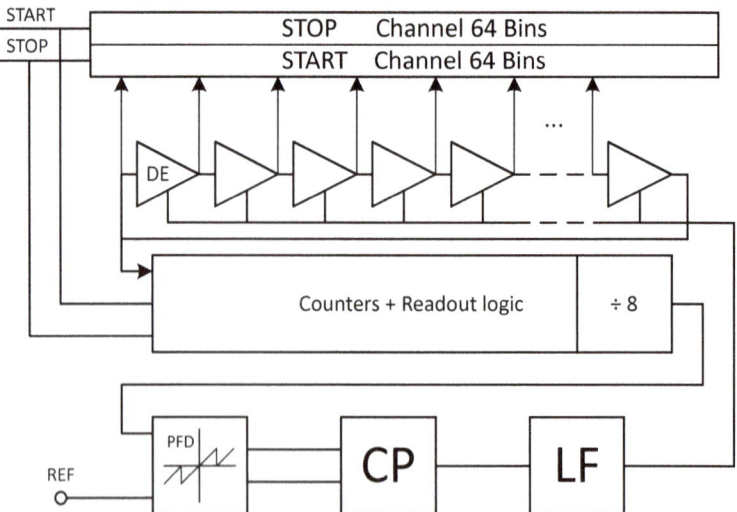

Figure 1. Radiation tolerant single shot time-to-digital converter (TDC) architecture based on a ring oscillator.

The TDC functionality is accomplished by having two sample channels, which sample the state of the VCO independently, these two channels correspond to the respective START and STOP input events. To further increase the dynamic range of the TDC a digital circuit is implemented which contains two counters of 12-bit (one for every channel), which extends the total dynamic range of the TDC to 4 µs. The divided feedback clock for the PLL is generated inside this digital circuit to save power. The entire digital block is designed with enclosed layout transistors (ELT) and synthesized with triple modular redundancy (TMR) to improve the resistance against total ionizing dose (TID) and single event effects (SEE) respectively. The entire chip can be configured digitally and read out, which reduces the number of pins drastically.

3. Circuit Implementation

The full TDC consisted of multiple parts: (1) ring oscillator based VCO, (2) phase-frequency detector, (3) charge pump, (4) loop filter, (5) two channels of C^2MOS sampling flip-flops (FF) and a (6) digital control block. In this section, some of these blocks and the implementation will be discussed separately.

3.1. VCO

The VCO used in this chip was based on a ring oscillator. The analog tuning of the delay element (DE) was done by changing the gate voltage of the P-type metal-oxide-semiconductor (PMOS) pair M5, M8 (Vc node) which changed the delay of the cell by limiting the current on the rising edge of the DE. The control voltage is inversely proportional to the delay of the DE and can tune the ring oscillator from 600 MHz up to 1.5 GHz. This voltage was controlled by the charge pump (CP) through the loop filter (LF). To reduce the gate delay of the DE, this implementation, has an extra feedforward input to cascade the DE's using the multipath approach [13], shown in Figure 2. For this ring oscillator, feedforward was foreseen to the third subsequent stage. This made the DE in this oscillator work 2.5 times faster than a ring oscillator without feedforward. Every increase in feedforward stage resulted in a gain of speed as shown in Figure 3. Nevertheless, this implies a serious layout restriction on the amount of stages to connect the feedforward path since routing complexity will increase drastically. Therefore, the feedforward path is only connected to three stages further. The layout of the cell is shown in Figure 4a. The pseudo differential property of the DE helps to have a signal with large voltage swings and steep edges. This decreased the effect of jitter and leads to cleaner sampling of the sense flip-flops (S-FF). The downside of this type of DE was the larger power consumption due to the feedforward path. It can be seen in Figure 3 that increasing the number of stages of the feedforward path, leads to an overall increase of the power consumption of the ring oscillator. This is because the feedforward devices M2 and M3 are drawing current for a longer time before the primairy path (M1, M4) is turned on.

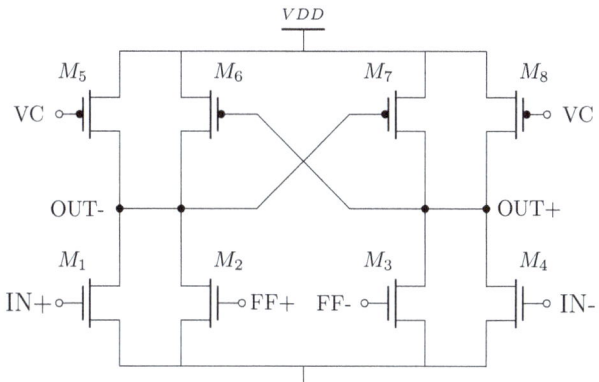

Figure 2. Schematic of the implemented delay element.

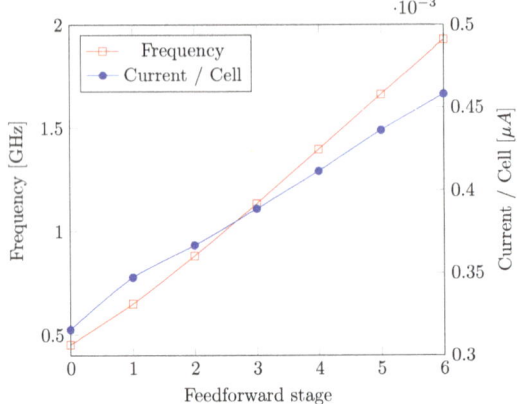

Figure 3. Simulation of the output frequency and power consumption of the ring oscillator with increasing number of feedforward stages.

By using 64 stages of the DE as shown in Figure 1, the ring oscillator can easily achieve a frequency of 1 GHz, with a phase noise of −114.3 dBc at 1 MHz shown in Figure 5. The Figure of merit (FOM) is 161.45 dB and did not change with increasing number of feedforward stages. The oscillator had 64 phases which results in a raw resolution of 15.6 ps. The entire ring oscillator was designed with enclosed layout transistors (ELT). This type of transistor has been proven to be more robust against TID effects up to a dose of 10 MGy [19]. ELTs were designed with an enclosed (circular) gate around the drain (Figure 4b). With this technique, the effects of charges captured in the STI were mitigated [20]. A practical restriction of using ELTs was the minimum gate width, which was larger compared to standard transistors because of the physical limitation of the drain contacts, by technology process rules. This also increased the power consumption of the cells.

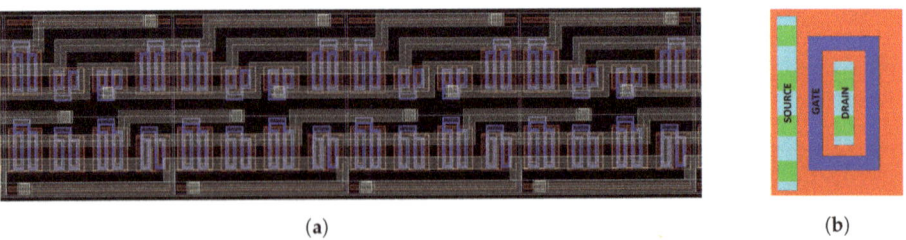

Figure 4. (a) Layout of the feedforward routing; (b) standard enclosed layout transistors (ELT) layout.

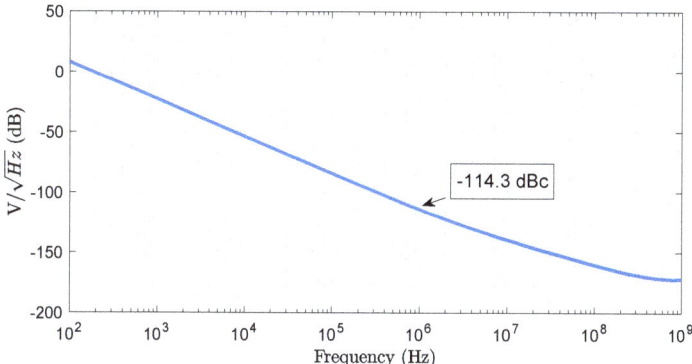

Figure 5. Simulated phase noise of the ring oscillator with a feedforward path of three stages.

3.2. Sampling Circuit

The VCO was sampled by two registers of sense flip-flops (S-FF) to quantize the phase of the ring oscillator. The S-FF was designed using the C^2MOS technique and was implemented in a pseudo differential way. As described in [21], flip-flops contributed significantly in the performance of the TDC due to the influence of the metastability of the flip-flop. The closer the changing input comes to the sampling event, the larger the propagation time. This can lead to bubble forming in the digital code [22]. Therefore, it was necessary to keep the metastable sampling window well below the raw resolution of the TDC. The designed S-FF shown in Figure 6, was therefore used in this design. The circuit consisted of two master–slave edge-triggered flip-flops which are used in a pseudo differential way with inverted cross connections to ensure the static behaviour of the S-FF.

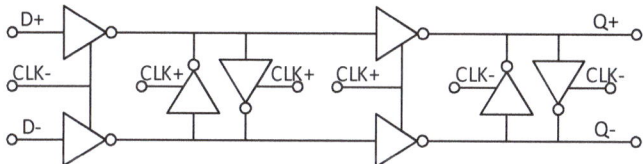

Figure 6. Schematic of the complementary metal oxide semiconductor (C^2MOS) pseudo differential sample flip-flop.

3.3. Digital Control Block

The digital back-end is a fully synthesized logic block which contains the counters to extend the dynamic range beyond the measurement intervals from the ring oscillator which is limited to 1 ns in this design. The digital core is a high-speed digital design which runs at 1 GHz clock speed and is fully triple modular redundant (TMR). This speed is on the edge of commercial 65 nm CMOS ELT cell libraries.

The readout logic saved the 32 bit circular thermometric data from both channels to be read out by the user to further decode the data off-line. The counters, to extend the dynamic range however, required some special attention. One 12 bit, TMR binary counter runs at the 1 GHz input clock coming form the VCO. This counter was incremented on the rising edge of the clock and the value of the counter is sampled by the start- or stop-signal. However, as shown in Figure 7, the output of this counter (C1) was unstable for a period of time after the rising edge of the input clock. This was due to the toggling of the logic. Firstly, if the start- or stop-signals sampled the value when the counter was unstable, the registers can become metastable. Secondly, due to an unknown delay in the registers and mismatch in the clock tree, the time for which C1 was still stable before it toggles was not exactly known. For these two reasons, it was expected that C1 is invalid from the clock edge until it has toggled and is fully stable. To overcome this issue, a second 12 bit register saved the C1 data upon the falling edge of the input clock such that this data is stable when C1 is unstable. Therefore either C1 or C2 will always contain a valid counter value. The selection of either C1 or C2, was based on the decoded 6 bit word coming from the sample registers. The MSB value of the decoded 6 bit word determined the phase of the start- or stop-signal relative to the VCO clock and can determine which counter is stable. For example, if the start signal occurred in the first half period of the VCO clock, the MSB of the fine code will be zero and C2 will be selected. Note that the counter should be constrained such that C1 was stable before half of the clock cycle, which places a constraint of 500 ps in this design. Concerning digital timing, it becomes challenging to meet the timing in the design, especially with respect to TID effects for which an additional timing overhead of 30% is included. The decoding in this design is done in an off-line way. The decoder accepts 32 bits from the start- and stop-register and should find the location on which a 1 to 0 transition occurs in the bit sequence. To overcome SEU errors and single bubbles, a "100" sequence was used to decode the raw TDC data. Finally, after the correct counter value has been selected, both 12 bit and 6 bit words were concatenated and a full time measurement with a dynamic range of 4 µs can be performed.

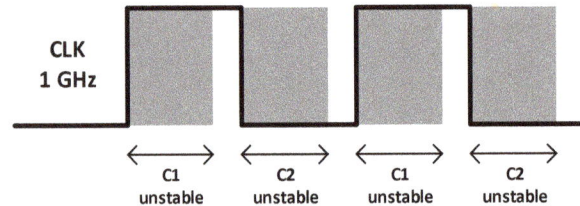

Figure 7. Stability of the reference counters.

4. Measurement Results

The TDC prototype was manufactured in 65 nm CMOS technology, with a die size of 0.6×0.52 mm^2. The macro picture is shown in Figure 8. To compare the performance of the chip before and after irradiation, the static INL and DNL were measured by performing a code density test, using a random hit generator which runs completely uncorrelated to the reference clock of the TDC [23]. The measured DNL and INL are shown in Figure 9 and are bound between $-0.42/+0.47$ LSB and $-0.71/+0.30$ LSB, respectively.

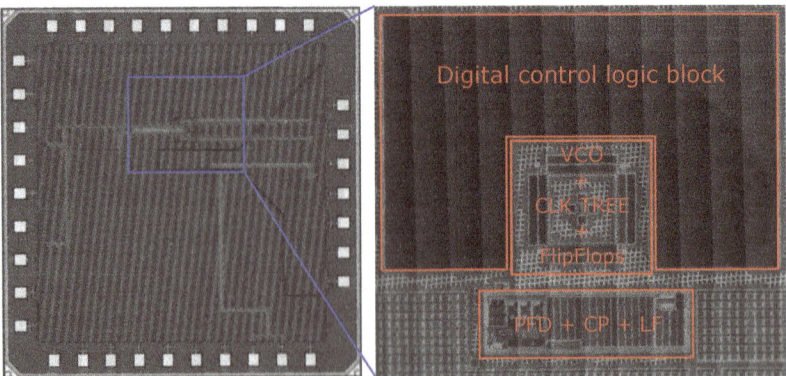

Figure 8. Photograph of the physical die.

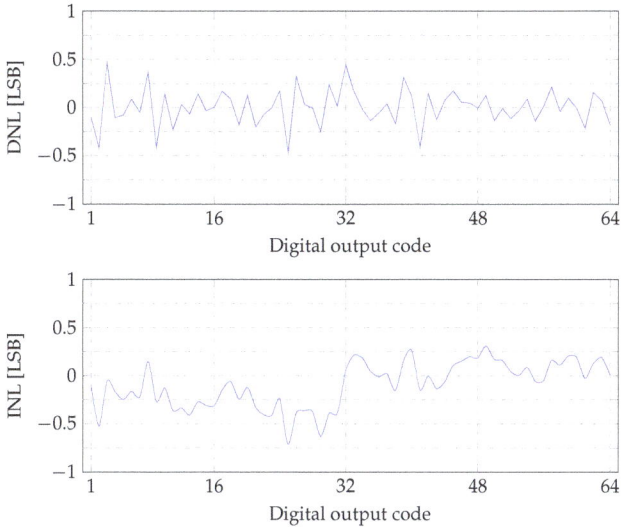

Figure 9. Measured differential non-linearity (DNL) and integral non-linearity (INL).

To test the performance of the TDC in terms of TID sensitivity, the TDC was placed under an X-ray beam coming from a 50 keV, 35 mA W-tube from Seifert (Figure 10). This resulted in a dose-rate of 54 kGy/h (SiO_2). During irradiation, a measurement of the TDC was conducted automatically every five minutes. In this case, there is no interruption of the irradiation which leads to a more precise measurement and more measurement points over time. Two samples were tested, both irradiated up to a dose of 2.5 MGy. During irradiation multiple measurements have been conducted, up to the point where the samples stopped working.

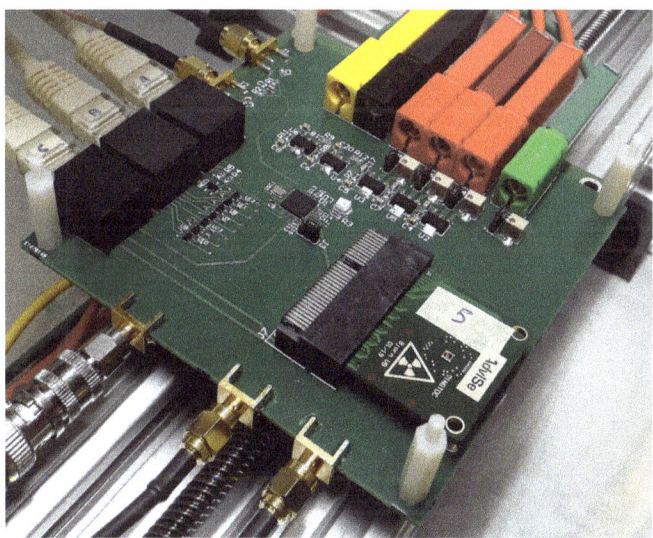

Figure 10. Picture of the used test setup for performing X-ray measurements.

First, a frequency sweep of the open loop ring oscillator was performed to measure the frequency degradation of the ring oscillator. The results of this measurement are shown in Figure 11.

Both measured samples only degraded 16% during the complete test. This results in a decreased VCO frequency from 1.5 GHz to 1.25 GHz, which is still more than the targeted frequency of 1 GHz. Therefore, the frequency degradation of the VCO is not the main cause of failing to lock to the phase of the reference signal.

In the second measurement, shown in Figure 12a, the in lock frequency of the PLL was measured. Here, a clear point of failure can be seen for both samples. The first sample reached a dose of 2.17 MGy and the second sample reached a dose of 2.52 MGy. The reason for the loss of lock was originating from the charge pump (CP) controlling the VCO through the loop filter. Although, the CP was designed to initialy deliver an equal up and down current, this also degrades under the influence of ionizing radiation. Known from [24], that PMOS devices degrade more compared to NMOS devices, the CP up and down current drift apart and becomes unbalanced. This combined with a decreasing set point for the VCO due to the decreasing oscillation frequency of the VCO causes the CP and LF to fail delivering the correct control voltage and therefore not able to lock to 1 GHz. Figure 12b, shows the measured current of the TDC while in a locked state. It can be seen that the overall current increased with dose, up to the point where the samples stopped working. To solve the issue of the unbalanced CP, a feedback loop can be implemented. This loop compensates one current source to be equal to the other.

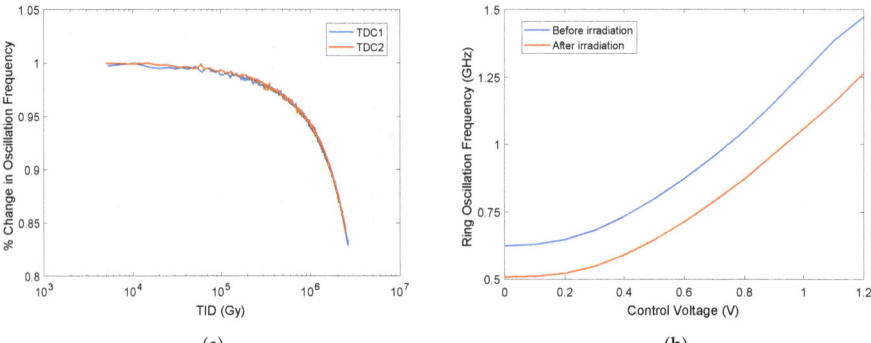

Figure 11. (a) Frequency degradation of the voltage controlled oscillator (VCO) in open-loop; (b) frequency sweep of the tunable ring oscillator before and after irradiation.

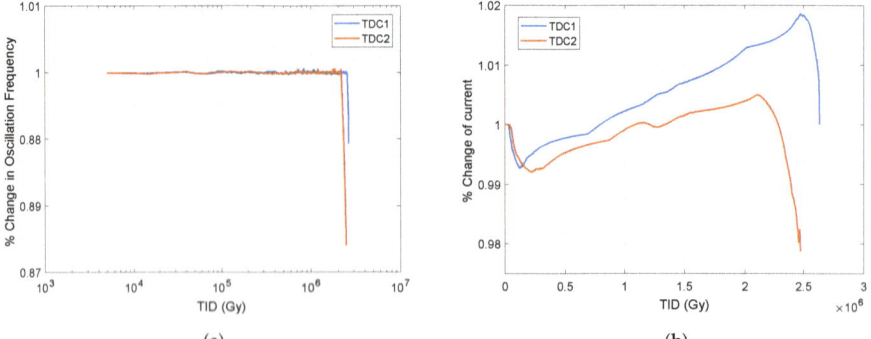

Figure 12. (a) Frequency degradation of the VCO in closed-loop; (b) percentage degradation of the VCO current.

Previous measurements gave an insight in the functional performance of the control loop and showed the point where it fails. The non-linearity of the TDC also gives a clear view of the performance of the TDC. In Figure 9, the INL and DNL of the TDC before irradiation are shown. During irradiation it is clear from Figure 13 that the DNL bounds are expanding with increasing dose. This was as expected since the mismatch between CMOS devices increased with dose [25].

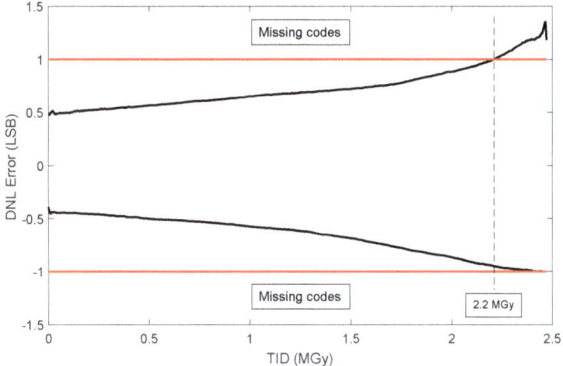

Figure 13. DNL degradation.

The limit of the DNL error that indicates if the TDC is still performing sufficiently, is $\pm 1\ LSB$. Beyond this point, missing codes can occur. For the measurement shown in Figure 13, this is up 2.2 MGy.

The performance of the proposed TDC is summarized and compared with the state-of-the-art TDCs in Table 1. The proposed TDC achieves the largest dynamic range and has a proven radiation tolerance of 2.2 MGy. Table 2 compares application related publication which report a radiation tolerance.

Table 1. Performance summary and comparison.

Reference	[26]	[27]	[13]	[22]	[28]	This Work
Technology (nm)	350	130	130	90	90	65
Technique	2-level DL	Vernier Ring	GRO	Pseudo Diff	Vernier GRO	Ring
Time resolution (ps)	24	8	6	17–21	6.4	15.6
Range (bit)	8	12	11	6	-	18
Power (mW)	50	7.5	2.2–21	6.9	4.32	134
DNL/INL (LSBrms)	0.55/1.5	-	-	0.7/0.7	-	0.22/0.34
Sample rate (MS/s)	160	15	50	26	250	200
Area (mm^2)	0.6	0.26	0.04	0.01	0.027	0.26
Radiation Tolerance	-	-	-	-	-	2.2 MGy

Table 2. Performance comparison with related application.

Reference	[29]	[11]	[30]	This Work
Technology (nm)	800	130	250	65
Technique	Pulse shrinking	$\Delta\Sigma$	DLL	Ring
Application	Space	LIDAR	High Energy Physics	High Energy Physics
RAW resolution (ps)	50	10.5	24	15.6
Range (bit)	11	11	21	18
Power (mW)	10	1.7	450 (multi channel)	134
DNL/INL (LSBrms)	x/0.45	-	0.21/2.1	0.22/0.34
Sample rate (MS/s)	1	50	8	200
Area (mm^2)	-	0.11	-	0.26
Radiation Tolerance	1 KGy	3.4 MGy	0.3 KGy	2.2 MGy

5. Conclusions

This work presents the evaluation of a single-shot TDC in terms of TID. The TDC has a measured resolution of 15.6 ps with a DNL and INL of 0.22 LSB rms and 0.34 LSB rms respectively. Two samples were irradiated and are able to reach a dose of 2.2 MGy before failing to meet specification due to an increased non-linearity error, originating from the increased mismatch in the sampling circuit. The reason for losing the locked state, was found to be originating from the difference in current drift between the up and down currents in the CP. The TDC was fabricated in 65 nm CMOS technology with an active area of 0.312 mm^2.

Author Contributions: Conceptualization, B.V.B. and J.P.; methodology, B.V.B. and J.P.; validation, B.V.B.; investigation, B.V.B. and J.P.; writing—original draft preparation, B.V.B., J.P. and P.L.; supervision, J.P. and P.L.

Funding: This research was funded by FWO.

Conflicts of Interest: The authors declare no conflict of interest. The funders had no role in the design of the study; in the collection, analyses, or interpretation of data; in the writing of the manuscript, or in the decision to publish the results.

References

1. Hu, C. Future CMOS Scaling and reliability. *Proc. IEEE* **1993**, *81*, 682–689. [CrossRef]
2. Jurgo, M.; Navickas, R. Comparison of TDC parameters in 65 nm and 0.13 μm CMOS. In Proceedings of the 2017 5th IEEE Workshop on Advances in Information, Electronic and Electrical Engineering (AIEEE), Riga, Latvia, 24–25 November 2017; pp. 1–4.
3. Jee, D.W.; Seo, Y.H.; Park, H.J.; Sim, J.Y. A 2 GHz fractional-N digital PLL with 1b noise shaping ΔΣ TDC. *IEEE J. Solid-State Circuits* **2012**, *47*, 875–883. [CrossRef]
4. Naraghi, S.; Courcy, M.; Flynn, M.P. A 9-bit, 14 μw and 0.06 mm2 pulse position modulation ADC in 90 nm digital CMOS. *IEEE J. Solid-State Circuits* **2010**, *45*, 1870–1880. [CrossRef]
5. Nose, K.; Kajita, M.; Mizuno, M. A 1-ps resolution jitter-measurement macro using interpolated jitter oversampling. *IEEE J. Solid-State Circuits* **2006**, *41*, 2911–2920. [CrossRef]
6. Dudek, P.; Szczepański, S.; Hatfield, J.V. A high-resolution CMOS time-to-digital converter utilizing a Vernier delay line. *IEEE J. Solid-State Circuits* **2000**, *25*, 240–247. [CrossRef]
7. Henzler, S.; Koeppe, S.; Kamp, W.; Mulatz, H.; Schmitt-Landsiedel, D. 90 nm 4.7 ps-Resolution 0.7-LSB single-shot precision and 19 pJ-per-shot local passive interpolation time-to-digital converter with on-chip characterization. In Proceedings of the IEEE International Solid-State Circuits Conference on Digest of Technical Papers, San Francisco, CA, USA, 3–7 February 2008.
8. Yao, C.; Jonsson, F.; Chen, J.; Zheng, L.R. A high-resolution Time-to-Digital Converter based on parallel delay elements. In Proceedings of the 2012 IEEE International Symposium on Circuits and Systems, Seoul, Korea, 20–23 May 2012; pp. 3158–3161.
9. Prinzie, J.; Steyaert, M.; Leroux, P. A Self-Calibrated Bang-Bang Phase Detector for Low-Offset Time Signal Processing. *IEEE Trans. Circuits Syst. II Express Briefs* **2016**, *63*, 453–457. [CrossRef]
10. Han, Y.; Rhee, W.; Wang, Z. A PVT-insensitive self-dithered TDC design by utilizing a ΔΣ DLL. In Proceedings of the 2012 IEEE 55th International Midwest Symposium on Circuits and Systems (MWSCAS), Boise, ID, USA, 5–8 August 2012.
11. Cao, Y.; De Cock, W.; Steyaert, M.; Leroux, P. 1-1-1 MASH ΔΣ time-to-digital converters with 6 ps resolution and third-order noise-shaping. *IEEE J. Solid-State Circuits* **2012**, *47*, 2093–2106. [CrossRef]
12. Gande, M.; Maghari, N.; Oh, T.; Moon, U.K. A 71dB dynamic range third-order ΔΣ TDC using charge-pump. In Proceedings of the IEEE Symposium on VLSI Circuits, Digest of Technical Papers, Honolulu, HI, USA, 13–15 June 2012.
13. Straayer, M.Z.; Perrott, M.H. A multi-path gated ring oscillator TDC with first-order noise shaping. *IEEE J. Solid-State Circuits* **2009**, *44*, 1089–1098. [CrossRef]
14. Elshazly, A.; Rao, S.; Young, B.; Hanumolu, P.K. A noise-shaping time-to-digital converter using switched-ring oscillators—Analysis, design, and measurement techniques. *IEEE J. Solid-State Circuits* **2014**, *49*, 1184–1197. [CrossRef]

15. Prinzie, J.; Christiansen, J.; Moreira, P.; Steyaert, M.; Leroux, P. Comparison of a 65 nm CMOS ring- and lc-oscillator based PLL in terms of TID and SEU sensitivity. *IEEE Trans. Nucl. Sci.* **2017**, *64*, 245–252. [CrossRef]
16. Moazedi, M.; Abrishamifar, A.; Sodagar, A.M. A highly-linear modified pseudo-differential current starved delay element with wide tuning range. In Proceedings of the 2011 19th Iranian Conference on Electrical Engineering, Tehran, Iran, 17–19 May 2011; p. 1.
17. Lacoe, R.C. Improving integrated circuit performance through the application of hardness-by-design methodology. *IEEE Trans. Nucl. Sci.* **2008**, *55*, 1903–1925. [CrossRef]
18. Prinzie, J. *Radiation Hardened CMOS Integrated Circuits for Time-Based Signal Processing*; Springer: Berlin/Heidelberg, Germany, 2017.
19. Faccio, F.; Michelis, S.; Cornale, D.; Paccagnella, A.; Gerardin, S. Radiation-Induced Short Channel (RISCE) and Narrow Channel (RINCE) Effects in 65 and 130 nm MOSFETs. *IEEE Trans. Nucl. Sci.* **2015**, *62*, 2933–2940. [CrossRef]
20. Johnston, A.H.; Swimm, R.T.; Allen, G.R.; Miyahira, T.F. Total dose effects in CMOS trench isolation regions. *IEEE Trans. Nucl. Sci.* **2009**, *56*, 1941–1949. [CrossRef]
21. Nikolić, B.; Oklobdžija, V.G.; Stojanović, V.; Jia, W.; Chiu, J.K.S.; Leung, M.M.T. Improved sense-amplifier-based flip-flop: Design and measurements. *IEEE J. Solid-State Circuits* **2000**, *35*, 876–884. [CrossRef]
22. Staszewski, R.B.; Vemulapalli, S.; Vallur, P.; Wallberg, J.; Balsara, P.T. 1.3 V 20 ps Time-to-Digital Converter for Frequency Synthesis in 90-nm CMOS. *IEEE Trans. Circuits Syst. II Express Briefs* **2006**, *53*, 220–224. [CrossRef]
23. Mota, M.; Christiansen, J. High-resolution time interpolator based on a delay locked loop and an RC delay line. *IEEE J. Solid-State Circuits* **1999**, *34*, 1360–1366. [CrossRef]
24. Barnaby, H.J. Total-ionizing-dose effects in modern CMOS technologies. *IEEE Trans. Nucl. Sci.* **2006**, *53*, 3103–3121. [CrossRef]
25. Verbeeck, J.; Leroux, P.; Steyaert, M. Radiation effects upon the mismatch of identically laid out transistor pairs. In Proceedings of the IEEE International Conference on Microelectronic Test Structures, Amsterdam, The Netherlands, 4–7 April 2011.
26. Hwang, C.S.; Chen, P.; Tsao, H.W. A high-precision time-to-digital converter using a two-level conversion scheme. *IEEE Trans. Nucl. Sci.* **2004**, *51*, 1349–1352. [CrossRef]
27. Yu, J.; Dai, F.F.; Jaeger, R.C. A 12-Bit vernier ring time-to-digital converter in 0.13 μ CMOS technology. *IEEE J. Solid-State Circuits* **2010**, *45*, 830–842. [CrossRef]
28. Lu, P.; Liscidini, A.; Andreani, P. A 3.6 mW, 90 nm CMOS gated-vernier time-to-digital converter with an equivalent resolution of 3.2 ps. *IEEE J. Solid-State Circuits* **2012**, *47*, 1626–1635. [CrossRef]
29. Karadamoglou, K.; Paschalidis, N.P.; Sarris, E.; Stamatopoulos, N.; Kottaras, G.; Paschalidis, V. An 11-bit high-resolution and adjustable-range CMOS time-to-digital converter for space science instruments. *IEEE J. Solid-State Circuits* **2004**, *39*, 214–222. [CrossRef]
30. Christiansen, J. *High Performance Time to Digital Converter*; CERN Microelectronics Group: Geneva, Switzerland, 2004.

© 2019 by the authors. Licensee MDPI, Basel, Switzerland. This article is an open access article distributed under the terms and conditions of the Creative Commons Attribution (CC BY) license (http://creativecommons.org/licenses/by/4.0/).

Article

Low-Power, Subthreshold Reference Circuits for the Space Environment: Evaluated with γ-rays, X-rays, Protons and Heavy Ions

Charalambos M. Andreou [1,*], Diego Miguel González-Castaño [2], Simone Gerardin [3], Marta Bagatin [3], Faustino Gómez Rodriguez [2], Alessandro Paccagnella [3], Alexander V. Prokofiev [4], Arto Javanainen [5,6], Ari Virtanen [5], Valentino Liberali [7], Cristiano Calligaro [8], Daniel Nahmad [9] and Julius Georgiou [1]

[1] Department of Electrical and Computer Engineering, University of Cyprus, Nicosia 1678, Cyprus; julio@ucy.ac.cy
[2] Radiation Physics Laboratory, Universidade de Santiago de Compostela, E-15705 Santiago de Compostela, Spain; diego.gonzalez@usc.es (D.M.G.-C.); faustino.gomez@usc.es (F.G.R.)
[3] Department of Information Engineering, University of Padova, 35122 Padova, Italy; simone.gerardin@dei.unipd.it (S.G.); marta.bagatin@dei.unipd.it (M.B.); alessandro.paccagnella@dei.unipd.it (A.P.)
[4] Department of Physics and Astronomy, Uppsala University, 75236 Uppsala, Sweden; alexander.prokofiev@physics.uu.se
[5] Department of Physics, University of Jyvaskyla, FI-40014 Jyvaskyla, Finland; arto.javanainen@jyu.fi (A.J.); ari.virtanen@phys.jyu.fi (A.V.)
[6] Department of Electrical Engineering and Computer Science, Vanderbilt University, Nashville, TN 37235, USA
[7] Department of Physics, Università degli Studi di Milano, 20133 Milano, Italy; valentino.liberali@unimi.it
[8] RedCat Devices, 20142 Milan, Italy; c.calligaro@redcatdevices.it
[9] R&D Department, Tower Semiconductor, Migdal Haemek 2310502, Israel; danielna@towersemi.com
* Correspondence: andreou.m.charalambos@ucy.ac.cy

Received: 16 April 2019; Accepted: 13 May 2019; Published: 21 May 2019

Abstract: The radiation tolerance of subthreshold reference circuits for space microelectronics is presented. The assessment is supported by measured results of total ionization dose and single event transient radiation-induced effects under γ-rays, X-rays, protons and heavy ions (silicon, krypton and xenon). A high total irradiation dose with different radiation sources was used to evaluate the proposed topologies for a wide range of applications operating in harsh environments similar to the space environment. The proposed custom designed integrated circuits (IC) circuits utilize only CMOS transistors, operating in the subthreshold regime, and poly-silicon resistors without using any external components such as compensation capacitors. The circuits are radiation hardened by design (RHBD) and they were fabricated using TowerJazz Semiconductor's 0.18 µm standard CMOS technology. The proposed voltage references are shown to be suitable for high-precision and low-power space applications. It is demonstrated that radiation hardened microelectronics operating in subthreshold regime are promising candidates for significantly reducing the size and cost of space missions due to reduced energy requirements.

Keywords: analog single-event transient (ASET); bandgap voltage reference (BGR); CMOS analog integrated circuits; gamma-rays; heavy-ions; ionization; protons; radiation hardening by design (RHBD); reference circuits; single-event effects (SEE); space electronics; total ionization dose (TID); voltage reference; X-rays

1. Introduction

Radiation-tolerant, high-accuracy, reference circuits are widely used in almost all circuits and systems that are intended for space applications. Analog and mixed-signal circuits and systems such as flash memories [1], ADCs, operational amplifiers, LDOs and DACs, require a stable and reliable reference voltage/current in order to perform within their specifications [2–5]. Any performance deviations of the reference voltage will consequently deteriorate the performance of all the subsequent circuits, leading to a malfunction or even failure of the overall system. Designing reference circuits that achieve a low temperature coefficient (TC) at a wide temperature range, whilst consuming little power, is challenging. Hence, when the supply voltage and power consumption specifications are very aggressive, the design has to operate within the subthreshold region, in which the non-linearities of the CMOS current components increase. Furthermore, in this region increased mismatch and process variations can be an issue. Nevertheless, it has been shown that it is possible to design and build an entire, mixed-signal, system-on-chip, operating predominantly in the subthreshold regime [6].

Traditional voltage reference circuits, such as the well known bandgap voltage reference (BGR), use the bipolar junction transistor (BJT) temperature dependence in order to generate a proportional to absolute temperature (PTAT) voltage, which is then utilized in order to produce a first-order temperature compensation scheme [2,7]. Subsequent approaches focus on partially canceling the BJT's base-emitter voltage non-linearities, in order to provide a higher-order, non-linear compensation [8–13]. The penalty of this approach is the higher design complexity and increased power consumption. More recently, low-power reference circuits utilize the metal oxide semiconductor (MOS) carrier mobility and threshold voltage temperature dependence in order to generate a first-order temperature-compensated reference voltage by summing a PTAT current and a complementary to absolute temperature (CTAT) current [14–17]. These circuits achieve low-power consumption but the TC is limited due to the non-linearities of MOS current components, which are higher when compared to the BJT ones.

Beyond the existing performance requirements of commercial applications, space microelectronics are required to be robust to the increased radiation levels of the space environment [5,18–27]. Hence, there is a lot of ongoing research activity to investigate the design of CMOS based analog, digital and mixed-signal, radiation tolerant circuits [28–45], including those that operate in the subthreshold regime [46–48].

Studies of the space industry have revealed that satellite/spacecraft size and cost can be significantly reduced by taking advantage of the inherent radiation hardness of modern CMOS commercial processes, in conjunction with radiation-hardening-by-design and low-power techniques [49,50]. Low-power is of major concern in space microelectronics, due to the isolation of the system and the limited available power.

One of the most promising solutions for achieving low-power consumption is to operate the devices in the subthreshold region. However, although designers can utilize well-known radiation hardening by design (RHBD) techniques, such as enclosed layout geometry transistors, it is not trivial to maintain good performance, when MOS transistors are biased in subthreshold. In this operating region, the transistor's drain-current is exponentially dependent on threshold voltage, therefore any deviations of the threshold voltage will severely impact the circuit's performance.

In this work, we've designed and characterized two custom subthreshold reference circuits [17,51]. In addition to achieving competitive performance for commercial applications, the proposed topologies are designed to be radiation tolerant so as to perform reliably in the space environment. The proposed reference circuits achieve high-order, non-linear curvature correction, which leads to an improved TC over a wide temperature range. In addition, the circuits perform reliably without failures and up to a certain extent with comparably low reference voltage variations when exposed to radiation such as γ-rays, X-rays, protons and heavy ions. The TC and TID performance are evaluated through fabricated silicon and experimental accelerated characterization results.

2. Radiation-Induced Effects in Subthreshold Circuits

Radiation-induced effects can be generally categorized into three kinds of radiation effects; those where the total ionization dose (TID) affects the devices properties, those where high-energy particles induce single event transients (SET) or device failures by dumping relatively large charges on critical nodes and those where the energetic particles cause displacement damage (DD) of the atomic lattice.

2.1. Total Ionization Dose Effects

When ionizing radiation impinges a material, such as Si and SiO_2, it loses energy (MeV/cm) which is absorbed by the material. The energy transfer, from high energy photons (i.e., γ-rays, X-rays) or charged particles (i.e., protons, electrons, α-particles, energetic heavy ions) towards the impinged material, is achieved through direct or indirect ionization mechanisms that generates electron–hole pairs. Ionizing radiation energy will extract electron–hole pairs from the material's atomic lattice. Some of the created electron–hole pairs will manage to recombine within a short time window; others, in the presence of an electric field, will escape recombination due to high mobility and will drift outside the gate oxide (within picoseconds) [52,53] towards the gate. This process will be triggered due to the gate electric field (assuming positive bias at the gate) or due to the built-in field. The holes (low mobility), that survived the recombination, will drift under the positive electric field of the gate towards the interface between gate oxide and silicon channel (Si/SiO_2 interface) [54,55], where charge trapping can occur. Other areas of CMOS processes, which are prone to charge trapping due to TID, are the shallow trench isolation and the deep trench isolation oxides.

In commercial CMOS processes, the gate oxide and isolation oxides (shallow trench isolation and deep trench isolation) which are structured by SiO_2 (insulator) are the most sensitive areas to be affected by ionizing radiation. The long-term charge trapping in the oxides will modify the electrical characteristics of the transistors and depends on total dose, dose rate, bias, time and temperature. The electrical characteristics that degrade include threshold voltage, carrier mobility, noise and leakage currents [41,56].

The threshold voltage shift (ΔV_{TH}) is proportional to the square of the oxide thickness (t_{ox}) up to a certain total dose as [57]:

$$\Delta V_{TH} \propto t_{ox}^2. \tag{1}$$

Above a certain dose, at which all the charge traps (oxide and interface states) are completely filled, the dependence becomes linear [57]:

$$\Delta V_{TH} \propto t_{ox} \tag{2}$$

The electrical characteristics that degrade will have different impact on a transistor/circuit operating in the subthreshold regime, compared to the same transistor/circuit designed in strong inversion regime. In order to identify the impact of radiation-induced effects in subthreshold regime, one has to explore the corresponding equations describing the MOS physics.

The threshold voltage shift (1) as well as the carriers mobility degradation (μ_{eff}) will modify the drain current (I_D) of an NMOS transistor operating in subthreshold/saturation such as:

$$I_D = K\mu C_{ox}(n-1)U_T^2 \exp\left(\frac{V_{GS} - V_{TH}}{nU_T}\right), \tag{3}$$

where K is the transistor's size aspect ratio W_{eff}/L_{eff}, μ is the mobility of carriers in the device channel, C_{ox} is the oxide capacitance per unit area, n is the subthreshold slope factor, $U_T = kT/q$ is the thermal voltage, V_{GS} is the gate-source voltage, and V_{TH} is the transistor threshold voltage. From (3) it can be deduced that the radiation-induced mobility degradation has the same impact on I_D in a transistor that operates in subthreshold regime compared to one that operates in strong inversion. However, the radiation-induced threshold voltage shift will impact I_D exponentially in a subthreshold MOS compared to a square law impact in a strong inversion MOS. In addition, the threshold voltage shift and the carriers mobility degradation will modify the transconductance (g_m) and the drain-source resistance (r_0) of a transistor operating in subthreshold/saturation through the I_D (neglecting the channel length modulation (λ)):

$$g_m = \frac{\delta I_D}{\delta V_{GS}} = \frac{I_D}{nU_T} \quad (4)$$

$$r_0 = \frac{1}{\lambda I_D}. \quad (5)$$

2.2. Displacement Damage Effects

High energy particles (~1 MeV), can induce crystal defects such as atomic lattice displacement (bulk damage), where atoms are displaced from their proper locations [58–61]. However, there is considerable DD even below (~1 MeV). This non-ionization effect is common when the impinging particle is electron, neutron or proton and can create Frenkel defects (vacancies or interstitials) [62]. The displacement damage will potentially degrade the minority carrier lifetime, the carrier mobility and the carrier concentration.

The non-ionizing energy loss (NIEL) which causes the displacement damage is mostly associated with bipolar transistors, whose operation depends on minority carrier lifetime [63]. However, the conduction of a MOS transistor operating in the subthreshold regime is due to the diffusion of minority carriers in the channel (caused by the lateral concentration gradient). Therefore, it is expected that displacement damage will affect the subthreshold MOS in a similar manner as a bipolar transistor.

2.3. Single Event Effects

Analog single-event transients (ASETs) are evanescent fluctuations of electrical charges in integrated circuits (IC). They may be observed when high-energy particles (alpha, protons and heavy ions), such as those found in the space environment (trapped particles in the Van Allen belts, solar energetic particles and galactic cosmic rays) [64], collide with analog ICs. When a high-energy particle penetrates the silicon substrate it ionizes the target material along its path. The ionized region is proximal to the ion path, generating a multitude of electron–hole pairs [65–68] in the vicinity of the ion track. Built-in electric fields or fields created by normal biasing conditions separate the pairs, leaving excess charge after the event. The excess charge injected at a sensitive circuit node can potentially disrupt the reliable functionality of the circuit, causing instantaneous or permanent failures.

Observable transients are most likely to occur when the impinging particles are heavy ions, such as silicon (Si), krypton (Kr) and xenon (Xe), which are high energy ions that have a high linear energy transfer (LET), and hence deposit more excess charge. The effect of the ASETs induced by these heavy ions on the desired signals, depends on the sensitivity of the particular analog circuit to the injected charge. The sensitivity is dependent on the circuit architecture, the devices' operating speed and the nominal operating voltage. Furthermore, as the technology nodes scale down, the decreased transistor geometries and thinner gate oxides, reduce the charge required to disrupt normal functionality, thus making the circuits more prone to ASETs. Thus in deep sub-micron technologies [69–71] ASETs are of major concern and impose critical issues for the microelectronic circuits reliability, while much ongoing research deals with characterizing the optimum circuit topologies, technology processes, devices and design approaches in order to mitigate ASETs in space applications [36,39,40,45,52,72–86].

3. Proposed Reference Circuits

In this work, two versions of low-power, subthreshold reference circuits were extensively assessed for the space environment. The design and analog performance tests for non-radiation environments was presented in [17,51,87]. This paper focuses on the radiation tolerance of two subthreshold topologies. The characterization experiments include measured results of the post-fabricated ICs in a wide temperature sweep, as well as measured results after exposing the circuits to γ-rays, X-rays, protons and heavy ions (silicon, krypton and xenon).

The designs were fabricated using TowerJazz Semiconductor's CMOS 0.18 µm technology. Both circuits utilize enclosed layout geometries [88] for all key NMOS transistors. These use enclosed layout geometry since they are prone to leakage currents from positive charges that are trapped in the deep-trench isolation structure and attract negative charges from the substrate, that form a parasitic channel along the edges of a planar layout geometry. The PMOS transistors are not prone to this effect, since their charge carriers are holes and trapped positive charges in oxide structures decrease the leakage current. The presented circuits are MOS-based voltage references, operating in subthreshold, which combine individually-linearized PTAT and CTAT currents. The proposed topologies incorporate two different types of polysilicon resistors. These, when combined with a subthreshold NMOS transistor, lead to a high-order curvature correction of the reference voltage, giving better performance across a wide range of temperatures. The two topologies have a high-impedance node at the drain of MP_3 transistor of the core structure. In addition, the reference voltage (VR)1 topology has an extra high-impedance node at the drain of MP_6 transistor.

Subthreshold operation of circuits and systems would be extremely beneficial in space microelectronics due to the limited energy sources. However, subthreshold circuits have to prove their reliability for such missions, where several radiation sources impact their electrical parameters. Therefore, this work investigates the resilience of subthreshold circuits with high total doses for a wide range of radiation sources. The circuits under evaluation are designed using the standard radiation-hardening-by-design (RHBD) techniques such as, extensive guard-rings, minimum gate extension to avoid leakage from the channel edge due to shallow trench isolation (STI), small layout and specialized circuit architectures.

The first reference circuit [51] is a low-power, wide-temperature-range topology which achieves a low TC over a temperature range of 190 °C, whilst being biased at a low supply voltage of 0.75 V and consuming only 4 µW of power. The circuit's topology is shown in Figure 1 and its layout in Figure 2a. This circuit occupies an area of 0.039 mm².

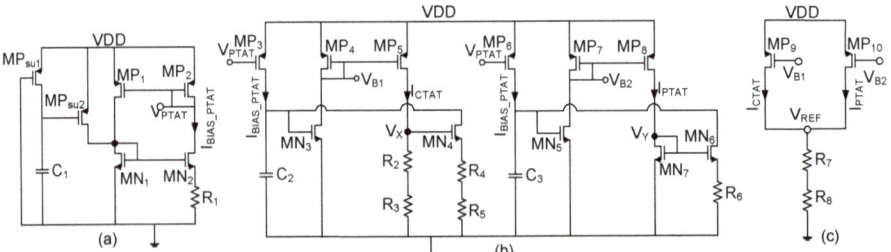

Figure 1. Schematic of the proposed voltage reference (VR1) [51]. (**a**) Proportional to absolute temperature (PTAT) current generator including a start up circuit, (**b**) main module utilizing the proposed novel method of high-order curvature correction of the reference voltage, (**c**) reference voltage output which sums the PTAT and complementary to absolute temperature (CTAT) curvature corrected currents.

The reference output voltage is generated by summing I_{PTAT} and I_{CTAT} currents across a resistance. This voltage is equal to:

$$V_{REF} = (I_{PTAT} + I_{CTAT}) \times R_{7,8} \tag{6}$$

where $R_{x,y} = R_x + R_y$. A detailed expression of the first reference voltage (VR1) can be expressed as:

$$V_{REF} = R_{7,8} K_{MN7} I_0 \exp\left(\frac{V_{GS7} - V_{TH}}{n U_T}\right)$$
$$+ \frac{R_{7,8}}{R_{2,3}} V_{GS4} + \frac{R_{7,8} R_{4,5}}{R_1 R_{2,3}} U_T \ln\left(\frac{K_{MN2}}{K_{MN1}}\right). \quad (7)$$

The measured post-trimmed TC at a bias voltage of 0.75 V is 15 ppm/°C for an extended temperature range of 190 °C (−60 °C to 130 °C) and is shown in Figure 3a.

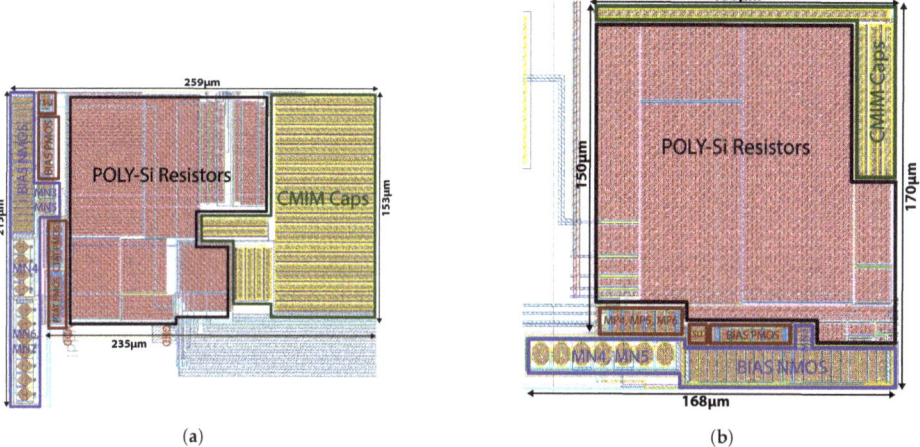

Figure 2. Layout of the proposed voltage references. (**a**) Layout of VR1 [51]; (**b**) Layout of VR2 [17].

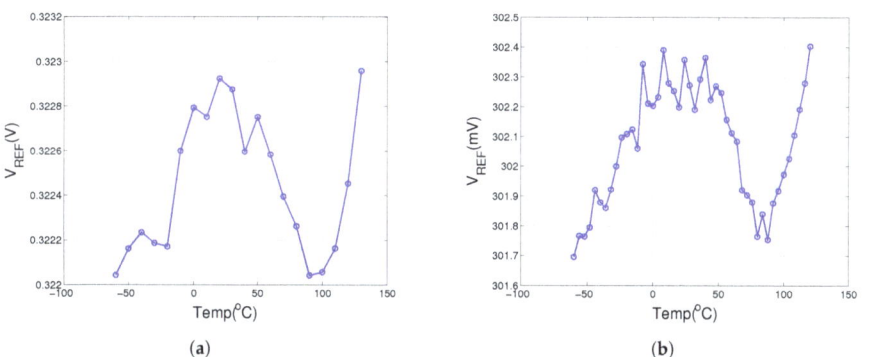

Figure 3. Measured temperature coefficient (TC) of the two reference circuits. (**a**) Measured TC of 15 ppm/°C of the first reference circuit; (**b**) measured TC of 12.9 ppm/°C of the second reference circuit.

The second reference circuit [17] is also a low-power, wide-temperature-range, curvature-compensated topology. The proposed topology achieves a temperature sensitivity of 12.9 ppm/°C for a temperature range of 180 °C (−60 °C to 120 °C) at a bias voltage of 0.7 V, whilst consuming 2.7 µW. It occupies an area of 0.023 mm².

The schematic of the proposed design is illustrated in Figure 4 while the layout is shown in Figure 2b. The topology consists of three main modules. A PTAT circuit, including the start-up circuit (MP_{su1}, MP_{su2}, C_1), is shown in Figure 4a, which generates a PTAT current for supplying the module of Figure 4b [15,89]. Figure 4b shows the core module, where both the linear and the non-linear compensation predominantly takes place.

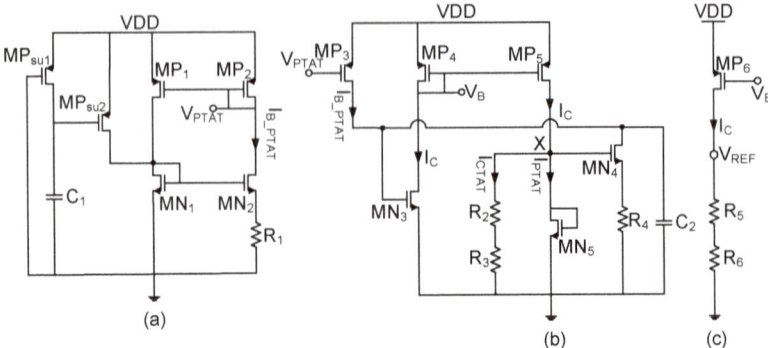

Figure 4. Schematic of the proposed voltage reference (VR2) [17]. (**a**) PTAT circuit including the start up (MPsu1, MPsu2,C1), (**b**) core module for implementing the high order compensation, (**c**) output stage to supply the reference voltage.

The reference voltage at the output of the proposed topology in Figure 4 can be expressed as:

$$V_{REF} = I_C \times R_{5,6}, \tag{8}$$

where $R_{x,y} = R_x + R_y$ and the current I_C consists of the currents through resistors R_2 and R_3, and the current through the transistor MN_5

The detailed equation for the second output reference voltage (VR2) is described by:

$$V_{REF} = \frac{V_{GS4} R_{5,6}}{R_{2,3}} + \frac{R_4 R_{5,6}}{R_1 R_{2,3}} \times U_T \ln\left(\frac{K_{MN2}}{K_{MN1}}\right) \\ + R_{5,6} K_{MN5} I_0 \exp\left(\frac{V_{GS5} - V_{TH}}{n U_T}\right), \tag{9}$$

where I_0 is:

$$I_0 = \mu C_{ox}(n-1) U_T^2, \tag{10}$$

where μ is the mobility of carriers in the device channel, C_{ox} is the oxide capacitance per unit area and n is the subthreshold slope factor.

4. Experimental Measurements on SET Irradiation Effects (Heavy Ions)

The proposed reference circuits were characterized at Radiation Effects Facility (RADEF) at the University of Jyvaskyla, Finland, for ASETs. Heavy ions (Si, Kr and Xe) from RADEF's standard 9.3 MeV/μm cocktail beam were used in order to provide different LET characteristics, so as to extract their cross-section (σ). The circuit irradiations were performed in air with a Kapton foil thickness of 25 μm and air thickness of 5 mm. During irradiation, the circuits were biased at their nominal supply voltages and the ASETs were recorded using a high sampling-rate oscilloscope (Agilent Technologies, Inc., Santa Clara, California, United States, DSO9104A 1 GHz/20 GS/s). The oscilloscope was set to record all the transient segments above a threshold trigger level (12 mV). This level was higher than the reference circuit noise floor and it ensures that stray electromagnetic fields at the testing facilities would not trigger the oscilloscope.

The RADEF's heavy ions cocktail provided Si ions with a LET(Si) of ~6.9 MeV·cm^2/mg, Kr ions with a LET(Si) of ~36.1 MeV·cm^2/mg and Xe ions with a LET(Si) of ~64.7 MeV·cm^2/mg. The charge deposited in the targeted material is greater at higher value of LET.

The cross-section, σ, is a metric to evaluate the resilience of the circuits under test when exposed to heavy-ions and can be expressed as:

$$\sigma = \frac{N_{ASET'S}}{\phi} \quad (\text{cm}^2), \tag{11}$$

where $N_{ASET'S}$ is the observed number of ASET events and ϕ is the uniform particle fluence (particles/cm^2).

The measured σ of the two voltage references are shown in Figure 5. The VR2 topology did not exhibit any sensitivity to Si ions, while an overall comparison shows that VR2 had less sensitivity compared to the VR1 circuit. The measured SET durations for the two circuits (VR1 and VR2) are shown in Figures 6–8 for silicon, krypton and xenon respectively. The VR2 circuit exhibits lower transients' duration compared to VR1.

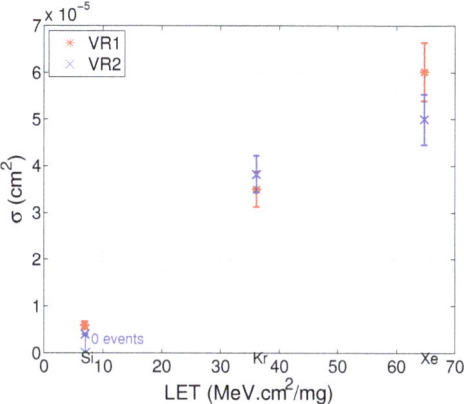

Figure 5. Measured Cross-Section of the two voltage reference circuits (VR$_1$ and VR$_2$) while exposed to silicon, krypton, and xenon ions.

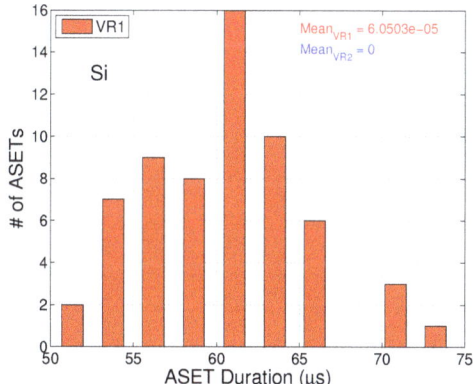

Figure 6. Measured number of ASETs versus ASETs duration for VR1 and VR2 circuits when exposed to Silicon.

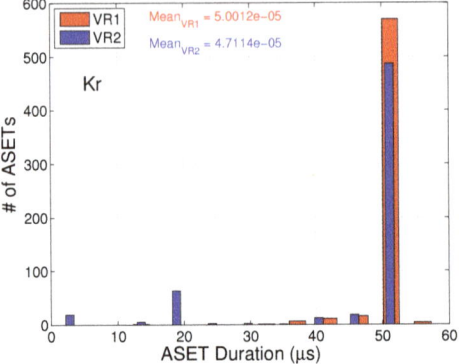

Figure 7. Measured number of ASETs versus ASETs duration for VR1 and VR2 circuits when exposed to Krypton.

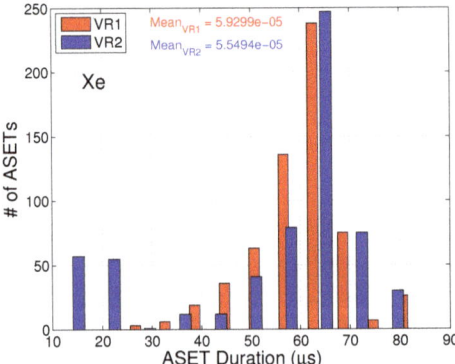

Figure 8. Measured number of ASETs versus ASETs duration for VR1 and VR2 circuits when exposed to Xenon.

The measured SET amplitudes for the two circuits (VR1 and VR2) are shown in Figures 9–11 for silicon, krypton and xenon respectively. The VR2 circuit exhibits smaller amplitudes compared to VR1.

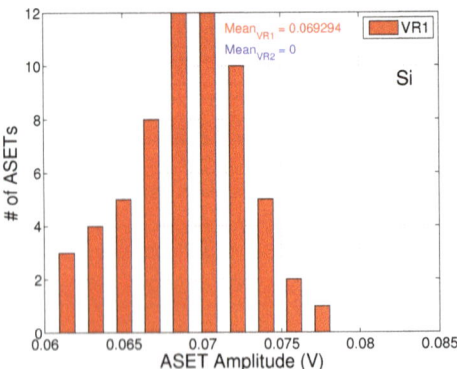

Figure 9. Measured number of ASETs versus ASETs peak amplitude for VR1 and VR2 circuits when exposed to Silicon.

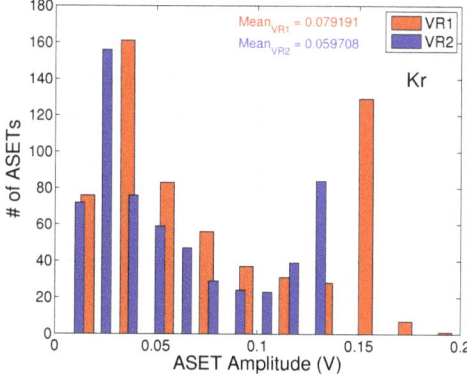

Figure 10. Measured number of ASETs versus ASETs peak amplitude for VR1 and VR2 circuits when exposed to Krypton.

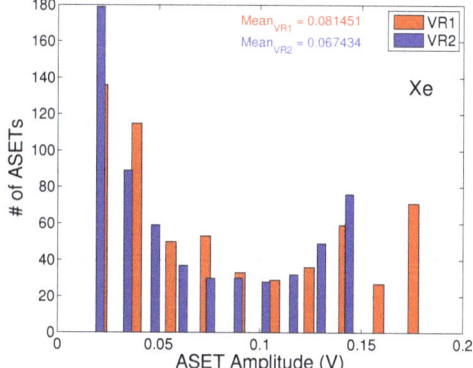

Figure 11. Measured number of ASETs versus ASET's peak amplitude for VR1 and VR2 circuits when exposed to Xenon.

The fact that VR2 outperforms VR1, in terms of σ, SETs' durations and amplitude, is possibly due to the smaller area of VR2 compared to VR1, since their circuit topology and operating conditions are very similar.

In general, the subthreshold circuits show somewhat higher sensitivity to heavy ions, in terms of σ and SETs' durations, compared to the strong inversion ones which were evaluated in previous work [90] by the authors. This is partly due to the lower supply voltage, which in turn needs less amount of deposited charge from the impinging ion in order to alter their nominal operating conditions of a particular circuit node. Furthermore, subthreshold circuits need more time to recover from a transient because of the lower current drive capability and the slower feedback. However, due to the limited bandwidth, longer transient durations are required in order to appear at the output, which gives subthreshold circuits an advantage. Part of the higher sensitivity could also be attributed to the larger silicon area of those topologies compared to the ones in [90]. The SETs' amplitudes in subthreshold and strong inversion circuits in [90] are comparable. This is probably due to the same reason described above, where the larger time constants of the subthreshold circuits tend to filter out some of the transients. An additional advantage of subthreshold circuits when exposed to heavy ions is that it is less probable to latch-up due to parasitic bipolar effect because of much smaller currents. This is a big advantage since parasitic bipolar effect could be fatal for a device and therefore an entire system.

5. Experimental Measurements on TID Irradiation Effects (γ-rays and X-rays)

In this section, a complete experimental characterization under γ-rays and X-rays TID induced effects was performed in order to evaluate the resilience of subthreshold reference circuit topologies against TID.

5.1. γ-ray Irradiation

The γ-ray irradiation was performed at the Radiation Physics Laboratory of the Universidade de Santiago de Compostela, using an AECL Theratron 780 Co-60 unit at room temperature. Dose rate and TID were monitored in real time by a 0.6 cm^3 therapy level ionization chamber connected to a reference class electrometer. The chamber and chips were positioned behind a 2 mm lead slab and 1 mm aluminium slab, to provide transient charge particle equilibrium in the gamma-ray field originating from the Co-60 source. Measured charge was converted to absorbed dose in air first, by employing the value of mean energy to produce a pair in air by a Co-60 beam (W_{Co60} = 33.97 J/C) and the mass of air enclosed in the chamber cavity with the appropriate correction due to ambient conditions. The correction factor between dose in air and dose in silicon inside the chips was calculated by Monte Carlo simulations (EGSnrc code) employing a realistic definition of geometries of the therapy unit, the ionization chamber and the chips.

Four chips from two different wafers (two chips from each wafer) were irradiated with γ-rays at a dose rate of 25 krad/h(Si) up to a total dose of 5.153 Mrad(Si), followed by room temperature annealing steps with the last annealing step measurement taken 25.97 h after the end of irradiation. The circuits under test were irradiated and measured at room temperature, with the supplies biased at nominal voltage during irradiation. The output voltages and current consumption of each circuit were measured at regular dose steps. The measured results of the relative output voltage are shown in Figure 12 for VR1 circuit and Figure 13 for VR2 circuit. Different wafers are expected to have some variation on device parameters, like threshold, voltage due to process variations. The measured results of the relative current consumption are shown in Figure 14 for VR1 circuit and Figure 15 for VR2 circuit.

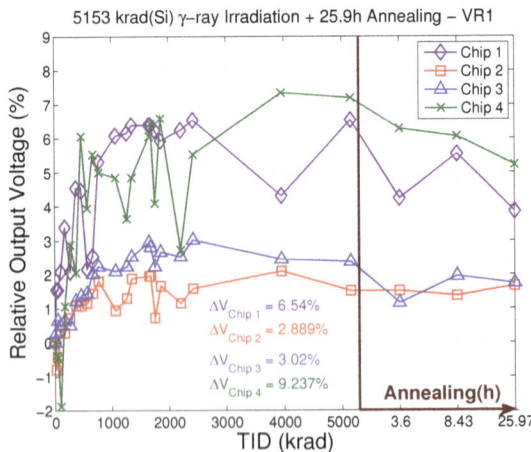

Figure 12. Relative output voltage (%) of the reference voltage (VR)1 when exposed to γ-rays at a dose rate of 25 krad/h(Si).

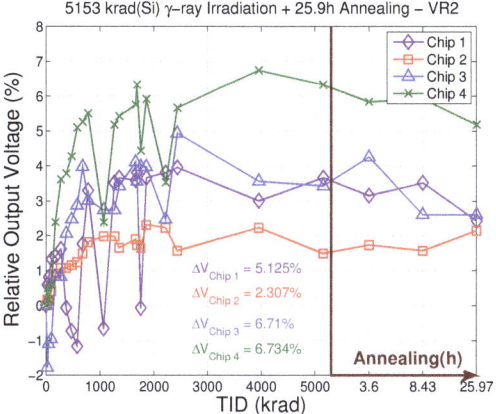

Figure 13. Relative output voltage (%) of the VR2 when exposed to γ-rays at a dose rate of 25 krad/h(Si).

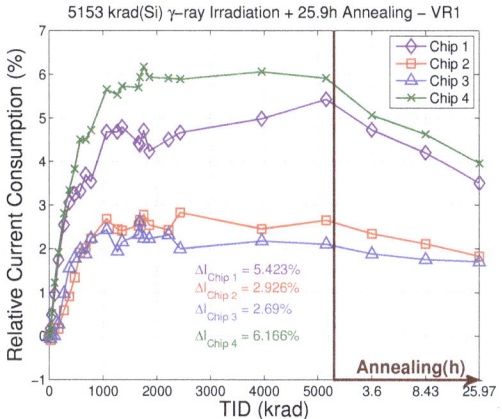

Figure 14. Relative current consumption (%) of the VR1 when exposed to γ-rays at a dose rate of 25 krad/h(Si).

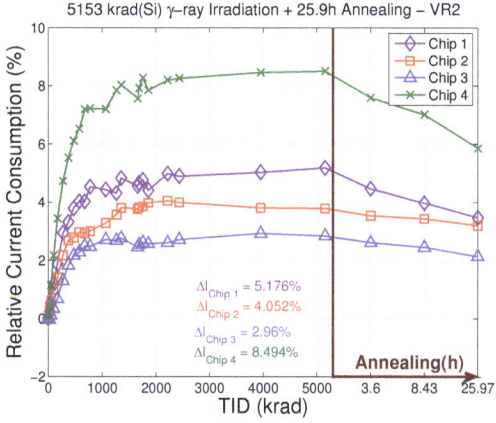

Figure 15. Relative current consumption (%) of the VR2 when exposed to γ-rays at a dose rate of 25 krad/h(Si).

The smallest output voltage variation is achieved by the VR2 circuit of chip 2 which is 2.307% and the highest by the VR1 circuit of chip 1 which is 9.237%. The two voltage references (VR1 and VR2) show comparable performance in terms of γ-rays irradiation TID effects, which for some of the chips is remarkable when considering the high total dose. The variation between different chips is due to the device mismatch within a circuit as well as due to process variation between different dies. Both circuits show a rising trend for total doses up to 1 Mrad, while for higher doses the output voltage remains almost unaffected. This could be explained by the competition between the interface state trapped charge and the oxide trapped charge. The interface state trapped charge occurs much slower and has a counterring effect in NMOS devices as opposed to the oxide trapped charge. Furthermore, when activated, interface trapped charge dominates the oxide trapped charge. The measured relative current consumption in Figures 14 and 15 show a similar trend with the relative output voltage variations. The current consumption variations are due to threshold voltage induced drain current variations as well as edge leakage currents due to STI trapped charge. During annealing at room temperature, the output voltage and current consumption partially recover. However, the interface trapped charge would need very high temperatures in order to anneal.

5.2. X-rays Irradiation

Two sessions of X-rays irradiation were performed. The first session was performed up to a total dose of 3.7 Mrad(Si) at the Radiation Physics Laboratory of the Universidade de Santiago de Compostela. The second session was performed up to a total dose of 80 Mrad(Si) at the Department of Information Engineering of the University of Padova. During both irradiation sessions all the circuits within the chips were biased at nominal supply voltage and their output reference voltages were measured at regular dose steps during irradiation.

5.2.1. X-ray Irradiation up to 3.17 Mrad(Si)

The irradiation was performed by a TW50 X-ray beam, delivered by an Oxford Instruments Neptune tube, with an added filtration by 1.1 mm of aluminum. X-ray beam outputs are generally measured by means of thin entrance window air filled ionization chambers. Conversion to absorbed dose to a relatively high atomic matter (high Z) medium like Silicon, surrounded by other higher than air Z materials, such as those present in a silicon chip, is not trivial. This is because the ratio of the absorbed dose to silicon and absorbed dose to air in the chamber varies sharply as a function of X-ray photon energy, due to enhanced photoelectric effect cross section. The beam output was characterized by measuring the X-ray spectrum and the exposure rate. The spectrum was measured with an AMPTEK CdTe scintillator connected to a MultiChannel Analyzer (MCA). The 50 keV tail was employed to calibrate the MCA in terms of photon energy. The exposure rate was measured with a PTW 23344 plate parallel X-ray chamber connected to an IBA DOSE-1 electrometer. The measured exposure rate was first converted to absorbed dose in air and then to absorbed dose in silicon in the area of the circuits under test. This was done by employing EGSnrc Monte Carlo code. The measured spectrum was used as the energy distribution of the Monte Carlo primary source. This spectrum was fine-tuned in order to reproduce experimental percent depth doses of the unfiltered 50 keV beam. A conversion factor was determined as the ratio of Monte Carlo absorbed dose in silicon inside the circuits under test and Monte Carlo absorbed dose to air inside the ionization chamber cavity.

In this first session, the two chips were irradiated with X-rays using a dose rate of 8.75 rad(Si)/s up to a total dose of 1.3 Mrad(Si) and then irradiated with a dose rate of 6.56 rad(Si)/s, up to a total dose of 3.173 Mrad(Si). The circuits were irradiated and measured at room temperature, with their supplies biased at nominal supply voltage. The output voltages as well as the current consumption were measured at regular dose steps during irradiation. Annealing steps at room temperature followed the irradiation, with the last annealing measurement taken 465 min after the end of irradiation. The measured results of the relative output voltage of the two chips of VR1 and VR2 reference circuits are shown in Figure 16 and their relative current consumption is shown in Figure 17.

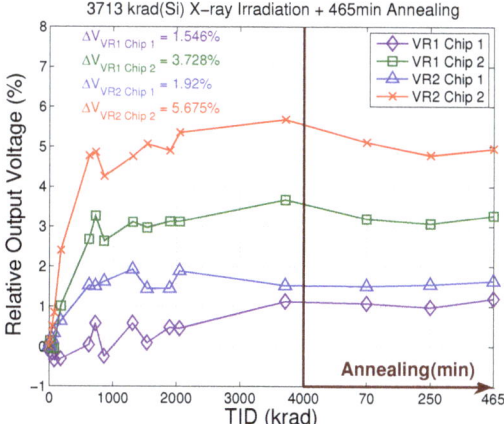

Figure 16. Relative output voltage (%) of the VR1 and VR2 when exposed to X-rays at a dose rate of 8.75 rad(Si)/s.

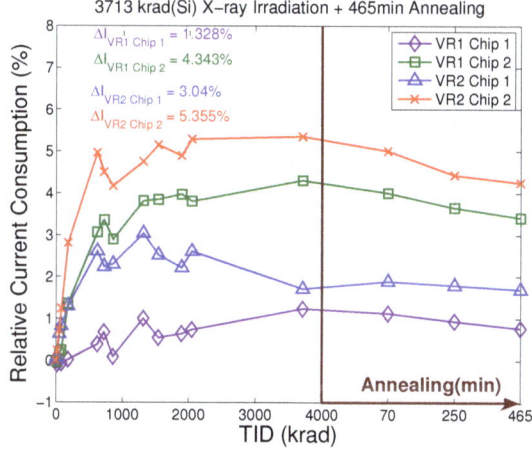

Figure 17. Relative current consumption (%) of the VR1 and VR2 when exposed to X-rays at a dose rate of 8.75 rad(Si)/s.

The smallest output voltage variation was achieved by the VR1 circuit of chip 1 which was 1.546% and the highest by the VR2 circuit of chip 2 which was 5.675%. The variation between different chips is again due to the process and mismatch variations between the circuits' devices. This can be justified by the fact that all the circuits of chip 1 demonstrated better performance than the ones of chip 2. Both circuits showed changing properties trend for total doses up to 800 krad, which then stabilizes at higher doses. This effect was in agreement with the γ-rays experiments and can be explained by the opposite effect of the interface state trapped charge and the oxide trapped charge on the device characteristics.

The measured relative current consumption in Figure 17 show almost identical trend with the relative output voltage variations. The current consumption variations can be attributed to the threshold voltage variation induced drain current variation as well as edge leakage currents due to STI trapped charge. During annealing at room temperature, the output voltage and current consumption of the VR1 and VR2 circuits of chip 2 (exhibited the highest TID variation) showed some recovering, while the corresponding circuits of chip 1 did not exhibit any significant change.

5.2.2. X-ray Irradiation up to 80 Mrad(Si)

In this session, a chip with the circuits under test was irradiated at a room temperature at the Department of Information Engineering of the University of Padova with 10-keV X-rays using a dose rate of 300 rad(Si)/s up to a total dose of 80 Mrad(Si). During irradiation both the circuits were biased at their nominal supply voltage. The output voltages were measured at regular dose steps during irradiation as well as during room temperature annealing, after the end of irradiation. The measured results for the relative output voltage are shown in Figure 18.

The two circuits showed similar performance to TID induced effects, where the smallest output voltage variation was achieved by the VR1 circuit which was 11.51% and the highest by the VR2 circuit which was 12.9%. The TID induced output voltage variations as shown in Figure 18 revealed an important outcome. Both circuits exhibited significant changes up to 10 Mrad, while for higher total doses they stabilized and then recovered significantly. This is in agreement with γ-rays and X-ray experiments and could again be explained by the competition between the interface state trapped charge and the oxide trapped charge. Another reason for this recovering during irradiation is the possible saturation of the oxide trapped charge first and then the interface trapped charge, which can be caused by the high total dose exposure. This can be supported by (1) and (2), where the irradiation induced rate of threshold voltage shift will reduce from square to linear dependence on oxide thickness. During annealing, at room temperature, the output voltage recovered at a higher rate.

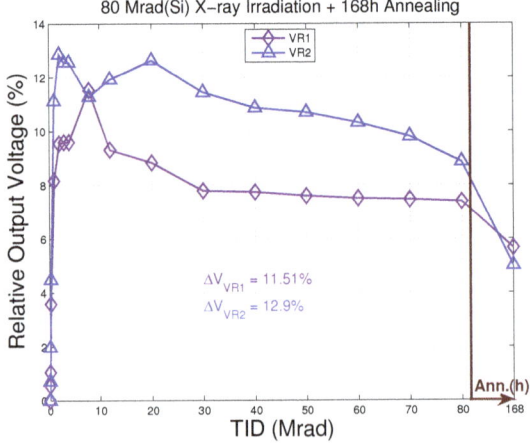

Figure 18. Relative output voltage (%) for the VR1 and VR2 when exposed to a 10 keV X-rays at a dose rate of 300 rad(Si)/s.

6. Experimental Measurements for TID and DD Irradiation Effects (Protons + X-rays and Protons)

The space radiation mixture is to a large extent composed from high-energy protons [91]. Therefore, in this section, the different topologies were irradiated with protons and X-rays and only-protons for the experimental characterization of DD/TID radiation-induced effects. These tests were required in order to classify the robustness of the subthreshold circuits in a more realistic scenario. It has to be noted that the effects induced by DD and TID interacted in a complicated fashion and were not simply additive [91].

6.1. Irradiation with Protons and X-rays

In this session the chip with the two circuits was irradiated with a 3 MeV proton beam in vacuum, with a flux of 10^9 p/cm²·s, up to a fluence of 1.47×10^{12} p/cm², corresponding to a total ionizing dose of 2 Mrad(Si). This was followed by three days of room temperature annealing. The circuits were unbiased during irradiation and they were biased just after the end of the irradiation in order to measure their output voltage. Then, they were remeasured again after one week (168 h) of room temperature annealing. The results of this irradiation session are shown in Figure 19 for the relative output voltage versus TID.

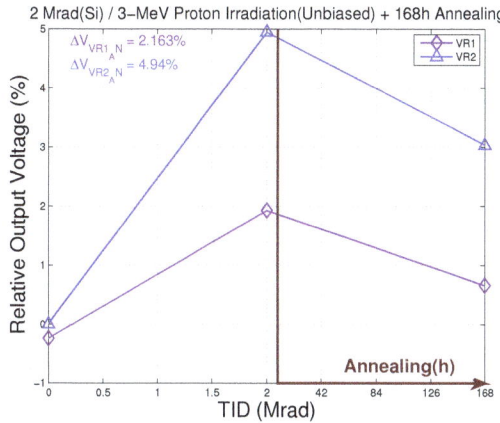

Figure 19. Relative output voltage (%) of the VR1 and VR2 when exposed to protons. The chips were irradiated unbiased with 3 MeV protons in vacuum, up to a fluence of 1.47×10^{12} p/cm² with a flux of 10^9 p/cm²·s. This corresponds to a total ionizing dose of 2 Mrad(Si), followed by three days of annealing.

The same chip, that was irradiated with a total dose of 2 Mrad(Si) of protons, was further irradiated with 10 keV X-rays after the proton exposure and annealing step. The X-ray irradiation and subsequent annealing were performed at room temperature, with all the circuits biased at the nominal supply voltage. A total dose of 78 Mrad(Si) was delivered through this X-ray irradiation session, so that a total dose of 80 Mrad(Si) was accumulated on the device, using a dose rate of 300 rad(Si)/s. The output voltages were measured at regular dose steps during irradiation at room temperature. Annealing steps followed the irradiation at room temperature, with the last annealing measurement taken one month (720 h) after the end of irradiation.

The results of the proton and subsequent X-ray irradiation are shown in Figure 20. The total dose reported on the X-axis of Figure 20 is the sum of the proton and the subsequent X-rays irradiation.

The VR1 circuit topology shows more resilience in comparison to the VR2 topology. The differences of the output voltage in Figure 20 when compared to the same irradiation dose of Figure 18 is possibly due to displacement damage that was induced from the proton irradiation on the subthreshold biased transistors. A 3 MeV proton beam with such a high fluence (1.47×10^{12} p/cm²) delivered both, TID as well as DD, for the equivalent dose of about two Mrad(Si). The conduction of a MOS transistor in sub-threshold regime was due to diffusion of minority carriers in the channel, where minority carrier lifetime was affected by DD in a similar manner as in bipolar transistors.

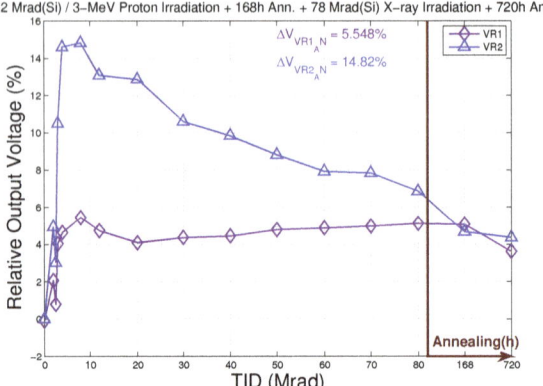

Figure 20. Relative output voltage (%) of the VR1 and VR2 when exposed to Protons and X-rays. The chips were irradiated unbiased with 3 MeV protons in vacuum, up to a total dose of 2 Mrad(Si), followed by three days of annealing. Then they were further irradiated with 10keV X-rays up to 78 Mrad(Si) with a dose rate of 300 rad(Si)/s. The dose reported on the x-axis is the sum of the proton and the subsequent X-ray irradiation which is 80 Mrad(Si).

6.2. Irradiation with Protons

In this irradiation session, another chip was irradiated in air with protons, at the Svedberg Laboratory in the University of Uppsala. The nominal primary proton energy was 180-MeV. In order to create a uniform proton field at the position of the chips under test, the primary proton beam was scattered by a Ta foil of 1.5 mm thickness. The chips were positioned at a distance of 200 cm from the foil. The average proton energy at the chips under test position amounted to 170.5-MeV. During irradiation, the incident proton beam was monitored by a telescope consisting of two scintillators, calibrated using a thin-film breakdown counter equipped with a fission foil. The telescope detected protons scattered by a stainless steel foil at the end of the vacuum pipe.

The chip was irradiated with the 170.5 MeV protons at the fluence of 3.9×10^{11} cm^{-2}. By the end of the experiment it had accumulated a total dose of 1400 krad(Si), at steps of 25 krad for low TID values, with wider steps for higher TID values, with an intermediate annealing step of 15 h. The irradiation and measurements were performed at room temperature with all the circuits biased at nominal supply voltage.

The output voltage of the circuits was measured at regular dose steps and is shown in Figure 21, where VR1 outperforms VR2. Both circuits show a considerable change for total doses up to 900 krad, which saturates at higher doses. This again agrees with the experimental sessions of γ-rays and X-rays and could be attributed to the same reasons explained above. The relative current consumption is shown in Figure 22 where the current consumption trend was identical with the relative output voltage variations trend. This is because the STI edge leakage current as well as the threshold voltage shift variations (gate oxide and interface states trapped charge) modify the current drained by the transistors and therefore the total current consumption of the circuits was modified accordingly. The current was increased mostly in the non-core NMOS transistors where planar layout was utilized. However, all transistors (including enclosed layout geometry ones) had small current variations due to threshold voltage shifts. During annealing both circuits show significant recovery in terms of output voltage as well as current consumption due to possible annealing of oxide trapped charge.

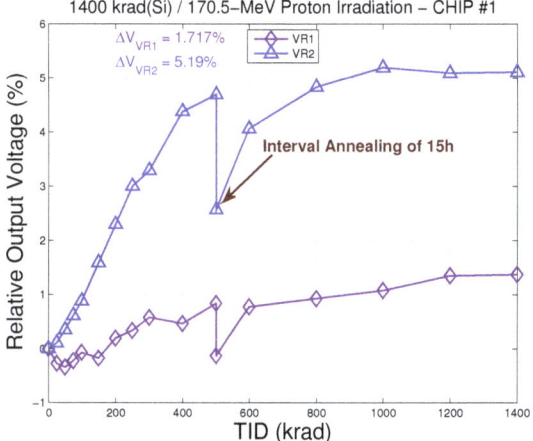

Figure 21. Relative output voltage change (%) of the VR1 and VR2 when exposed to 170.5-MeV Protons irradiation, accumulating a total dose of 1400 krad(Si) with a fluence of 3.9×10^{11} cm^{-2}. An interval annealing step at room temperature was performed at 500 krad(Si).

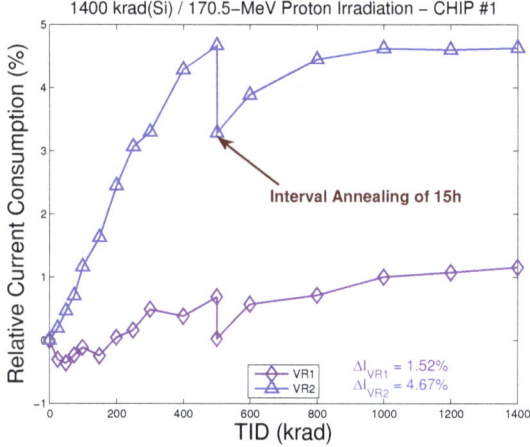

Figure 22. Relative change in current consumption (%) of the VR1 and VR2 when exposed to 170.5 MeV Proton irradiation, accumulating a total dose of 1400 krad(Si) with a fluence of 3.9×10^{11} cm^{-2}. An interval annealing step at room temperature was performed at 500 krad(Si).

7. Discussion on Subthreshold Radiation Effects

In this paper we have presented the results of radiation tests involving two subthreshold circuits. When comparing TID radiation-induced effects between subthreshold and strong inversion circuits, the major difference is the dependence of the drain current to threshold voltage shift. This dependence is exponential in subthreshold circuits as opposed to square-law dependence in strong inversion circuits. This is the major disadvantage of subthreshold circuits in radiation environment, however this disadvantage is rapidly diminishing in deep sub-micron semiconductor technologies, where according to (1) and (2) the oxide thickness reduction will diminish the radiation-induced threshold voltage shift. Thus, in deep sub-micron technologies, subthreshold circuits will potentially emerge as an attractive and promising solution for space microelectronics. The leakage currents due to trapped charge at the STI oxides will have more relative impact in subthreshold circuits compared to strong inversion ones, since the actual current through the channel could be in the order of the STI induced leakage currents. However, this disadvantage, which concerns only NMOS devices, can be remedied by using enclosed layout geometry transistors at the layout level. On the other hand, the low-voltage operation of subthreshold circuits applies lower electric fields across the oxides. This will reduce the rate of electron–hole separation and increase the probability of recombination. Therefore, this induces lower trapped charge in the oxides and hence lower will be the radiation-induced threshold voltage shift and leakage current.

SETs usually originate from particle strikes which traverse reverse-biased pn-junctions or areas with strong electric fields. Built-in electric fields or fields created by normal biasing conditions separate the pairs, leaving excess charge after the event. This is particularly a problem at the transistor's drain terminal, especially with the deep submicron technologies, where the generated plasma of e-h pairs drifts apart because of the high electric fields across the depletion region. When comparing SET radiation-induced effects between subthreshold and strong inversion circuits, the major difference is the supply voltage. The gate oxide thickness (depends on the type of device that is selected) and the supply voltage, both affect the amount of energy needed in order to alter the transistor's normal operating conditions. Therefore, the lower supply voltage of subthreshold circuits will require less charge and therefore less energy from the impinging ion in order to alter its nominal state, which makes them more vulnerable compared to strong inversion circuits. On the other hand, the advantage of low-voltage operation of subthreshold circuits is the lower electric field across the pn-junctions which reduces the amount of separated e-h pairs as well as the charge collected at the impinged node. An additional advantage of subthreshold circuits is that it is not possible to form a BJT through the substrate (parasitic bipolar effect) after a heavy ion strike [46].

An important advantage of subthreshold regime, which applies in both TID and SET effects, is that the drain-source voltage for saturation is 4 kT ≈ 104 mV which, in contrast with strong inversion regime, is very low and independent of the gate-source voltage and threshold voltage. Therefore, in contrast with strong inversion regime, it is not easy to get a subthreshold transistor out of saturation region due to irradiation-induced effects. This has been proven throughout all the experimental sessions in this work, where there was not any complete functional failure observed, even in very high TID irradiation or heavy ion strikes.

8. Conclusions

A comprehensive evaluation of two subthreshold voltage reference circuits with respect to their resilience to SEE, TID and TID/DD was performed. The evaluation is supported by measured results with γ-rays, X-rays, protons and heavy ions. The high total doses applied in this range of experiments provide a complete evaluation of subthreshold circuits in the whole range of space applications, radiation physics instruments and medical applications.

The fact that VR2 outperforms VR1, in terms of σ, and SETs' durations and amplitude, is due to the smaller area of VR2 compared to VR1. The circuit topology and operating conditions of the two circuits are very similar, since VR2 combines two feedback loops within a single branch.

The critical nodes that affect the output voltage are those that generate I_{CTAT} and I_{PTAT}, namely nodes V_X and V_Y in VR1 and node X in VR2. The reason that the VR1 circuit outperforms the VR2 circuit in the TID experiments is because VR1's critical nodes have a slightly higher impedance than that of VR2, given that in VR2 the I_{CTAT} and I_{PTAT} appear in parallel at a single node X. Therefore if the TID reduced the threshold of MN_4 the change in voltage at node X will create a greater change in current at node X, given the lower impedance, when compared to the equivalent critical nodes in VR1. In VR1 the threshold change will influence MN_6 and MN_4, which influence the voltage at the higher impedance nodes V_X and V_Y, thus leading to smaller relative change in the output current and consequently reference voltage.

In general, the subthreshold reference circuits show promising performance for space applications, especially in high total doses where they stop deviating or partially recover towards their nominal performance. It is also important that they do not show any signs of collapse or functional failure in any of the experiments, even in uncommonly high total ionization doses or heavy ion strikes. In addition, as expalined above, they will benefit from more advanced technology nodes with thinner gate oxides. These conclusions, along with their main advantage of low-power consumption, make subthreshold circuits candidates for future space missions due to significantly reducing the size, cost and power requirements of space applications. Therefore, there is still room to explore more in the future in terms of different types of circuits and devices.

Author Contributions: Conceptualization, analysis, original draft and revised manuscript preparation: C.M.A. and J.G.; design and execution of the irradiation experiments: C.M.A., D.M.G.-C., S.G., M.B., F.G.R., A.P., A.V.P., A.J., A.V., V.L., C.C., D.N. and J.G.

Funding: This work was supported by: European Union SkyFlash Project 262890 FP7-SPACE-2010-1 (SPA.2010.2.2-01 Space technologies), European Space Agency (ESA/ESTEC Contract 4000111630/14/NL/PA) and, Academy of Finland under the Finnish Centre of Excellence Programme 2012–2017 (Project No 2513553, Nuclear- and Accelerator-Based Physics).

Conflicts of Interest: The authors declare no conflict of interest. The funders had no role in the design of the study; in the collection, analyses, or interpretation of data; in the writing of the manuscript, or in the decision to publish the results.

References

1. Arbat, A.; Calligaro, C.; Dayan, V.; Pikhay, E.; Roizin, Y. SkyFlash EC project: Architecture for a 1Mbit S-Flash for space applications. In Proceedings of the 2012 19th IEEE International Conference on Electronics, Circuits and Systems (ICECS), Seville, Spain, 9–12 December 2012; pp. 617–620.
2. Tsividis, Y.P. Accurate analysis of temperature effects in I_c-V_{be} characteristics with application to bandgap reference sources. *IEEE J. Solid-State Circuits* **1980**, *15*, 1076–1084. [CrossRef]
3. Malcovati, P.; Maloberti, F.; Fiocchi, C.; Pruzzi, M. Curvature-compensated BiCMOS bandgap with 1-V supply voltage. *IEEE J. Solid-State Circuits* **2001**, *36*, 1076–1081. [CrossRef]
4. Rincon-Mora, G.A.; Allen, P.E. A 1.1-V current-mode and piecewise-linear curvature-corrected bandgap reference. *IEEE J. Solid-State Circuits* **1998**, *33*, 1551–1554. [CrossRef]
5. Moreira, P. Radiation Effects on the "CERN_bandgap" Circuit. Available online: http://proj-qpll.web.cern.ch/proj-qpll/images/bandgapRadEffects.pdf (accessed on 16 May 2019)
6. Georgiou, J.; Toumazou, C. A 126 µW cochlear chip for a totally implantable system. *IEEE J. Solid-State Circuits* **2005**, *40*, 430–443. [CrossRef]
7. Banba, H.; Shiga, H.; Umezawa, A.; Miyaba, T.; Tanzawa, T.; Atsumi, S.; Sakui, K. A CMOS bandgap reference circuit with sub-1-V operation. *IEEE J. Solid-State Circuits* **1999**, *34*, 670–674. [CrossRef]
8. Basyurt, P.B.; Bonizzoni, E.; Aksin, D.Y.; Maloberti, F. A 0.4-V Supply Curvature-Corrected Reference Generator with 84.5-ppm/°C Average Temperature Coefficient within −40 °C to 130 °C. *IEEE Trans. Circuits Syst. II Express Briefs* **2017**, *64*, 362–366. [CrossRef]
9. Huang, Y.; Zhu, L.; Kong, F.; Cheung, C.; Najafizadeh, L. BiCMOS-Based Compensation: Toward Fully Curvature-Corrected Bandgap Reference Circuits. *IEEE Trans. Circuits Syst. I Reg. Pap.* **2017**, *64*, 1210–1223. [CrossRef]
10. Kamath, U.; Cullen, E.; Jennings, J.; Cical, I.; Walsh, D.; Lim, P.; Farley, B.; Staszewski, R.B. A 1 V Bandgap Reference in 7-nm FinFET with a Programmable Temperature Coefficient and an Inaccuracy of ±0.2% from −45 °C to 125 °C. In Proceedings of the ESSCIRC 2018 IEEE 44th European Solid State Circuits Conference (ESSCIRC), Dresden, Germany, 3–6 September 2018; pp. 78–81. [CrossRef]
11. Ma, B.; Yu, F. A Novel 1.2-V 4.5-ppm/°C Curvature-Compensated CMOS Bandgap Reference. *IEEE Trans. Circuits Syst. I Reg. Pap.* **2014**, *61*, 1026–1035. [CrossRef]
12. Chen, H.M.; Lee, C.C.; Jheng, S.H.; Chen, W.C.; Lee, B.Y. A Sub-1 ppm/ °C Precision Bandgap Reference with Adjusted-Temperature-Curvature Compensation. *IEEE Trans. Circuits Syst. I Reg. Pap.* **2017**, *64*, 1308–1317. [CrossRef]
13. Luong, P.; Christoffersen, C.; Rossi-Aicardi, C.; Dualibe, C. Nanopower, Sub-1 V, CMOS Voltage References with Digitally-Trimmable Temperature Coefficients. *IEEE Trans. Circuits Syst. I Reg. Pap.* **2017**, *64*, 787–798. [CrossRef]

14. De Vita, G.; Iannaccone, G. A Sub-1-V, 10 ppm/°C, Nanopower Voltage Reference Generator. *IEEE J. Solid-State Circuits* **2007**, *42*, 1536–1542. [CrossRef]
15. Giustolisi, G.; Palumbo, G.; Criscione, M.; Cutri, F. A low-voltage low-power voltage reference based on subthreshold MOSFETs. *IEEE J. Solid-State Circuits* **2003**, *38*, 151–154. [CrossRef]
16. Seok, M.; Kim, G.; Blaauw, D.; Sylvester, D. A Portable 2-Transistor Picowatt Temperature-Compensated Voltage Reference Operating at 0.5 V. *IEEE J. Solid-State Circuits* **2012**, *47*, 2534–2545. [CrossRef]
17. Andreou, C.M.; Georgiou, J. A 0.7 V, 2.7 μW, 12.9 ppm/°C over 180 °C CMOS subthreshold voltage reference. *Int. J. Circuit Theory Appl.* **2017**, *45*, 1349–1368. [CrossRef]
18. Liu, F.; Yang, F.; Wang, H.; Xiang, X.; Zhou, X.; Hu, S.; Lin, Z.; Bermak, A.; Tang, F. Radiation Hardened CMOS Negative Voltage Reference for Aerospace Application. *IEEE Trans. Nucl. Sci.* **2017**, *64*, 2505–2510. [CrossRef]
19. Cardoso, A.; Chakraborty, P.; Karaulac, N.; Fleischhauer, D.; Lourenco, N.; Fleetwood, Z.; Omprakash, A.; England, T.; Jung, S.; Najafizadeh, L.; et al. Single-Event Transient and Total Dose Response of Precision Voltage Reference Circuits Designed in a 90-nm SiGe BiCMOS Technology. *IEEE Trans. Nucl. Sci.* **2014**, *61*, 3210–3217. [CrossRef]
20. Gromov, V.; Annema, A.; Kluit, R.; Visschers, J.; Timmer, P. A Radiation Hard Bandgap Reference Circuit in a Standard 0.13 μm CMOS Technology. *IEEE Trans. Nucl. Sci.* **2007**, *54*, 2727–2733. [CrossRef]
21. McCue, B.; Blalock, B.; Britton, C.; Potts, J.; Kemerling, J.; Isihara, K.; Leines, M. A Wide Temperature, Radiation Tolerant, CMOS-Compatible Precision Voltage Referencefor Extreme Radiation Environment Instrumentation Systems. *IEEE Trans. Nucl. Sci.* **2013**, *60*, 2272–2279. [CrossRef]
22. Najafizadeh, L.; Sutton, A.; Diestelhorst, R.; Bellini, M.; Jun, B.; Cressler, J.; Marshall, P.; Marshall, C. A Comparison of the Effects of X-ray and Proton Irradiation on the Performance of SiGe Precision Voltage References. *IEEE Trans. Nucl. Sci.* **2007**, *54*, 2238–2244. [CrossRef]
23. Cao, Y.; De Cock, W.; Steyaert, M.; Leroux, P. A 4.5 MGy TID-Tolerant CMOS Bandgap Reference Circuit Using a Dynamic Base Leakage Compensation Technique. *IEEE Trans. Nucl. Sci.* **2013**, *60*, 2819–2824. [CrossRef]
24. Abare, W.; Brueggeman, F.; Pease, R.; Krieg, J.; Simons, M. Comparative analysis of low dose-rate, accelerated, and standard cobalt-60 radiation response data for a low-dropout voltage regulator and a voltage reference. In Proceedings of the 2002 IEEE Radiation Effects Data Workshop, Phoenix, AZ, USA, 15–19 July 2002; pp. 177–180.
25. McClure, S.; Gorelick, J.; Pease, R.; Rax, B.; Ladbury, R. Total dose performance of radiation hardened voltage regulators and references. In Proceedings of the 2001 IEEE Radiation Effects Data Workshop, Vancouver, BC, Canada, 16–20 July 2001; pp. 1–5.
26. Piccin, Y.; Lapuyade, H.; Deval, Y.; Morche, C.; Seyler, J.Y.; Goutti, F. Radiation-Hardening Technique for Voltage Reference Circuit in a Standard 130 nm CMOS Technology. *IEEE Trans. Nucl. Sci.* **2014**, *61*, 967–974. [CrossRef]
27. Andreou, C.M.; Paccagnella, A.; Gonzalez-Castano, D.M.; Gomez, F.; Liberali, V.; Prokofiev, A.V.; Calligaro, C.; Javanainen, A.; Virtanen, A.; Nahmad, D.; et al. A Subthreshold, Low-Power, RHBD Reference Circuit, for Earth Observation and Communication Satellites. In Proceedings of the IEEE International Symposium on Circuits and Systems (ISCAS), Lisbon, Portugal, 24–27 May 2015.
28. Martin, H.; Martin-Holgado, P.; Morilla, Y.; Entrena, L.; San-Millan, E. Total Ionizing Dose Effects on a Delay-Based Physical Unclonable Function Implemented in FPGAs. *Electronics* **2018**, *7*, 163. [CrossRef]
29. Prinzie, J.; Christiansen, J.; Moreira, P.; Steyaert, M.; Leroux, P. Comparison of a 65 nm CMOS Ring- and LC-Oscillator Based PLL in Terms of TID and SEU Sensitivity. *IEEE Trans. Nucl. Sci.* **2017**, *64*, 245–252. [CrossRef]
30. Prinzie, J.; Steyaert, M.; Leroux, P. *Radiation Effects in CMOS Technology*. In *Radiation Hardened CMOS Integrated Circuits for Time-Based Signal Processing*; Analog Circuits and Signal Processing; Springer: Cham, Switzerland, 2018. [CrossRef]
31. Quinn, H.; Black, D.; Robinson, W.; Buchner, S. Fault Simulation and Emulation Tools to Augment Radiation-Hardness Assurance Testing. *IEEE Trans. Nucl. Sci.* **2013**, *60*, 2119–2142. [CrossRef]
32. Quinn, H. Challenges in Testing Complex Systems. *IEEE Trans. Nucl. Sci.* **2014**, *61*, 766–786. [CrossRef]
33. Esqueda, I.; Barnaby, H.; Holbert, K.; El-Mamouni, F.; Schrimpf, R. Modeling of Ionizing Radiation-Induced Degradation in Multiple Gate Field Effect Transistors. *IEEE Trans. Nucl. Sci.* **2011**, *58*, 499–505. [CrossRef]

34. Manghisoni, M.; Ratti, L.; Re, V.; Speziali, V.; Traversi, G.; Candelori, A. Comparison of ionizing radiation effects in 0.18 and 0.25 μm CMOS technologies for analog applications. *IEEE Trans. Nucl. Sci.* **2003**, *50*, 1827–1833. [CrossRef]
35. Cochran, D.; Buchner, S.; Irwin, T.; Kniffin, S.; Ladbury, R.; Palor, C.; LaBel, K.; Marshall, C.; Reed, R.; Sanders, A.; et al. Current total ionizing dose results and displacement damage results for candidate spacecraft electronics for NASA. In Proceedings of the 2004 IEEE Radiation Effects Data Workshop, Atlanta, GA, USA, 22 July 2004; pp. 19–25.
36. Atkinson, N. System-Level Radiation Hardening of Low-Voltage Analog/Mixed-Signal Circuits. Ph.D. Thesis, Vanderbilt University, Nashville, TN, USA, 2013.
37. Najafizadeh, L.; Phillips, S.; Moen, K.; Diestelhorst, R.; Bellini, M.; Saha, P.; Cressler, J.; Vizkelethy, G.; Turowski, M.; Raman, A.; et al. Single Event Transient Response of SiGe Voltage References and Its Impact on the Performance of Analog and Mixed-Signal Circuits. *IEEE Trans. Nucl. Sci.* **2009**, *56*, 3469–3476. [CrossRef]
38. Gerardin, S.; Bagatin, M.; Paccagnella, A.; Grurmann, K.; Gliem, F.; Oldham, T.; Irom, F.; Nguyen, D. Radiation Effects in Flash Memories. *IEEE Trans. Nucl. Sci.* **2013**, *60*, 1953–1969. [CrossRef]
39. Roig, F.; Dusseau, L.; Khachatrian, A.; Roche, N.H.; Privat, A.; Vaille, J.R.; Boch, J.; Warner, J.; Saigne, F.; Buchner, S.; et al. Modeling and Investigations on TID-ASETs Synergistic Effect in LM124 Operational Amplifier From Three Different Manufacturers. *IEEE Trans. Nucl. Sci.* **2013**, *60*, 4430–4438. [CrossRef]
40. Anelli, G.; Campbell, M.; Delmastro, M.; Faccio, F.; Floria, S.; Giraldo, A.; Heijne, E.; Jarron, P.; Kloukinas, K.; Marchioro, A.; et al. Radiation tolerant VLSI circuits in standard deep submicron CMOS technologies for the LHC experiments: practical design aspects. *IEEE Trans. Nucl. Sci.* **1999**, *46*, 1690–1696. [CrossRef]
41. Ferlet-Cavrois, V. Electronic radiation hardening—Radiation Hardness Assurance and Technology Demonstration Activities. In Proceedings of the JUICE Instrument Workshop, Darmstadt, Germany, 9–11 November 2011.
42. Gerardin, S.; Bagatin, M.; Paccagnella, A.; Ferlet-Cavrois, V. Degradation of Sub 40-nm NAND Flash Memories Under Total Dose Irradiation. *IEEE Trans. Nucl. Sci.* **2012**, *59*, 2952–2958. [CrossRef]
43. Freitag, R.; Brown, D.; Dozier, C. Evidence for two types of radiation-induced trapped positive charge. *IEEE Trans. Nucl. Sci.* **1994**, *41*, 1828–1834. [CrossRef]
44. Gaillardin, M.; Goiffon, V.; Marcandella, C.; Girard, S.; Martinez, M.; Paillet, P.; Magnan, P.; Estribeau, M. Radiation Effects in CMOS Isolation Oxides: Differences and Similarities With Thermal Oxides. *IEEE Trans. Nucl. Sci.* **2013**, *60*, 2623–2629. [CrossRef]
45. Roig, F.; Dusseau, L.; Ribeiro, P.; Auriel, G.; Roche, N.H.; Privat, A.; Vaille, J.R.; Boch, J.; Saigne, F.; Marec, R.; et al. The Role of Feedback Resistors and TID Effects in the ASET Response of a High Speed Current Feedback Amplifier. *IEEE Trans. Nucl. Sci.* **2014**, *61*, 3201–3209. [CrossRef]
46. Casey, M.; Armstrong, S.; Arora, R.; King, M.; Ahlbin, J.; Francis, S.; Bhuva, B.; McMorrow, D.; Hughes, H.; McMarr, P.; et al. Effect of Total Ionizing Dose on a Bulk 130 nm Ring Oscillator Operating at Ultra-Low Power. *IEEE Trans. Nucl. Sci.* **2009**, *56*, 3262–3266. [CrossRef]
47. Casey, M.; Amusan, O.; Nation, S.; Loveless, T.; Balasubramanian, A.; Bhuva, B.; Reed, R.; McMorrow, D.; Weller, R.; Alles, M.; et al. Single-Event Effects on Combinational Logic Circuits Operating at Ultra-Low Power. *IEEE Trans. Nucl. Sci.* **2008**, *55*, 3342–3346. [CrossRef]
48. Tai-Hua, C.; Jinhui, C.; Clark, L.; Knudsen, J.; Samson, G. Ultra-Low Power Radiation Hardened by Design Memory Circuits. *IEEE Trans. Nucl. Sci.* **2007**, *54*, 2004–2011.
49. Maki, G.K.; Yeh, P.S. *Radiation Tolerant Ultra Low Power CMOS Microelectronics: Technology Development Status*; NASA: Washington, DC, USA, 2003.
50. Xapsos, M.; Summers, G.; Jackson, E. Enhanced total ionizing dose tolerance of bulk CMOS transistors fabricated for ultra-low power applications. *IEEE Trans. Nucl. Sci.* **1999**, *46*, 1697–1701. [CrossRef]
51. Andreou, C.M.; Georgiou, J. A 0.75-V, 4-μW, 15-ppm/°C, 190 °C temperature range, voltage reference. *Int. J. Circuit Theory Appl.* **2016**, *44*, 1029–1038. [CrossRef]
52. Oldham, T.; McLean, F. Total ionizing dose effects in MOS oxides and devices. *IEEE Trans. Nucl. Sci.* **2003**, *50*, 483–499. [CrossRef]
53. Schwank, J.; Shaneyfelt, M.; Fleetwood, D.; Felix, J.; Dodd, P.; Paillet, P.; Ferlet-Cavrois, V. Radiation Effects in MOS Oxides. *IEEE Trans. Nucl. Sci.* **2008**, *55*, 1833–1853. [CrossRef]

54. Paccagnella, A. Radiation response and reliability of oxides used in advanced processes. Short Course. In Proceedings of the Nuclear and Space Radiation Effects Conference, Monterey, CA, USA, 21–25 July 2003.
55. Ivan Sanchez Esqueda, I. Modeling of Total Ionizing Dose Effects in Advanced Complementary Metal-Oxide-Semiconductor Technologies. Ph.D. Thesis, Arizona State University, Phoenix, AZ, USA, 2011.
56. Faccio, F.; Cervelli, G. Radiation-induced edge effects in deep submicron CMOS transistors. *IEEE Trans. Nucl. Sci.* **2005**, *52*, 2413–2420. [CrossRef]
57. Lacoe, R. CMOS scaling, design principles and hardening-by-design methodologies. Short Course. In Proceedings of the Nuclear and Space Radiation Effects Conference, Monterey, CA, USA, 21–25 July 2003.
58. Fourches, N. *Total Dose and Dose Rate Effects on Some Current Semiconducting Devices, Current Topics in Ionizing Radiation Research*; InTech: London, UK, 2012.
59. Virmontois, C.; Goiffon, V.; Magnan, P.; Girard, S.; Inguimbert, C.; Petit, S.; Rolland, G.; Saint-Pe, O. Displacement Damage Effects Due to Neutron and Proton Irradiations on CMOS Image Sensors Manufactured in Deep Submicron Technology. *IEEE Trans. Nucl. Sci.* **2010**, *57*, 3101–3108. [CrossRef]
60. Puchner, H.; Whatley, M.; Tausch, J. Neutron displacement damage of an integrated bandgap voltage reference of a 90 nm technology CMOS based SRAM. In Proceedings of the 2011 12th European Conference on Radiation and Its Effects on Components and Systems (RADECS), Sevilla, Spain, 19–23 September 2011; pp. 812–814.
61. Srour, J.; Marshall, C.; Marshall, P. Review of displacement damage effects in silicon devices. *IEEE Trans. Nucl. Sci.* **2003**, *50*, 653–670. [CrossRef]
62. Maurer, R.; Fraeman, M.; Martin, M.; Roth, D. Harsh Environments: Space Radiation Environment, Effects, and Mitigation. *Johns Hopkins APL Tech. Dig.* **2008**, *28*, 17–29.
63. Buchner, S.; Baze, M. Single-Event Transients in Fast Electronic Circuits. Short Course. In Proceedings of the Nuclear and Space Radiation Effects Conference, Vancouver, BC, Canada, 16–20 July 2001.
64. Stassinopoulos, E.; Raymond, J.P. The space radiation environment for electronics. *Proc. IEEE* **1988**, *76*, 1423–1442. [CrossRef]
65. Atkinson, N.M.; Ahlbin, J.R.; Witulski, A.F.; Gaspard, N.J.; Holman, W.T.; Bhuva, B.L.; Zhang, E.X.; Chen, L.; Massengill, L.W. Effect of Transistor Density and Charge Sharing on Single-Event Transients in 90-nm Bulk CMOS. *IEEE Trans. Nucl. Sci.* **2011**, *58*, 2578–2584. [CrossRef]
66. Buchner, S.; McMorrow, D. Single-Event Transients in Bipolar Linear Integrated Circuits. *IEEE Trans. Nucl. Sci.* **2006**, *53*, 3079–3102. [CrossRef]
67. Ahlbin, J.R.; Member, S.; Gadlage, M.J.; Atkinson, N.M.; Narasimham, B.; Bhuva, B.L.; Member, S.; Witulski, A.F.; Holman, W.T.; Eaton, P.H.; et al. Effect of Multiple-Transistor Charge Collection on Single-Event Transient Pulse Widths. *IEEE Trans. Device. Mater. Reliab.* **2011**, *11*, 401–406. [CrossRef]
68. Javanainen, A.; Member, S.; Schwank, J.R.; Shaneyfelt, M.R.; Harboe-sørensen, R.; Virtanen, A.; Kettunen, H.; Dalton, S.M.; Dodd, P.E.; Member, S.; et al. Heavy-Ion Induced Charge Yield in MOSFETs. *IEEE Trans. Nucl. Sci.* **2009**, *56*, 3367–3371. [CrossRef]
69. Baumann, R.C. Single event effects in advanced CMOS technology. In Proceedings of the IEEE Nuclear and Space Radiation Effects Conference Short Course Text, Seattle, WA, USA, 11–15 July 2005; pp. 1–59.
70. Ahlbin, J.R. Characterization of the Mechanisms Affecting Single-Event Transients in Sub-100 Nm Technologies. Ph.D. Thesis, Vanderbilt University, Nashville, TN, USA, 2012.
71. Wang, T. Study of Single-Event Transient Effects on Analog Circuits. Ph.D. Thesis, University of Saskatchewan, Saskatoon, SK, Canada, 2011.
72. Ferlet-Cavrois, V.; Massengill, L.; Gouker, P. Single Event Transients in Digital CMOS—A Review. *IEEE Trans. Nucl. Sci.* **2013**, *60*, 1767–1790. [CrossRef]
73. Ferlet-Cavrois, V.; Paillet, P.; McMorrow, D.; Fel, N.; Baggio, J.; Girard, S.; Duhamel, O.; Melinger, J.; Gaillardin, M.; Schwank, J.; et al. New Insights Into Single Event Transient Propagation in Chains of Inverters—Evidence for Propagation-Induced Pulse Broadening. *IEEE Trans. Nucl. Sci.* **2007**, *54*, 2338–2346. [CrossRef]
74. Adell, P.; Schrimpf, R.; Barnaby, H.; Marec, R.; Chatry, C.; Calvel, P.; Barillot, C.; Mion, O. Analysis of single-event transients in analog circuits. *IEEE Trans. Nucl. Sci.* **2000**, *47*, 2616–2623. [CrossRef]
75. Chen, C.H.; Knag, P.; Zhang, Z. Characterization of Heavy-Ion-Induced Single-Event Effects in 65 nm Bulk CMOS ASIC Test Chips. *IEEE Trans. Nucl. Sci.* **2014**, *61*, 2694–2701. [CrossRef]

76. Hofbauer, M.; Schweiger, K.; Zimmermann, H.; Giesen, U.; Langner, F.; Schmid, U.; Steininger, A. Supply Voltage Dependent On-Chip Single-Event Transient Pulse Shape Measurements in 90-nm Bulk CMOS Under Alpha Irradiation. *IEEE Trans. Nucl. Sci.* **2013**, *60*, 2640–2646. [CrossRef]
77. Reed, R.A. Fundamental Mechanisms for Single Particle-Induced Soft Errors. In Proceedings of the IEEE NSREC Short Course 2008, Tucson, AZ, USA, 14–18 July 2008.
78. Gerardin, S.; Bagatin, M.; Paccagnella, A.; Visconti, A.; Bonanomi, M.; Pellizzer, F.; Vela, M.; Ferlet-Cavrois, V. Single Event Effects in 90-nm Phase Change Memories. *IEEE Trans. Nucl. Sci.* **2011**, *58*, 2755–2760. [CrossRef]
79. Schwank, J.; Shaneyfelt, M.; Dodd, P. Radiation Hardness Assurance Testing of Microelectronic Devices and Integrated Circuits: Test Guideline for Proton and Heavy Ion Single-Event Effects. *IEEE Trans. Nucl. Sci.* **2013**, *60*, 2101–2118. [CrossRef]
80. Adell, P.; Schrimpf, R.; Cirba, C.; Holman, W.; Zhu, X.; Barnaby, H.; Mion, O. Single event transient effects in a voltage reference. *Microelectron. Reliab.* **2005**, *45*, 355–359. [CrossRef]
81. Irom, F.; Miyahira, T.; Adell, P.; Laird, J.; Conder, B.; Pouget, V.; Essely, F. Investigation of Single-Event Transients in Linear Voltage Regulators. *IEEE Trans. Nucl. Sci.* **2008**, *55*, 3352–3359. [CrossRef]
82. Savage, M.; Titus, J.; Turflinger, T.; Pease, R.; Poivey, C. A comprehensive analog single-event transient analysis methodology. *IEEE Trans. Nucl. Sci.* **2004**, *51*, 3546–3552. [CrossRef]
83. Poivey, C.; Howard, J.; Buchner, S.; LaBel, K.; Forney, J.; Kim, H.; Assad, A. Development of a test methodology for single-event transients (SETs) in linear devices. *IEEE Trans. Nucl. Sci.* **2001**, *48*, 2180–2186. [CrossRef]
84. Fleetwood, D.; Rodgers, M.; Tsetseris, L.; Zhou, X.; Batyrev, I.; Wang, S.; Schrimpf, R.; Pantelides, S. Effects of device aging on microelectronics radiation response and reliability. *Microelectron. Reliab.* **2007**, *47*, 1075–1085. [CrossRef]
85. Chen, W.; Varanasi, N.; Pouget, V.; Barnaby, H.; Vermeire, B.; Adell, P.; Copani, T.; Fouillat, P. Impact of VCO Topology on SET Induced Frequency Response. *IEEE Trans. Nucl. Sci.* **2007**, *54*, 2500–2505. [CrossRef]
86. Ferlet-Cavrois, V.; McMorrow, D.; Kobayashi, D.; Fel, N.; Melinger, J.; Schwank, J.; Gaillardin, M.; Pouget, V.; Essely, F.; Baggio, J.; et al. A New Technique for SET Pulse Width Measurement in Chains of Inverters Using Pulsed Laser Irradiation. *IEEE Trans. Nucl. Sci.* **2009**, *56*, 2014–2020. [CrossRef]
87. Andreou, C.M.; Georgiou, J. An all-subthreshold, 0.75 V supply, 2 ppm/°C, CMOS Voltage Reference. In Proceedings of the 2013 IEEE International Symposium on Circuits and Systems (ISCAS2013), Beijing, China, 19–23 May 2013; pp. 1476–1479. [CrossRef]
88. Bucher, M.; Nikolaou, A.; Papadopoulou, A.; Makris, N.; Chevas, L.; Borghello, G.; Koch, H.D.; Faccio, F. Total ionizing dose effects on analog performance of 65 nm bulk CMOS with enclosed-gate and standard layout. In Proceedings of the 2018 IEEE International Conference on Microelectronic Test Structures (ICMTS), Austin, TX, USA, 19–22 March 2018; pp. 166–170.
89. Vittoz, E.; Fellrath, J. CMOS analog integrated circuits based on weak inversion operations. *IEEE J. Solid-State Circuits* **1977**, *12*, 224–231. [CrossRef]
90. Andreou, C.M.; Javanainen, A.; Rominski, A.; Virtanen, A.; Liberali, V.; Calligaro, C.; Prokofiev, A.V.; Gerardin, S.; Bagatin, M.; Paccagnella, A.; et al. Single Event Transients and Pulse Quenching Effects in Bandgap Reference Topologies for Space Applications. *IEEE Trans. Nucl. Sci.* **2016**, *63*, 2950–2961. [CrossRef]
91. Schrimpf, R. Physics and Hardness Assurance for Bipolar Technologies. Short Course. In Proceedings of the Nuclear and Space Radiation Effects Conference, Vancouver, BC, Canada, 16–20 July 2001.

 © 2019 by the authors. Licensee MDPI, Basel, Switzerland. This article is an open access article distributed under the terms and conditions of the Creative Commons Attribution (CC BY) license (http://creativecommons.org/licenses/by/4.0/).

Article

Single Event Transients in CMOS Ring Oscillators

Jeffrey Prinzie * and Valentijn De Smedt

Department of Electrical Engineering (ESAT), KU Leuven, 3000 Leuven, Belgium; valentijn.desmedt@kuleuven.be
* Correspondence: jeffrey.prinzie@kuleuven.be

Received: 28 April 2019; Accepted: 29 May 2019; Published: 1 June 2019

Abstract: In this paper, a time-variant analysis is made on Single-Event Transients (SETs) in integrated CMOS ring oscillators. The Impulse Sensitive Function (ISF) of the oscillator is used to analyze the impact of the relative moment when a particle hits the circuit. The analysis is based on simulations and verified experimentally with a Two-Photon Absorption (TPA) laser setup. The experiments are done using a 65 nm CMOS test chip.

Keywords: Single-Event Upsets (SEUs); radiation effects; Ring Oscillators; Impulse Sensitive Function; Radiation Hardening by Design

1. Introduction

Integrated, high-speed clock generation circuits are essential blocks in nearly all modern silicon systems. A wide variety of circuits and architectures is available in the literature. Most design choices depend on the desired quality and frequency of the generated clock signal and the reference clock. The vast majority of high-speed clock generators employ a Phase Locked Loop (PLL) [1] that ensures a fixed frequency multiplication and a known phase relationship between the on-chip high-speed oscillator and off-chip reference clock. However, other architectures such as Delay Locked Loops (DLLs) [2] and Multiplying DLLs (MDLLs) [3] are gaining more interest over the past years.

The quality of a synthesized clock mainly depends on the quality of the reference clock and the on-chip oscillator [4]. While the latter is in the hand of a designer, the former is usually not. Therefore, the study of integrated CMOS oscillators has been an interesting research topic for the past decades. Two main types of oscillators are commonly used: LC-tank oscillators and ring oscillators [5]. LC-oscillators rely on a resonant tank of an inductor and a capacitor which resonates at a frequency $\omega = 1/\sqrt{LC}$. They are known to exhibit superior phase noise and jitter performance and are widely used in low jitter clock synthesis, down to 100 fs RMS and low phase noise local oscillators for wireless communication links where out-of-band phase noise can limit the performance in the case of a strong interferer [6,7]. While the performance and power efficiency of an LC-oscillator is excellent, the main downside is its large area and limited tuning range. Typically, the inductor occupies more than 150×150 μm^2. Secondly, since the quality factor of the inductor peaks in the GHz frequency range but falls for lower frequencies, such oscillators are rarely used below several 100 MHz without the use of a divider. When such extremely low noise levels are not mandatory, integrated ring oscillators can prove their usage. In terms of area usage, ring oscillators can be as small as a few tens of digital gates [8]. They rely on the total delay of a closed loop of digital delay cells. Ring oscillators find their application on digital systems such as microprocessors, complex SoCs and serial communication links.

Today's most advanced electronic systems also find their application in harsh environments containing ionizing radiation. Examples of applications are space systems such as satellites and deep space probes [9], high-energy physics experiments such as the ATLAS [10] and CMS detectors [11,12] at the Large Hadron Collider (LHC) at CERN. However, terrestrial applications such as autonomous airplanes, cars and high-reliable computing systems in data centers are also affected by ionizing radiation. Charged particles that impact a silicon chip can generate charges in the silicon. This occurs nearly instantly when a particle crosses the circuit. When this occurs near the active source and drain junctions of a transistor, these charges can be collected by the junctions and injected in the circuit. In digital circuits, these Single Event Effects (SEEs) are Single Event Transients (SETs) or Single-Event Upsets (SEUs). The former is only a temporal error while the latter remains erroneous. SEEs also strongly impact the oscillator in a clock generator. In particular, SEEs generate phase transients in the clock that can cause errors in synchronous logic clocked by the clock generator. Especially when timing is critical, phase jumps as large as 20 ps can be catastrophic for the reliability of a digital platform. Therefore, a solid understanding of the basic mechanisms of SEEs in CMOS ring oscillators is essential to give more insight in hardening and protection methods of these blocks.

This paper is organized as follows. Section 2 introduces a time-dependent model for the sensitivity of an oscillator to current impulses, which is applied to estimate its response to SEEs. In Section 3, experiments are shown that prove the time-dependent radiation effects using a two-photo laser absorption setup. Finally, conclusions are drawn.

2. Time Dependent Effects in Oscillators

A commonly used theory to understand translation of white and colored noise sources in oscillator circuits is the Linear-Time-Variant (LTV) noise theory of Hajimiri [13–15]. In this theory, noise is modeled as a current source injecting pulses on the different nodes of the oscillator, as depicted in Figure 1. This image shows an abstract (high-level) representation of a ring oscillator where a current impulse is injected at one particular node, which is used in the analysis below. The shown oscillator has N odd amount of stages for the analysis. This current pulse is causing phase steps in the oscillator, which are integrated over time, resulting in an uncertainty on the phase, also called phase noise. The impact of the injected pulses is weighted by the so-called Impulse-Sensitivity-Function (ISF), a dimensionless and frequency-independent function determining the sensitivity of the oscillator node to the injected noise. The instantaneous value of the ISF is a direct measure for the noise-to-phase transfer function of the oscillator [16]:

$$h_\phi(t, \tau) = \frac{\Gamma(\omega_0 \cdot \tau)}{q_{max}} \cdot u(t - \tau) \quad (1)$$

where t is the time, τ is the moment of impact of the current pulse, $h_\phi(t, \tau)$ is the current-to-phase impulse response, $\Gamma(\omega_0 \cdot \tau)$ is the ISF at time τ, q_{max} is the charge displacement during an oscillator cycle (proportional to the amplitude) and $u(t - \tau)$ is the unit step function. Since the oscillator is assumed to be a LTV system, the superposition principle can be applied to calculate the impact of a infinite series of pulses:

$$\phi(t) = \int_{-\infty}^{\infty} h_\phi(t, \tau) \cdot i(\tau) d\tau = \int_{-\infty}^{t} \frac{\Gamma(\omega_0 \cdot \tau)}{q_{max}} \cdot i(\tau) d\tau \quad (2)$$

where $\phi(t)$ represents the integrated phase deviation compared to the ideal oscillator and $i(t)$ is the injected noise current over time. In [13], it is shown that application of the LTV approach to a noise spectrum results in the typical $1/f^\alpha$ noise shape around the oscillator carrier. In this article, the ISF is used to calculate the impact of current pulses caused by a particle strike on a four-stage ring oscillator.

In Figure 2, a waveform is shown where a disturbance is injected at two different moments in time. In Figure 2a, a current impulse is injected in the maximal saturated region of the oscillator. Since the current does not change in the saturated shape of the waveform, the ISF for current injections in this region is approximately zero. However, as shown in Figure 2b, when the current is injected during the signal transition times, the total phase error is maximal. In general, the ISF is proportional to the slew rate of the waveform, which is large during transition.

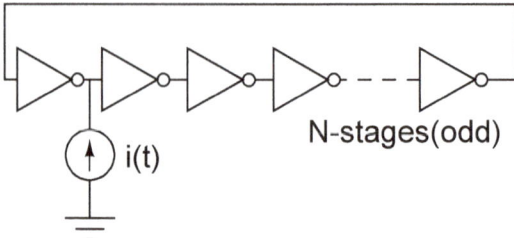

Figure 1. Current pulse injection in a ring oscillator with arbitrary number of N stages.

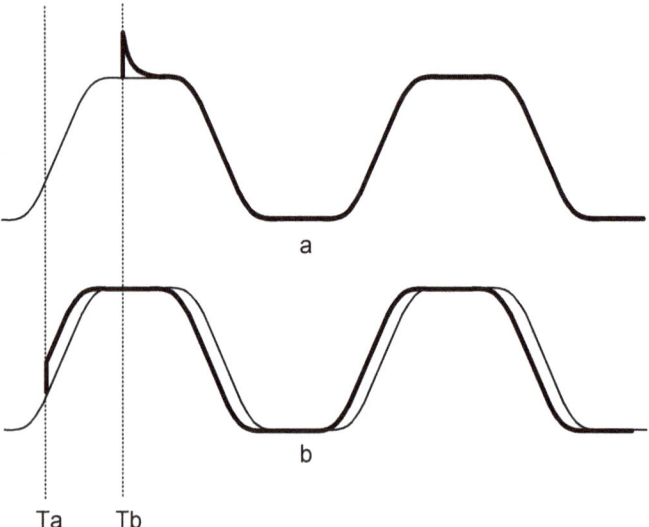

Figure 2. Time-dependent effect in a ring oscillator waveform: (**a**) Injection in saturation region. (**b**) Injection during transition.

2.1. Calculation of the ISF

To calculate the ISF of an oscillator, different methods are described in the literature [13]. The first method is based on circuit simulations where a current pulse is injected at different moments spread along the oscillator period. By calculating the induced phase shift, the ISF can easily be calculated. However, more analytical methods are also available, making use of the state space description of an oscillator. In a ring oscillator, the state variables f_i can be considered to be the node voltages at the output of each stage. After all, a state variable is an independent memory state in the system: a current through an inductor or a voltage across a (parasitic) capacitance [15]. In this case, the ISF for a pulse injected at the output node of stage i of an n-stage oscillator can be written as:

$$\Gamma_i(\omega \cdot t) = \frac{f'_i}{\sum_{j=1}^{n} f'^2_j}. \tag{3}$$

This shows that the ISF is low (zero) when the node voltage is constant and high during the transients. In [16], the ISF of a ring oscillator is approximated by a flat line equal to zero, with a triangular (alternating positive and negative) pulse at each transient of the considered oscillator node. Although this is a piecewise linearized approximation, the results for a typical statistical noise source are satisfying.

The approximation, however, only holds for oscillators of which the node voltages are strongly saturating against the supply rails with sharp transients in between. The oscillator considered in this article is based on a differential Maneatis cell, resulting a much smoother and non-saturating waveforms and therefore also a smoother shape of the ISF.

2.2. Particle Strikes as a Noise Source

Similar to noise, charged particles ionize the silicon substrate resulting in free charges in the substrate. In the region of the source and drain junctions of the transistors, these charges can be collected in the junctions by the strong electric field in the depletion region. As such, a part of the generated charges is injected in the circuit nodes. An exact calculation of the impact is a tedious and complex task: analytical methods are based on approximations and are therefore often inaccurate; numerical TCAD simulations, on the other hand, are considered to present the most accurate results but are time-consuming and sensitive to doping and geometry inaccuracies. The most common issue here is that many of the technology parameters are unknown to designers. A well accepted model to assess charge injection in analog circuits is the double exponential current shape [17]:

$$i(t) = Q \cdot \frac{e^{-a \cdot t} - e^{-b \cdot t}}{b - a} \quad (4)$$

where Q is the total collected charge and a and b are technology dependent time constants. In the case where the time constants are significantly faster than the overall circuit dynamics, the injected current can be simplified as an impulse current:

$$i(t) \approx Q \cdot \delta(t) \quad (5)$$

For a particle strike at time τ, the resulting phase error is equal to (using Equation (1)):

$$\phi(t) \approx \int_{-\infty}^{\infty} h_\phi(t, \tau) \cdot Q \cdot \delta(\tau) d\tau = \frac{Q \cdot \Gamma(\omega_0 \cdot \tau)}{q_{max}} \cdot u(t - \tau) \quad (6)$$

The previous analysis only considered a single node in the oscillator. Practically, an oscillator is built with N stages where N is odd for single ended oscillators and N can be either even or odd for differential oscillators. An oscillator will oscillate at a frequency:

$$f = \frac{1}{2 \cdot N \cdot T_d} \quad (7)$$

in which T_d is the gate delay per stage and N is the number of stages. The signal waveform is therefore shifted between two successive stages by T_d. As a consequence, the ISF is also shifted between successive stages. The current-to-phase impulse response for all successive stages in an inverting ring oscillator is therefore:

$$h_\phi(t, \tau)[i] = \frac{\Gamma(\omega_0 \cdot \tau - i/N)}{q_{max}} \cdot u(t - \tau) \quad (8)$$

where i = [0...(N − 1)] represents each stage. This is elaborately discussed in [14]. The time-shifting is experimentally shown in Section 3 by measuring the ISF at different stages in the experimental design. The phase shift of all stages is considered to be of significant interest for phase noise analysis and noise folding due to common noise sources, such as supply or substrate noise. However, since radiation effects are only impacting one node simultaneously (if the cells are sufficiently large), each stage can be represented by the same ISF and the phase shift can be ignored.

3. Experiments

3.1. Experimental Circuit Description

To quantify the magnitude of the ISF experimentally, a ring oscillator was designed and prototyped in a 65 nm CMOS technology with standard-Vt devices an a core voltage of 1.2 V. The oscillator had four differential voltage controlled delay stages and oscillates from 1.5 GHz to 3.2 GHz with a nominal frequency of 2 GHz (further assumed in all experiments). The circuit schematic of the delay cells is shown in Figure 3a and was designed for these experiments based on a well known and frequently used Maneatis delay cell [18]. A PMOS equivalent circuit of the Maneatis delay cell was designed to reduce 1/f noise. In addition, the bias block from [18] was used. The transistor sizes were chosen to meet the target frequency of 2 GHz in this technology. The delay through the ring was adjusted using the bias voltage of M3 (biasp), which regulated the current through the cells. The bias voltage of the NMOS load was adjusted to keep the oscillation amplitude relatively constant. These voltages were generated by a bias generator, as shown in Figure 3b, which was shared by all stages. M4 converted the VCO input tuning voltage to a current that was mirrored by M3. The right branch was a replica of the delay cell, which stabilized the oscillation amplitude. A bypass resistor ensured that a non-zero current flowed when the tuning voltage was equal to zero (or below V_{th} of M4) to prevent a failure in oscillation. The layout of a single delay cell is shown in Figure 4. PMOS and NMOS devices were isolated with two p- and n-guard rings to reduce the probability of latch-up. Devices M1a and M1b had identical finger widths and shared the same drain and source voltages. Therefore, when considering the charge collection after a particle strike, the drain nodes of M1a and M1b behaved in exactly the same fashion. The full layout of the ring oscillator is shown in Figure 5. The left part of the layout consisted of the bias circuitry. The right part was the four-stage differential ring oscillator. The red dots indicate the locations where the laser was focused and charges were injected during the experiments. In each stage, both M2 and M1a were studied. For the reason explained above, M1b is not reported since its results were identical. Further circuit details on the design and electrical measurements of the ring-oscillator are reported in [19].

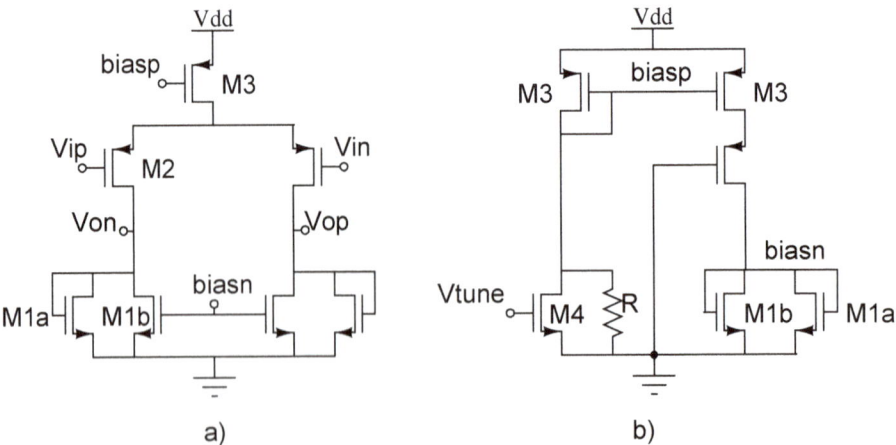

Figure 3. VCO circuit diagram: (a) VCO delay stage; and (b) VCO common bias circuit.

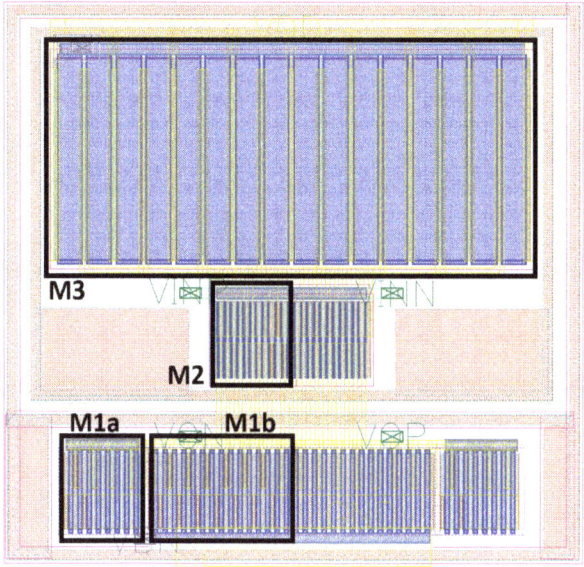

Figure 4. VCO delay cell layout.

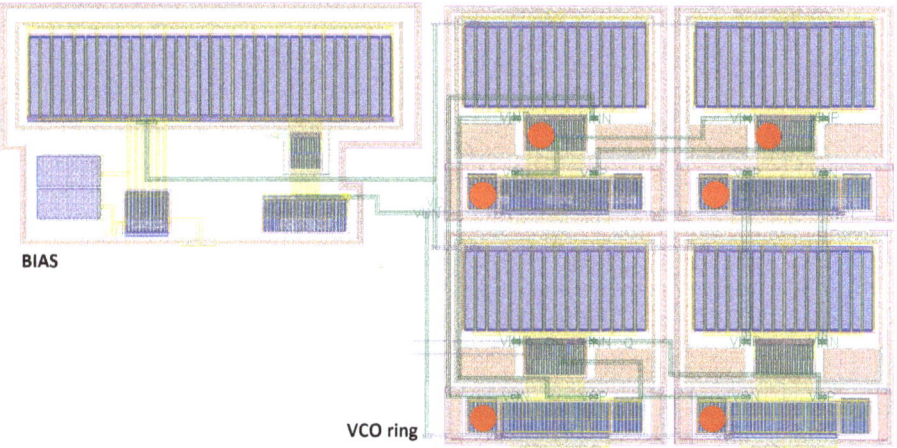

Figure 5. Full VCO delay cell layout with indicated laser points.

3.2. Simulation Results

To qualify the measurements, a simulation was performed to correlate with the simulated data. The ring oscillator was simulated using a periodic steady state (PSS) analysis, using a shooting engine. The PSS resolved a periodic behavior of the circuit and determined the harmonic content of the waveform [20]. This technique was commonly used to analyze the performance of an oscillator. The signal waveforms at the internal nodes of the oscillator are shown in Figure 6a, respectively, Q1–Q4 of V_{op} on each delay stage. The phase of the four waveforms was distributed in the interval 0-π. This was slightly different from what is traditionally expected from a ring oscillator. However, since the number of stages was even, an inversion was made by crossing two oscillator waveforms in the loop. Otherwise, the oscillator would fail to oscillate. As a consequence, at the oscillation frequency, the total phase delay of all stages should only satisfy $\phi_{loop}(f_0) = \pi$. The same phase distribution was visible in the results and measurements of the ISF.

The simulation of the ISF was done using the pulse projection vector (PPV) method, available in PSS simulators [21,22]. This method estimated the amount of phase deviation that originated from a disturbance at a particular node, which represented the ISF of the oscillation node. An alternative method would be to run several transient simulations and measuring the phase error, resulting from narrow current pulses, injected at various successive moments in time. While the latter was similar to the experimental setup, it required relatively large post processing and less accuracy than the former method. The results of the simulations of the ISF are shown in Figure 6b. As expected, the ISF of each successive stage was shifted with respect to the preceding stage, similar to the phase deviation in the signal waveform. As discussed above, the oscillator was mostly sensitive to current impulses, which were injected near the steep edges of the waveform.

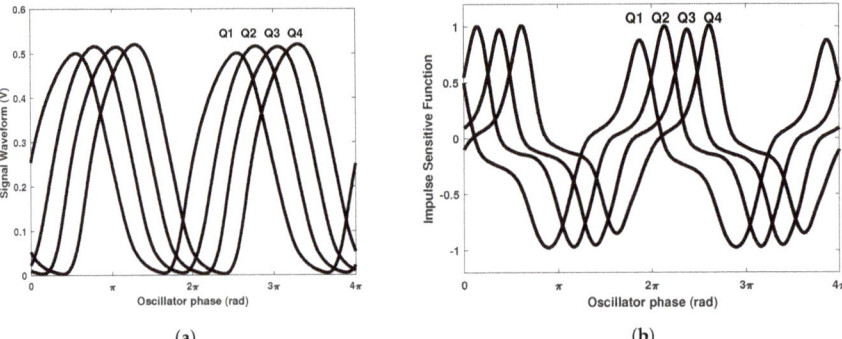

(a)   (b)

Figure 6. Simulated results of the ring oscillator circuit. (**a**) Simulated Waveforms of the prototyped oscillator; (**b**) Simulated ISF of the prototyped oscillator.

3.3. Experimental Measurement Setup

To experimentally prove the time dependency of the phase errors to the moment when the charge was collected, a setup was used with a two photon absorption (TPA) laser. An abstract representation of the setup is shown in Figure 7. A femto-second laser pulse was generated at a pulse rate of 100 Hz, which was generated in the laser source and was not synchronized to the electrical setup. The laser beam was focused on the chip and locally generated free carriers near the focal point of the beam by means of the non-linear two photon absorption mechanism. This allowed accurately generating charges only locally in the silicon substrate with spot sizes of < 1 µm. To measure the ISF of the oscillator, either the laser clock needed to be synchronized to the oscillator or the arrival time of the laser pulse needed to be measured. Practically, the latter was preferred since the laser arrival time could be accurately measured by extracting part of the laser beam and detecting it with a photo detector, which converted the laser pulse to an electrical signal. The accuracy of the arrival time detection was limited by the intrinsic noise of the detector but a jitter of less than 2 ps could be achieved. The setup was based on a statistically random sampling of different arrival times of the laser beam on the chip. Since the laser clock was asynchronous to the oscillator, the pulse could arrive at any moment in time. Both the oscillator waveform and the photo diode signal were captured by a high speed sampling oscilloscope. The scope was triggered by the laser clock, which indicated an occurrence of a pulse. The relative phase of the pulse to the oscillator phase was extracted by post-processing by measuring the time difference between the photo detector signal pulse and the oscillator edge. This calculation provided the injection time, relative to the oscillator zero-crossing, as well as the X-data point of the sampled ISF. The vertical value of the ISF was the total phase error that was caused by the laser pulse. This was also measured by comparing the phase of the oscillator before and after pulse injection.

A picture of the experimental setup is shown in Figure 8. Figure 8a shows the test board of the chip with the focusing lens of the laser. The laser beam was injected vertically and focused on the substrate of the chip. As shown in Figure 7, a second bidirectional splitter was used to visualize the substrate of the chip with an infrared camera. A snapshot of the layout that was investigated is shown in Figure 8b.

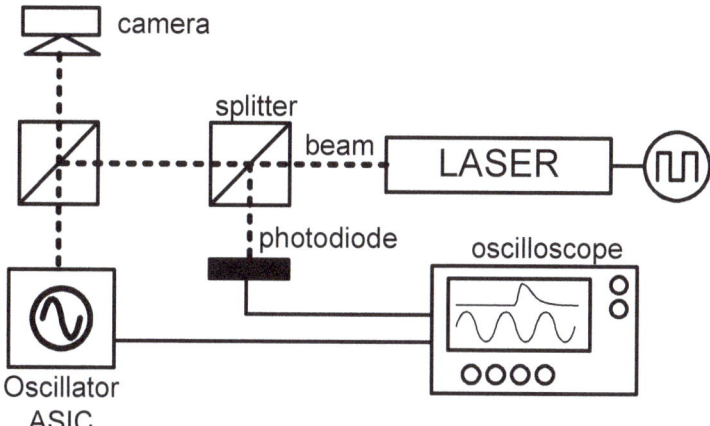

Figure 7. Measurement Setup.

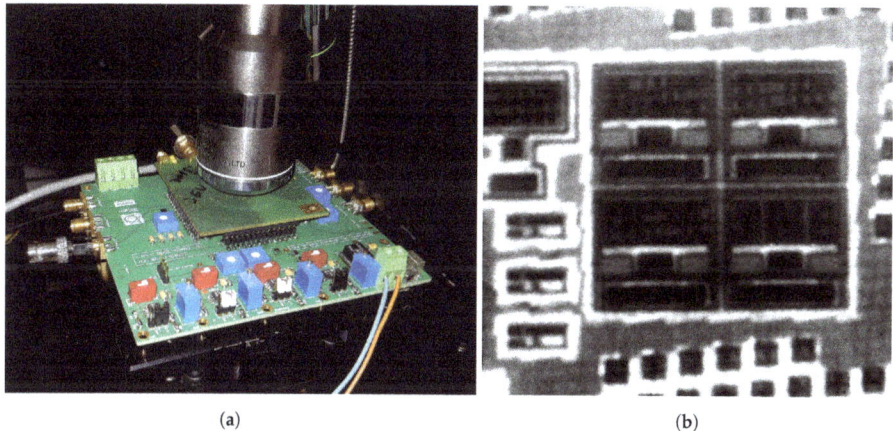

Figure 8. Experimental laser facility setup. (**a**) Photograph of the experimental laser setup; (**b**) Picture of the ring oscillator layout during laser experiments.

3.4. Experimental Results

The results of an injection campaign on the NMOS device M1a of the first stage is shown in Figure 9. The X-axis shows the moment when the pulse was injected, relative to the oscillator zero-crossing. The Y-axis shows the phase error caused by the pulse. Each pulse provided a single point in the scatter plot. Although the waveform of the ISF was periodic in 2π, two periods are shown in Figure 9 to improve readability. However, the data in the interval 2–4π were identical to the first period. In total, 500 pulses were injected in the oscillator. The measurement time was not limited by the laser pulse frequency but by the processing time of the sampling oscilloscope to save the data upon trigger. The scatter plot was used to fit a periodic function with eight harmonics and qA overlaid to the data points. The measurements clearly indicated that the phase error was highly time-dependent as estimated beforehand. The fit of the scatter plot achieved an $R^2 = 0.88$, which was of sufficient quality to analyze and compare the ISF from the fitted curves. This could also be observed well from the plot.

To further analyze the periodic behavior of the oscillator for injected charges, the same analysis was made on all four stages of the oscillator. Although the absolute arrival time of the laser with respect to the oscillator could not be measured, a relative comparison could be made. This occurred since several delays in the setup were not accurately known. Firstly, the relative delay in the laser beam between the pulse arriving at the photo diode and the one actually arriving at the chip was not known. Secondly, several coaxial SMA cables were used to measure the signals on the oscilloscope. Both delays, however, were static and identical for all measurements and were considered as a bias for our setup. The results of injections at all four stages (M1a) is shown in Figure 10. Each ISF was measured by the method described sbove. For improved readability, only the fitted curves are shown here, representing the ISF of the oscillator node. As expected and similar to the simulations, the ISF of the successive stages was phase shifted due to the delays of the oscillator at its oscillation frequency. From these measurements, it became clear the the presented theory could be accurately applied to investigate the impact of SEEs to ring oscillators. A careful observation of the four ISF waveforms indicated that the shapes of the four successive stages were not identical. This was due to a change in the laser focus, which was difficult to control accurately across a wide area, such as this device. If the device was not planar or was slightly tilted, the vertical focus of the laser changes with position and the charges were generated at different depths, resulting in deformed effects. However, a manual refocus was done for each measurement to match the collected charges at all nodes as accurately as practically achievable.

Both experiments shown above only present results from charges that were collected by the NMOS devices, the junctions of which could only drain charge from the oscillator nodes to the substrate. However, if charges were collected by the PMOS device M2, these junctions could only supply charges from the supply (nwell) to the output node. Therefore, the charge injection of PMOS and NMOS was opposite and the measured phase shift was inverted as well. Figure 11 shows both ISF waveforms of the first stage when the laser was focused on both the PMOS and NMOS device in the oscillator. These results clearly indicate the inversion of the phase error due to a reversed current flow.

One node which was not addressed in the results was the drain of M3. This is a common-mode net in the delay cells. Therefore, it is expected that this node does not contribute to a direct phase error since the effects on both differential nodes cancel. However, one could expect a frequency error since the common mode voltage could impact the delay of the stage. However, this effect was experimentally negligible and is therefore not discussed.

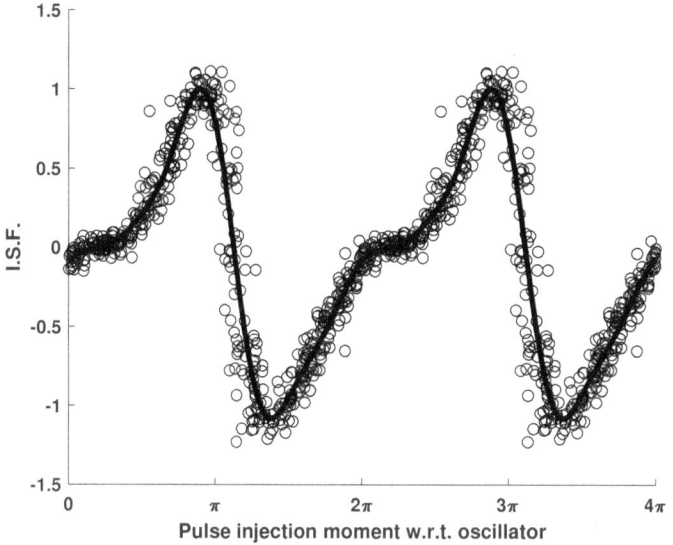

Figure 9. Measured ISF with collected data points ($R^2 = 0.88$).

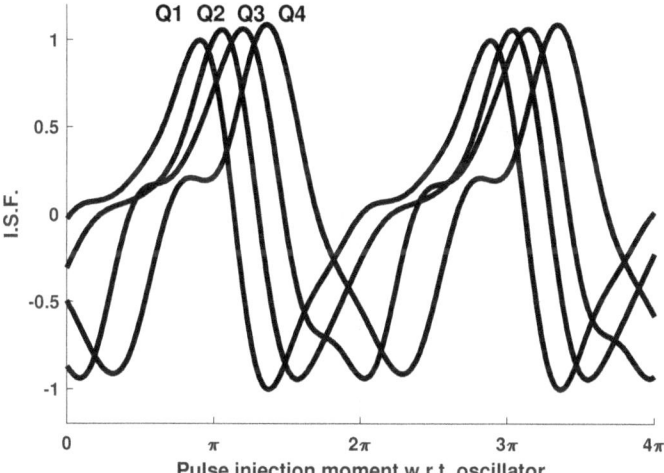

Figure 10. Measured ISF at four stages.

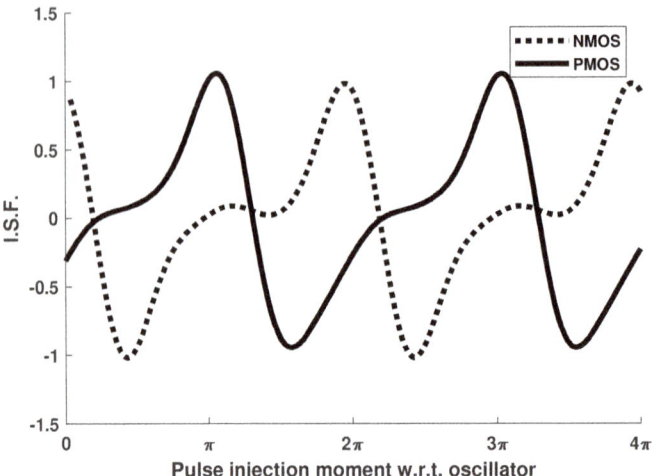

Figure 11. Comparison between NMOS-PMOS sensitivity of a ring oscillator stage.

4. Conclusions

This paper presents a time-dependent analysis on Single-Event Transients on CMOS ring oscillators. The analysis is based on the periodic Impulse Sensitive Function (ISF) of an oscillator, which represents the phase error due to an impulse current that predicts that Single-Event Transients are time-dependent. It was shown that the oscillator phase error due to radiation depends on the moment when charges are injected to the circuit, relative to the phase of the oscillator. To prove the theoretical analysis, a test chip was manufactured in a 65 nm CMOS technology to measure the time-dependent effects. A measurement campaign is presented that, for the first time, proved this theory experimentally in the time-domain and verified the proposed analysis in practice. The experiments were done with a Two-Photon Laser setup to inject charges in the silicon devices with a measurement accuracy better than 2 ps. The measurements were done based on statistical random sampling of the pulse arrival time with respect to the oscillator waveform. The measurements showed that the ring oscillator's phase error depends on the moment when charges are generated in the devices, which correlate with the signal waveform. It was also observed that the ISF of different stages is phase shifted, which was predicted by the theoretical analysis. We can therefore conclude that the theoretical models can be applied to calculate Single-Event Transients in CMOS ring oscillators. where possible.

Author Contributions: Conceptualization, J.P. and V.D.S.; methodology, J.P. and V.D.S.; validation, J.P. and V.D.S.; investigation, J.P. and V.D.S.; and writing—original draft preparation, J.P. and V.D.S.

Funding: This research was funded by FWO.

Conflicts of Interest: The authors declare no conflict of interest. The founding sponsors had no role in the design of the study; in the collection, analyses, or interpretation of data; in the writing of the manuscript, and in the decision to publish the results.

References

1. Young, I.A.; Greason, J.K.; Wong, K.L. A PLL clock generator with 5 to 110 MHz of lock range for microprocessors. *IEEE J. Solid-State Circuits* **1992**, *27*, 1599–1607. [CrossRef]
2. Jung, D.; An, Y.; Ryu, K.; Park, J.; Jung, S. All-Digital Fast-Locking Delay-Locked Loop Using a Cyclic-Locking Loop for DRAM. *IEEE Trans. Circuits Syst. II* **2015**, *62*, 1023–1027. [CrossRef]

3. Yang, S.; Yin, J.; Mak, P.; Martins, R.P. A 0.0056 mm2 -249-dB-FoM All-Digital MDLL Using a Block-Sharing Offset-Free Frequency-Tracking Loop and Dual Multiplexed-Ring VCOs. *IEEE J. Solid-State Circuits* **2019**, *54*, 88–98. [CrossRef]
4. McNeill, J.A. Jitter in ring oscillators. *IEEE J. Solid-State Circuits* **1997**, *32*, 870–879. [CrossRef]
5. Prinzie, J.; Christiansen, J.; Moreira, P.; Steyaert, M.; Leroux, P. Comparison of a 65 nm CMOS Ring- and LC-Oscillator Based PLL in Terms of TID and SEU Sensitivity. *IEEE Trans. Nucl. Sci.* **2017**, *64*, 245–252. [CrossRef]
6. Yang, Z.; Chen, R.Y. High-Performance Low-Cost Dual 15 GHz/30 GHz CMOS LC Voltage-Controlled Oscillator. *IEEE Microwave Wirel. Compon. Lett.* **2016**, *26*, 714–716. [CrossRef]
7. Lu, J.; Wang, N.; Chang, M.F. A single-LC-tank 5–10 GHz quadrature local oscillator for cognitive radio applications. In Proceedings of the 2011 IEEE Radio Frequency Integrated Circuits Symposium, Baltimore, MD, USA, 5–7 June 2011; pp. 1–4. [CrossRef]
8. Buhr, S.; Kreißig, M.; Ellinger, F. Low Power 16 Phase Ring Oscillator and PLL for Use in sub-ns Time Synchronization over Ethernet. In Proceedings of the 2018 25th IEEE International Conference on Electronics, Circuits and Systems (ICECS), Bordeaux, France, 9–12 December 2018; pp. 53–56. [CrossRef]
9. Bravhar, K.; Martins, V.; Santos, L.; Codinachs, D.M. BRAVE NG-MEDIUM FPGA reconfiguration through SpaceWire: example use case and performance analysis. In Proceedings of the 2018 IEEE NASA/ESA Conference on Adaptive Hardware and Systems (AHS), Edinburgh, UK, 6–9 August 2018; pp. 135–141.
10. Kuppambatti, J.; Ban, J.; Andeen, T.; Brown, R.; Carbone, R.; Kinget, P.; Brooijmans, G.; Sippach, W. A radiation-hard dual-channel 12-bit 40 MS/s ADC prototype for the ATLAS liquid argon calorimeter readout electronics upgrade at the CERN LHC. *Nucl. Instrum. Methods Phys. Res. Sect. A* **2017**, *855*, 38–46. [CrossRef]
11. Hansen, M. CMS ECAL electronics developments for HL-LHC. *J. Instrum.* **2015**, *10*, C03028. [CrossRef]
12. Prinzie, J.; Christiansen, J.; Moreira, P.; Steyaert, M.; Leroux, P. A 2.56-GHz SEU Radiation Hard LC-Tank VCO for High-Speed Communication Links in 65-nm CMOS Technology. *IEEE Trans. Nucl. Sci.* **2018**, *65*, 407–412. [CrossRef]
13. Hajimiri, A.; Lee, T.H. A general theory of phase noise in electrical oscillators. *IEEE J. Solid-State Circuits* **1998**, *33*, 179–194. [CrossRef]
14. Lee, T.H.; Hajimiri, A. Oscillator phase noise: A tutorial. *IEEE J. Solid-State Circuits* **2000**, *35*, 326–336. [CrossRef]
15. De Smedt, V.; Gielen, G.; Dehaene, W. *Temperature-And Supply Voltage-Independent Time References for Wireless Sensor Networks*; Springer: Cham, Switzerland, 2015.
16. Hajimiri, A.; Limotyrakis, S.; Lee, T.H. Jitter and phase noise in ring oscillators. *IEEE J. Solid-State Circuits* **1999**, *34*, 790–804. [CrossRef]
17. Black, D.A.; Robinson, W.H.; Wilcox, I.Z.; Limbrick, D.B.; Black, J.D. Modeling of Single Event Transients With Dual Double-Exponential Current Sources: Implications for Logic Cell Characterization. *IEEE Trans. Nucl. Sci.* **2015**, *62*, 1540–1549. [CrossRef]
18. Maneatis, J.G. Low-jitter process-independent DLL and PLL based on self-biased techniques. *IEEE J. Solid-State Circuits* **1996**, *31*, 1723–1732. [CrossRef]
19. Prinzie, J.; Steyaert, M.; Leroux, P. *Radiation Hardened CMOS Integrated Circuits for Time-Based Signal Processing*; Springer: Cham, Switzerland, 2018.
20. Kundert, K. Predicting the phase noise and jitter of PLL-based frequency synthesizers. In *Phase-Locking in High-Performance Systems: From Devices to Architectures*; John Wiley & Sons, Inc.: New York, NY, USA, 2003; pp. 46–69.
21. Vanassche, P.; Gielen, G.; Gielen, G.; Gielen, G.; Sansen, W. On the difference between two widely publicized methods for analyzing oscillator phase behavior. In Proceedings of the 2002 IEEE/ACM International Conference on Computer-Aided Design, San Jose, CA, USA, 10–14 November 2002; pp. 229–233.
22. Levantino, S.; Maffezzoni, P.; Pepe, F.; Bonfanti, A.; Samori, C.; Lacaita, A.L. Efficient Calculation of the Impulse Sensitivity Function in Oscillators. *IEEE Trans. Circuits Syst. II* **2012**, *59*, 628–632. [CrossRef]

© 2019 by the authors. Licensee MDPI, Basel, Switzerland. This article is an open access article distributed under the terms and conditions of the Creative Commons Attribution (CC BY) license (http://creativecommons.org/licenses/by/4.0/).

Article

RHBD Techniques to Mitigate SEU and SET in CMOS Frequency Synthesizers

V. Díez-Acereda, Sunil L. Khemchandani, J. del Pino and S. Mateos-Angulo *

Institute of Applied Microelectronics (IUMA), Department of Electronic and Automatic Engineering, University of Las Palmas de Gran Canaria (ULPGC), 35001 Las Palmas de Gran Canaria, Spain; vdiez@iuma.ulpgc.es (V.D.-A.); sunil.lalchand@ulpgc.es (S.L.K.); jpino@iuma.ulpgc.es (J.d.P.)
* Correspondence: smateos@iuma.ulpgc.es; Tel.: +34-928-457329

Received: 30 April 2019; Accepted: 14 June 2019; Published: 19 June 2019

Abstract: This paper presents a thorough study of radiation effects on a frequency synthesizer designed in a 0.18 µm CMOS technology. In CMOS devices, the effect of a high energy particle impact can be modeled by a current pulse connected to the drain of the transistors. The effects of SET (single event transient) and SEU (single event upset) were analyzed connecting current pulses to the drains of all the transistors and analyzing the amplitude variations and phase shifts obtained at the output nodes. Following this procedure, the most sensitive circuits were detected. This paper proposes a combination of radiation hardening-by-design techniques (RHBD) such as resistor–capacitor (RC) filtering or local circuit-redundancy to mitigate the effects of radiation. The proposed modifications make the frequency synthesizer more robust against radiation.

Keywords: single event transient (SET); single event opset (SEU); radiation-hardening-by-design (RHBD); frequency synthesizers; voltage controlled oscillator (VCO); frequency divider by two; CMOS

1. Introduction

Wireless sensor networks (WSN) are used in a large number of applications due to their known properties such as low cost, low power consumption, small size, flexibility, etc. This has been possible thanks to the use of complementary metal-oxide-semiconductor (CMOS) technologies that, although responsible for most of the mentioned advantages, have the disadvantage that they are very sensitive to environmental radiation. This limits the use of WSN in sectors such as aeronautics, space or nuclear power plants.

When a high-energy particle hits a semiconductor device, it produces a single event effect (SEE) [1,2]. This can be reduced with shielding materials, but at the expense of an increase in the weight of the sensor node, which prevents its application in sectors where weight is important, such as aeronautics or satellites. One way to solve these problems is to make electronic components and systems resistant to damage or malfunction caused by ionizing radiation, a technique commonly called radiation hardening-by-design (RHBD) [3].

There are different types of SEEs, which are classified depending on the effects they can produce on the circuits [2]. This paper focuses on single event transient (SET) and single event upset (SEU). SETs occur when a particle impacts near or through a PN junction creating a transient pulse [4]. This temporary voltage or current disturbance at a circuit node produces amplitude variations and phase shifts that worsen the signal-to-noise ratio and could cause a change in the circuit state. SEUs occur when a particle passes through a sensitive node of a storage element, such as a flip-flop, causing changes in the stored content [5]. SEUs are especially harmful in frequency synthesizers. As shown in Figure 1, they are based on frequency dividers and a change in the state of one of the flip-flops results in a count error, changing the output frequency [6].

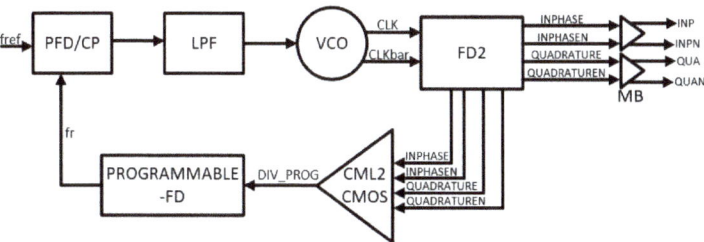

Figure 1. Frequency synthesizer block diagram.

In [5–8], the effects of space radiation on a phase-locked loop (PLL) and frequency synthesizer were studied, but no RHBD techniques were presented. In [9], a comparison of the effect of radiation on an LC oscillator and a ring oscillator within a PLL was done. It was observed that LC oscillators are more vulnerable than ring oscillators, but no RHBD technique was implemented.

Zhang et al. [10] proposed a RHBD technique to enhance the radiation tolerance of an LC tank oscillator. This technique consists of adding a coupled capacitor to accelerate the current pulse discharge in the bias transistor. In addition, two AC coupled capacitors are added between the varactors and the LC tank to block the voltage distortion.

In [11], a set-hardened-by-design charge pump (SET-HBD-CP) to improve SET tolerance of the CP in a PLL was proposed. The SET-HBD-CP approach consists of a basic CP, a reference circuit and a radiation-hardened circuit. When an impact occurs, the radiation-hardened circuit will work and provide the compensation current so as to improve the additional charge of the struck node. Improved results for the recovery time, phase shift and disturbance in the PLL were obtained. For example, when a particle impacts with a linear energy transfer (LET) of 1 pC/μm on the output node of the charge pump, the maximum recovery time, voltage perturbation and phase shift improvement are 72%, 93.7% and 91.8%, respectively.

In [12], an SEU tolerant frequency divider was proposed. This RHBD approach detects the SEU-induced errors via counting the number of rising clock edges and corrects the errors via resetting the faulty frequency divider to a proper state.

In [13], several RHBD techniques for a low-jitter PLL in 130 nm partially depleted-silicon-on-insulator (PD-SOI) process were presented. For the CP, a stacked RHBD technique based on low mismatch current was implemented. The RHBD voltage controlled oscillator (VCO) technique was based on a current compensation scheme and a triple modular redundancy (TMR) technique. In addition, the TMR technique was also implemented in the programmable frequency divider (Programmable-FD) to make the circuit robust to radiation.

In this paper, a thorough study of SETs and SEUs effects in a frequency synthesizer is presented. Section 2 describes the design and operation of the frequency synthesizer that will serve as the basis for our study and, in Section 3, the radiation analysis is performed. On the basis of this study, Section 4 describes the RHBD techniques that are implemented to increase the robustness of the most vulnerable circuits. Finally, some conclusions are drawn in Section 5.

2. Architecture of the Frequency Synthesizer

The frequency synthesizer was designed to fulfill the specifications of the IEEE 802.15.4 standard when a zero-IF receiver architecture is used. In this case, a frequency range from 2405 MHz to 2480 MHz has to be covered with a channel spacing of 5 MHz and the phase noise has to be −102 dBc/Hz @ 3.5 MHz [14].

2.1. Phase-Frequency Detector (PFD) and Charge Pump (CP)

A conventional PFD was designed. This circuit employs a sequential logic and responds to the rising edges of the two inputs [15]. Typically, the PFD outputs are connected to a CP that consists of two switched current sources that charge or discharge the loop filter according to two logical signal inputs [16].

2.2. Low Pass Filter (LPF)

The loop filter is a passive three-pole filter. This comprises a second order filter section and a RC section, providing an extra pole to assist the attenuation of the side-bands at multiples of the comparison frequency that may appear [17].

2.3. Voltage Controlled Oscillator (VCO)

The VCO was implemented as an LC oscillator in CMOS configuration where all the tank components are integrated on-chip. This architecture provides higher transconductance for a given bias current, which results in faster switching and low sensitivity to ion impacts [18].

Figure 2a shows the schematic of the VCO. The core is composed by the cross-coupled pair transistors (M1–M4) to obtain the negative resistance and the LC tank. To control the oscillation frequency of the VCO, a voltage (VTune) is applied to the MOS varactors (C1 and C2). The output signals are CLK and CLKbar and the current source (IRef) and the transistors M5 and M6 are used to bias the oscillator.

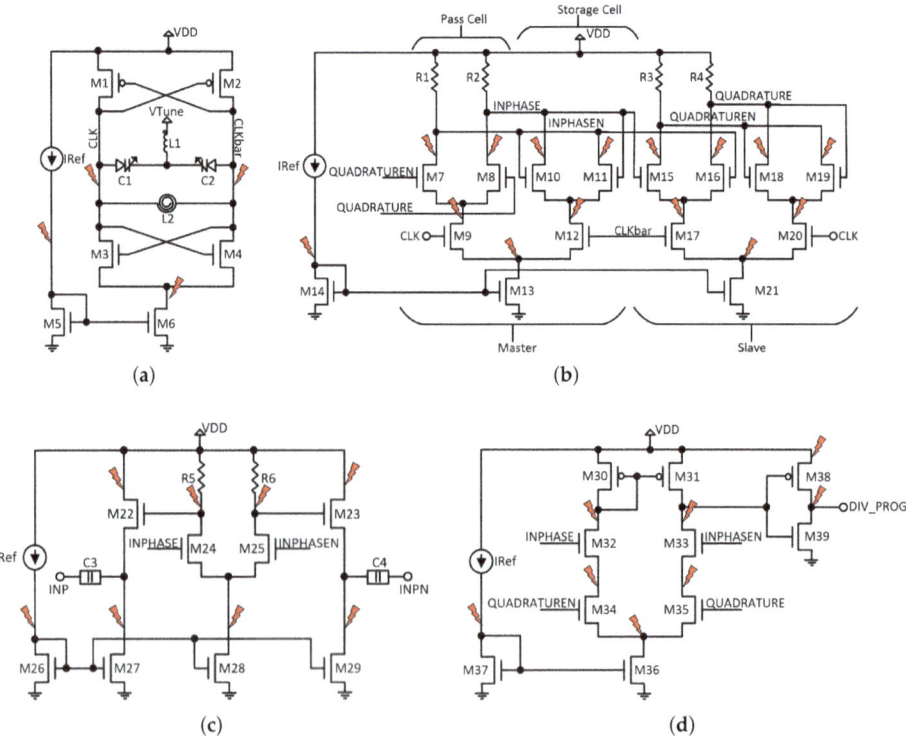

Figure 2. Schematic of the frequency synthesizer most vulnerable circuits. (**a**) Voltage controlled oscillator (VCO); (**b**) Frequency divider by two (FD2); (**c**) Mixer buffer (MB); (**d**) Current-mode logic to CMOS (CML2CMOS) converter.

2.4. Frequency Divider by Two (FD2)

The VCO output is connected to the FD2 in order to generate the in-phase (I) and quadrature (Q) signals necessary for the O-QPSK modulation [14]. The CML (current-mode logic) configuration is chosen to obtain the I-Q signals.

Figure 2b shows the schematic of the FD2 with the CML structure, which consists of two D flip-flops in master–slave configuration. It is composed by a sampling stage (M7, M8, M15 and M16), and a hold stage (M10, M11, M18 and M19) whose objective is to maintain the voltage at the output node. The current source (IRef); the transistors M13, M14 and M21; and the pull-up resistances are used to bias the circuit. The control signals, CLK and CLKbar, come from the VCO output. The differential output signals are in-phase signals (INPHASE and INPHASEN) and quadrature signals (QUADRATURE and QUADRATUREN).

2.5. Mixer Buffer (MB) and CML to CMOS Converter (CML2CMOS Converter)

The MB for the in-phase signal is shown in Figure 2c. The same schematic is used for the quadrature signal. The MB has a differential input stage and a source follower output stage.

To drive the Programmable-FD, a CML2CMOS converter in a D2SE (Differential to Single-Ended) configuration is used [19]. The schematic of this circuit is shown in Figure 2d. The control signals of the differential stages are in-phase signals (INPHASE and INPHASEN). The quadrature signals (QUADRATURE and QUADRATUREN) are added to avoid overloading the output. The current source IRef and transistors M36 and M37 are used to bias the circuit. At the output, an inverter stage formed by M38 and M39 is used to obtain a digital signal to drive the Programmable-FD.

2.6. Programmable Frequency Divider (Programmable-FD)

Figure 3 shows a block diagram of the Programmable-FD. It is a conventional dual-modulus frequency divider composed by a dual modulus prescaler (P/P + 1), and two programmable counters (Np and A). The prescaler divides by 2/3 the output of the CML2CMOS converter, and Np and A divide by 256 and 8, respectively.

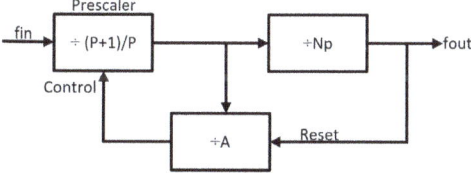

Figure 3. Programmable frequency divider (Programmable-FD) block diagram.

3. SET and SEU Analysis in the Frequency Synthesizer

3.1. Impacts Modeling

In CMOS devices, the most affected areas by an ion impact are the reverse bias junctions. This corresponds to the n-p and p-n junctions between the drain and substrate in NMOS and PMOS transistors, respectively [4]. The effect of an ion impact can be modeled by a current pulse connected to the drain of the transistor. This current pulse is generally described as a double exponential with the following expression [20,21]:

$$I_{SET} = \frac{Q}{(t_f - t_r)} \times (e^{\frac{-t}{t_f}} - e^{\frac{-t}{t_r}}), \qquad (1)$$

where Q is the collected charge, while t_f and t_r are the fall and rise times, respectively. Typically, t_f and t_r are of the order of hundreds of picoseconds and tens of picoseconds, respectively [22,23], and Q goes from several hundreds of fC to about 1200 fC. Table 1 shows the values used in this study, which are the expected numbers for a 180 nm CMOS process [22].

Table 1. Current pulses.

t_r (ps)	t_f (ps)	Q (fC)	LET, d = 2 µm (MeV·cm^2·mg^{-1})
50	200	300	14.47
50	200	525	25.31
100	400	501	24.16
100	400	801	38.62
100	400	990	47.74
100	400	1200	57.86

The LET is defined as the energy that loses the particle until it reaches rest. It can be calculated by the following expression:

$$LET = \frac{Q \times 3.6}{(e \times \rho_{Si} \times d)}, \quad (2)$$

where e is the electron charge, ρ_{Si} is the silicon density, d is the sensitive depth of the charge collection and the constant 3.6 corresponds to the energy required in eV to create an electron–hole pair in silicon [20,22]. Figure 4 shows the time domain waveforms of the current pulses. An impact of a charged particle in the n-p and p-n junctions between the drain and substrate in NMOS and PMOS transistors generates electron–hole pairs, which results in a current peak. This current slowly decays as these extra electron–hole pairs recombine. Therefore, the width of the pulse is strictly related with the number of electron–hole pairs generated and how fast they recombine [24].

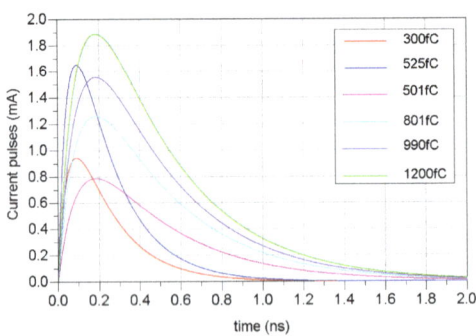

Figure 4. Current pulse shapes used for single event transient (SET) prediction.

3.2. SET and SEU Analysis

To perform the analysis of the radiation effects on every transistor of the frequency synthesizer, a current pulse was connected to its drain resulting in an amplitude variation and a phase shift of the output signal. Figure 5 compares the output signal for the cases when there is an impact (red) and when there is no impact (blue). The recovery time is defined as the time it takes the signal amplitude to return to its value without impact with an error of less than 5%. This was done for all transistors of the synthesizer. It is important to note that, since the width of the pulses covered several periods of a 5 GHz signal, there was no appreciable difference between impacting at the peaks or at the zero crossings of the signals [25].

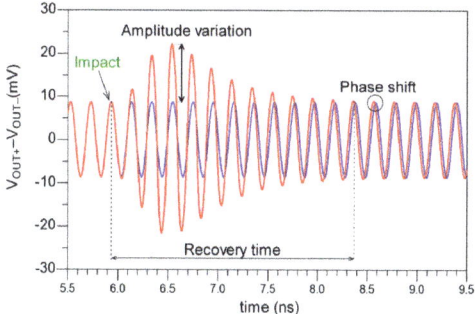

Figure 5. The blue signal is the output without an impact on any transistor and the red signal is the signal when there is an impact on a transistor.

Low frequency circuits such as the PFD, the CP and the LPF were practically unaffected by the current pulses since their operating frequency is 2.5 MHz. This means that the period of the signal was 0.4 µs, which was at least one order of magnitude bigger than the pulse widths considered in this study (in the order of a few nanoseconds). For the same reason, high frequency circuits such as the VCO, the FD2, the MB, the CML2CMOS converter and the Programmable-FD were more affected. Figure 2 shows the schematics of those circuits with the current pulses, represented as rays, connected to the drain of the transistors.

Figure 6 shows the phase shift of one of the frequency synthesizer outputs (INP) due to impacts on the transistors of the VCO, the CML2CMOS converter, the FD2 and the MB. The other frequency synthesizer outputs had a similar behavior. As shown in Figure 6, there were circuits that were very vulnerable to impacts and others that were practically invulnerable. For example, impacts on the MB were not affected since the phase changes were practically negligible (see Figure 6a). However, other circuits such as the VCO and the CML2CMOS converter were more affected by the impacts. In the case of the VCO, the current mirror transistor M5 was the most sensitive component since at high energy a high phase shift was produced [25]. This can be seen in Figure 6b. Figure 6c shows that, in the CML2CMOS converter, the most vulnerable transistor was M35 since at certain LET values the phase shift was also large. It should be taken into account that a 180° phase shift is equivalent to a SEU that can be propagated to other circuits.

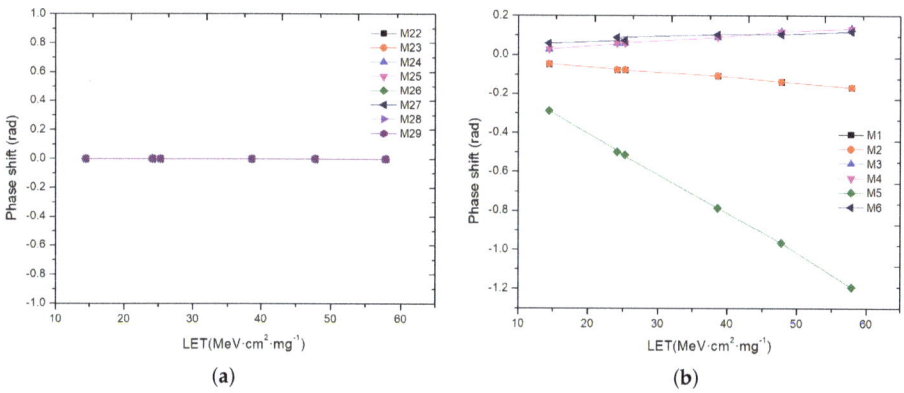

Figure 6. Cont.

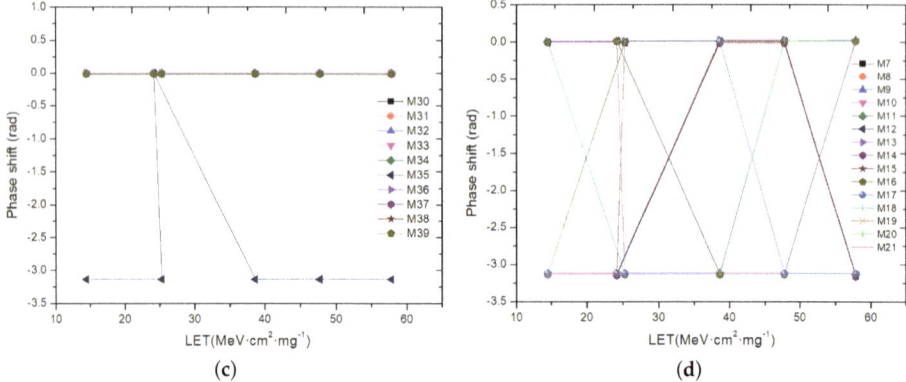

Figure 6. INP output phase shift due to impacts on the transistors. (**a**) MB; (**b**) VCO; (**c**) CML2CMOS converter; (**d**) FD2.

The FD2 was extremely sensitive to impacts. Figure 6d shows that an ion impact on any transistor produced a 180° phase shift for different LET values. As stated above, this produced a SEU at the divider output that can be propagated to other parts of the circuit, causing an incorrect output.

Figure 7 shows the output of the Programmable-FD when impacts were applied each nanosecond in the prescaler. As seen in the figure, amplitude variations were observed at the output of this circuit but no changes in the state of the divider was observed.

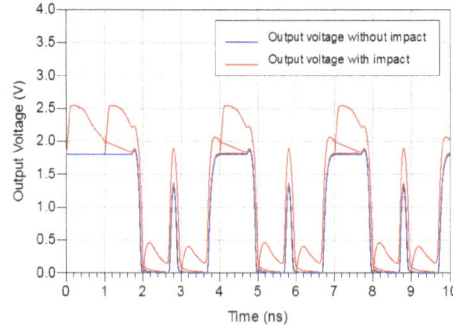

Figure 7. The blue signal is the output of the Programmable-FD without any impact and the red signal is when impacts were applied each nanosecond.

4. RHBD Design

In this section, the most sensitive circuits are redesigned using RHBD techniques.

4.1. VCO

The simulation results shown in Section 3 indicate that the VCO was considerably more sensitive to strikes at the biasing network of the circuit (transistors M5 and M6) [26]. Therefore, the RHBD technique proposed in this paper is focused on this node. This mitigation technique should improve the SET performance of the VCO without affecting its RF performance.

Resistor–capacitor (RC) filtering was implemented to achieve a RHBD VCO. Figure 8a shows the schematic of the redesigned VCO. Two resistors, R1 and R2 (both with a value of 4.7 kΩ), and the capacitor Cg (1 pF) were added to increase the time constant (τ = RC) on the gates of transistors M5 and M6, resulting in a much lower output voltage deviation due to the SET. A drawback of this technique is that the phase noise of the oscillator no longer met the standard specification (−102 dBc/Hz @3.5 MHz) due to the thermal noise of resistors R1 and R2. To improve the phase noise, a common technique used is to increase the amplitude of oscillation [27]. To do this, a capacitive divider consisting of four capacitors and two resistors were added to the N and P cross-coupled pair transistors. This divider reduced the voltage in the drains of M1–M4, thus increasing the maximum output swing at the drains of these transistors.

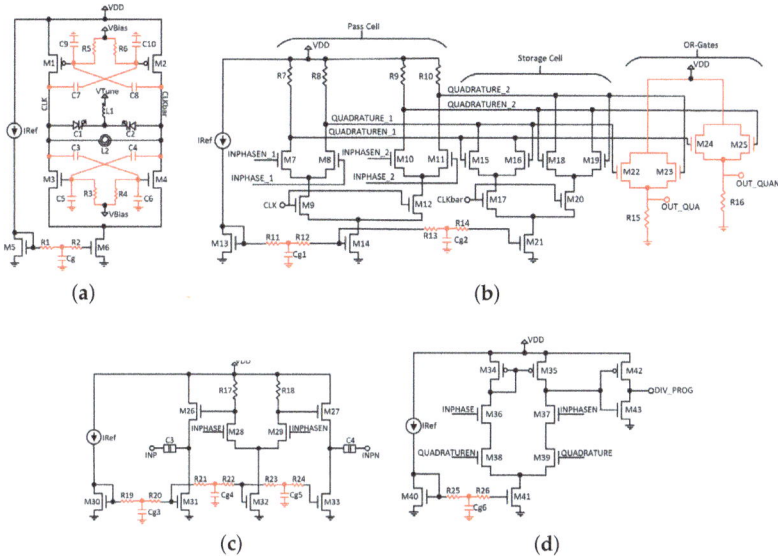

Figure 8. Radiation hardening-by-design (RHBD) schematics. (**a**) VCO; (**b**) Slave stage of the FD2; (**c**) MB; (**d**) CML2CMOS converter.

Figure 9 shows the phase noise of the VCO before and after RHBD design. As seen in the figure, the VCO met the specifications of the standard with a phase noise that was even better than the original design.

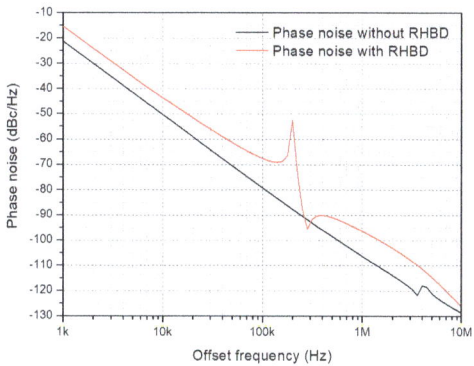

Figure 9. VCO phase noise.

Figure 10a shows the INP output phase shift due to impacts on the RHBD VCO transistors. The phase shifts were reduced almost to zero, which means a reduction of 50.3%. The recovery time decreased by approximately 81%, needing only 2 ns to stabilize, as shown in Figure 10b.

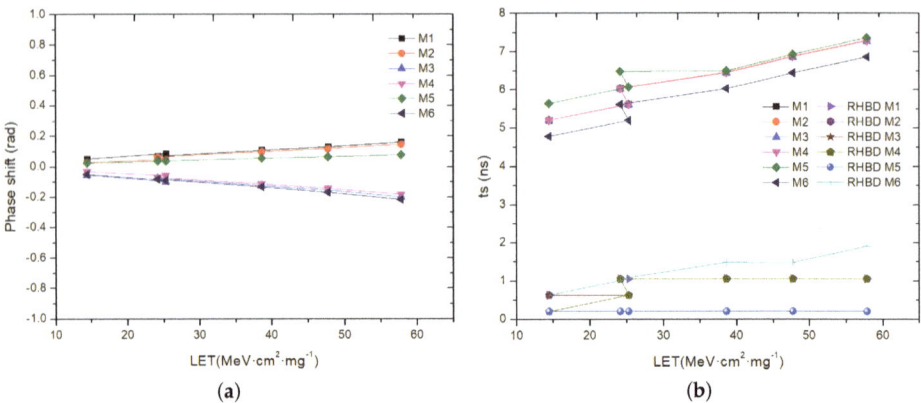

Figure 10. INP output of the RHBD VCO results. (**a**) Phase shift; (**b**) Recovery time.

4.2. FD2

Two RHBD circuit-redundancy techniques were employed in the FD2: Dual interlocked storage cell (DICE) [28] and gated feedback cell (GFC) [29]. In both techniques, the layout should be carefully implemented and the critical nodes should be spatially separated.

The FD2 schematic shown in Figure 2b is based on a standard master–slave D-flip-flop and suffers from an increased vulnerability to SEU due to cross-coupling at the transistor-level required for the storage cell functionality. Local redundancy mitigates its SEU sensitivity with only a moderate increase in power consumption and circuit complexity. Figure 8b shows the schematic of the master stage of the D-flip flop with local redundancy. The gate and the drain of the transistors in the storage cell are not connected to the same differential pair in the pass cell, thus achieving effective decoupling of the gate and drain terminals of the transistors in the storage cell and, as a consequence, reducing its SEU sensitivity. This technique is commonly referred as dual interlocked storage cell.

The other technique, called gated feedback cell, is also shown in Figure 8b applied to a single latch of the divider by two. The latch outputs are connected to a pair of OR gates that hold the circuit stable when an impact occurs, since the output of a two-input OR gate changes state only when both inputs change their state from high to low or from low to high [29]. The OR operation comprises a pair of source followers (M22–M25) that helps to transmit the correct logic value to the storage cell inputs even when one of the OR gate inputs is in error due to an ion strike. Resistances R15 and R16, each of 55 kΩ, were also included.

The OR-gate-based feedback to the storage cell inputs, in addition to local redundancy, is expected to offer a high SEU immunity. However, the technique of increasing the time constant in the bias circuits was also used. Resistors R11–R14 (each of 4.7 kΩ) and capacitors Cg1 and Cg2 (1 pF each) were added to increase the time constant ($\tau = RC$) in the gate terminal of transistors M13, M14 and M21, which are part of the bias circuit network.

As shown in Figure 11, the SEU effect produced in the FD2 was mitigated with an improvement of 74.52%. Figure 12 shows the recovery time before and after using the RHBD techniques. As seen in the figure, an improvement of 32.05% for the recovery time was achieved.

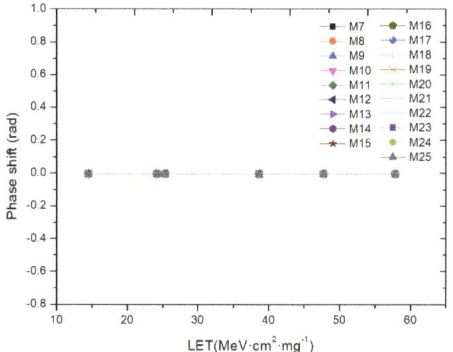

Figure 11. INP output of the RHBD FD2 for the phase shift results.

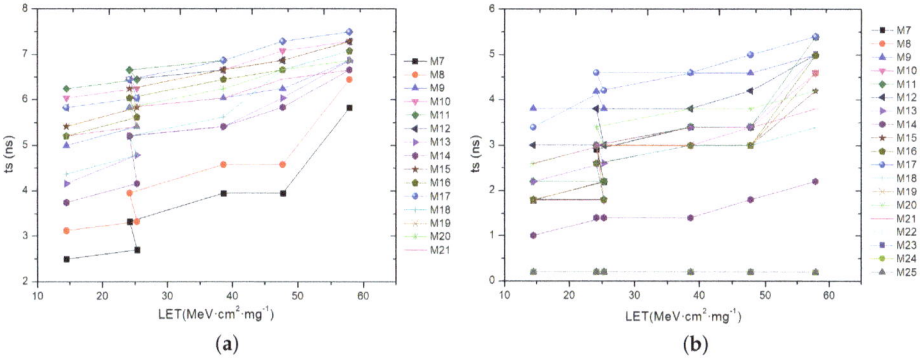

Figure 12. INP output recovery time of the FD2. (**a**) Before RHBD; (**b**) After RHBD.

4.3. CML2CMOS Converter and MB

Similar to the previous circuits, the most sensitive nodes of the CML2CMOS converter and the MB were those of the biasing networks. The same RHBD technique based on increasing the time constant was used. The resistances and capacitors included had the same values as for the VCO and the FD2, 4.7 kΩ and 1 pF, respectively. Figure 8c,d shows the RHBD schematics.

Figure 13a shows that the phase shift of the CML2CMOS converter was almost zero for all transistors, disappearing the SEU in this circuit. Figure 13b shows a comparison of the recovery time before and after the RHBD design. As seen in the figure, after applying the RHBD technique, the recovery time was reduced by 3 ns, obtaining an overall improvement of 78.83%.

As mentioned in Section 3, impacts in the MB did not produce 180° phase shifts at the output. As shown in Figure 14a, the results are very similar before (see Figure 6a) and after RHBD design. Figure 14b shows the recovery time of the INP output before and after RHBD design. The recovery time was almost zero, thus an overall improvement of 77.9% was obtained for this circuit.

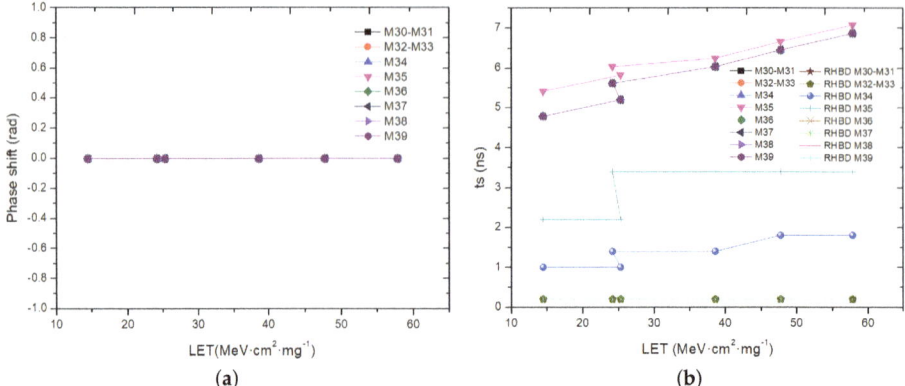

Figure 13. INP output of the RHBD CML2CMOS converter results. (**a**) Phase shift; (**b**) Recovery time.

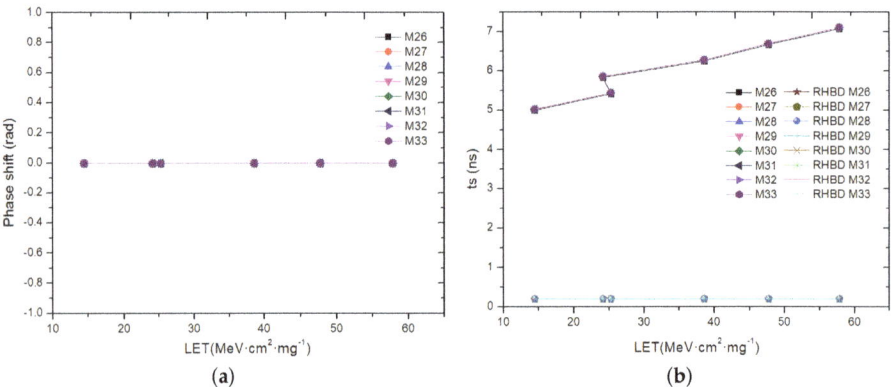

Figure 14. INP output of the RHBD MB results. (**a**) Phase shift; (**b**) Recovery time.

5. Conclusions

This paper presents a comprehensive study of the effects of SETs and SEUs on a frequency synthesizer for the IEEE 802.15.4 standard. The blocks that work at low frequencies, such as the PFD, the CP and the LPF, are not affected by ion impacts. However, high frequency circuits such as the Programmable-FD, the VCO, the FD2, the CML2CMOS converter and the MB are more vulnerable. In the Programmable-FD, amplitude variations of the output signal occur, but there is no change in the state of the divider, thus no RHBD techniques were applied. This is not the case in the rest of the high frequency circuits, thus RHBD techniques were implemented on them. The VCO's radiation tolerance was improved using resistor–capacitor (RC) filtering and a capacitive divider was introduced to improve the degraded phase noise. The combination of both techniques resulted in a substantial improvement on the VCO performance, reducing by approximately 50% the output phase displacement and by 81% the recovery time. RC filtering of the bias circuits was also used in the FD2, the CML2CMOS converter and the MB. In addition, local circuit-redundancy hardening techniques were employed in the flip flops of the FD2. The proposed modifications make the frequency synthesizer more robust against radiation: SEU effects were fully mitigated and the SETs were reduced considerably. Furthermore, the power consumption of the PLL was increased from 18.5 mW to 21.2 mW due to the local circuit-redundancy technique.

Author Contributions: All authors contributed to the study of the radiation in the frequency synthesizer and the writing process of the paper.

Funding: This research was funded by the Spanish Ministry of Economy and Competitiveness grant number TEC2015- 71072-C03-01; by the Spanish Ministry of Science, Innovation and Universities grant number RTI2018-099189-B-C22; by the Canary Agency for Research, Innovation and Information Society (ACIISI) of the Canary Islands Government grant number ProID2017010067; and the "Programa Predoctoral de Formación del Personal Investigador" of the ULPGC.

Acknowledgments: The authors would like to thank Raquel León-Martín for her contributions on this paper.

Conflicts of Interest: The authors declare no conflict of interest.

Abbreviations

The following abbreviations are used in this manuscript:

CML	Current-Mode Logic
CML2CMOS converter	CML to CMOS converter
CMOS	Complementary Metal-Oxide-Semiconductor
CP	Charge Pump
DICE	Dual Interlocked Storage Cell
D2SE	Differential to Single-Ended
FD2	Frequency Divider by 2
GFC	Gated Feedback Cell
IEEE	The Institute of Electrical and Electronics Engineers
LET	Linear Energy Transfer
LPF	Low Pass Filter
MB	Mixer Buffer
O-QPSK	Offset Quadrature Phase Shift Keying
PD-SOI	Partially Depleted-Silicon-on-Insulator
PFD	Phase Frequency Detector
Programmable-FD	Programmable Frequency Divider
PLL	Phase-Locked Loop
RC	Resistor–Capacitor
RF	Radio Frequency
RHBD	Radiation Hardening-By-Design
SEE	Single Event Effect
SET	Single Event Transient
SET-HBD-CP	SET-Hardened-By-Design Charge Pump
SEU	Single Event Upset
TMR	Triple Modular Redundancy
VCO	Voltage Controlled Oscillator
WSN	Wireless Sensor Networks

References

1. Bagatin, M.; Gerardin, S. *Ionizing Radiation Effects in Electronics: From Memories To Imagers*; CRC Press: Boca Raton, FL, USA, 2015; p. 391.
2. Gaillard, R. Single Event Effects: Mechanisms and Classification. In *Soft Errors in Modern Electronic Systems. Frontiers in Electronic Testing*; Springer: Boston, MA, USA, 2011; pp. 27–54. [CrossRef]
3. Kerns, S.E.; Shafer, B.D. The Design of Radiation-Hardened ICS for Space: A Compendium of Approaches. *Proc. IEEE* **1988**, *76*, 1470–1509. [CrossRef]
4. Wang, T. Study of Single-Event Transient Effects on Analog Circuits. Ph.D. Thesis, University of Saskatchewan, Saskatoon, SK, Canada, 2011.
5. Dayaratna, L.; Seehra, S.; Bogorad, A.; Ramos, L. Single event upset characteristics of some digital integrated frequency synthesizers. In Proceedings of the 1999 IEEE Radiation Effects Data Workshop. Workshop Record. Held in conjunction with IEEE Nuclear and Space Radiation Effects Conference (Cat. No.99TH8463), Norfolk, VA, USA, 12–16 June 1999; pp. 46–52. [CrossRef]

6. Chung, H.H.; Chen, W.; Bakkaloglu, B.; Barnaby, H.J.; Vermeire, B.; Kiaei, S. Analysis of Single Events Effects on Monolithic PLL Frequency Synthesizers. *IEEE Trans. Nucl. Sci.* **2006**, *53*, 3539–3543. [CrossRef]
7. Sotskov, D.I.; Elesin, V.V.; Kuznetsov, A.G.; Nazarova, G.N.; Chukov, G.V.; Boychenko, D.V.; Telets, V.A.; Usachev, N.A. Total Ionizing Dose Effects in Phase-Locked Loop ICs and Frequency Synthesizers. In Proceedings of the 2015 15th European Conference on Radiation and Its Effects on Components and Systems (RADECS), Moscow, Russia, 14–18 September 2015; pp. 1–3. [CrossRef]
8. Chen, Z.; Ding, D.; Dong, Y.; Shan, Y.; Zhou, S.; Hu, Y.; Zheng, Y.; Peng, C.; Chen, R. Study of Total-Ionizing-Dose Effects on a Single-Event-Hardened Phase-Locked Loop. *IEEE Trans. Nucl. Sci.* **2018**, *65*, 997–1004. [CrossRef]
9. Prinzie, J.; Christiansen, J.; Moreira, P.; Steyaert, M.; Leroux, P. Comparison of a 65 nm CMOS ring- and lc-oscillator based PLL in terms of TID and SEU sensitivity. *IEEE Trans. Nucl. Sci.* **2017**, *64*, 245–252. [CrossRef]
10. Zhang, Z.; Chen, L.; Djahanshahi, H. A Hardened-By-Design Technique for LC-Tank Voltage Controlled Oscillator. In Proceedings of the 2018 IEEE Canadian Conference on Electrical & Computer Engineering (CCECE), Quebec City, QC, Canada, 13–16 May 2018; pp. 1–4. [CrossRef]
11. Zhou, Q.; Zhang, C.; Tan, J.; Zhu, L.; Wang, L.; Luo, W. A SET-hardened phase-locked loop. In Proceedings of the 2017 IEEE 3rd Information Technology and Mechatronics Engineering Conference (ITOEC), Chongqing, China, 3–5 October 2017; pp. 196–199. [CrossRef]
12. Li, N.; She, X. Single event transient tolerant frequency divider. *IET Comput. Digit. Tech.* **2014**, *8*, 140–147. [CrossRef]
13. Chen, Z.; Lin, M.; Ding, D.; Zheng, Y.; Sang, Z.; Zou, S. Analysis of Single-Event Effects in a Radiation-Hardened Low-Jitter PLL Under Heavy Ion and Pulsed Laser Irradiation. *IEEE Trans. Nucl. Sci.* **2017**, *64*, 106–112. [CrossRef]
14. Standards Committee of the IEEE Computer Society. *IEEE Standard for Local and Metropolitan Area Networks—Part 15.4: Low-Rate Wireless Personal Area Networks (WPANs)*; IEEE Std 802.15.4-2011; Standards Committee of the IEEE Computer Society: New York, NY, USA, 2006.
15. Weste, N.H.E.; Harris, D.M. *CMOS VLSI Design: A Circuits and Systems Perspective*; Addison Wesley: Boston, MA, USA, 2011; p. 838.
16. Razavi, B. *Design of Analog CMOS Integrated Circuits*; McGraw-Hill: New York, NY, USA, 2001; p. 684.
17. Fujitsu Microelectronics America Inc. *Super PLL Application Guide: Integer PLL Loop Filter Design*; Fujitsu Microelectronics America Inc.: San Jose, CA, USA, 2002; pp. 22–24.
18. Hajimiri, A.; Lee, T. Design issues in CMOS differential LC oscillators. *IEEE J. Solid-State Circuits* **1999**, *34*, 717–724. [CrossRef]
19. Senjaliya, C.; Neelakantan, U. 180 nm CMOS process based L-band CML to CMOS converter. *IJARIIE* **2018**, *4*, 3697–3700.
20. Portela-Garcia, M.; Lopez-Ongil, C.; Garcia-Valderas, M.; Entrena, L.; Thys, G.; Redant, S. Assessing SET sensitivity of a PLL. In Proceedings of the Design of Circuits and Integrated Systems, Madrid, Spain, 26–28 November 2014; pp. 1–6. [CrossRef]
21. Garg, R.; Khatri, S.P. Analysis and Design of Resilient VLSI Circuits. Ph.D. Thesis, Texas A&M University, TX, USA, 2010. [CrossRef]
22. Lochner, S.; Deppe, H. Radiation studies on the UMC 180 nm CMOS process at GSI. In Proceedings of the 2009 European Conference on Radiation and Its Effects on Components and Systems, Bruges, Belgium, 14–18 September 2009; pp. 614–616. [CrossRef]
23. Eaton, P.; Benedetto, J.; Mavis, D.; Avery, K.; Sibley, M.; Gadlage, M.; Turflinger, T. Single event transient pulsewidth measurements using a variable temporal latch technique. *IEEE Trans. Nucl. Sci.* **2004**, *51*, 3365–3368. [CrossRef]
24. Mateos-Angulo, S.; Rodríguez, R.; del Pino, J.; González, B.; Khemchandani, S.L. Single event effects analysis and charge collection mechanisms on AlGaN/GaN HEMTs. *Semicond. Sci. Technol.* **2019**, *34*, 035029. [CrossRef]
25. González Ramírez, D.; Lalchand Khemchandani, S.; del Pino, J.; Mayor-Duarte, D.; San Miguel-Montesdeoca, M.; Mateos-Angulo, S. Single event transients mitigation techniques for CMOS integrated VCOs. *Microelectron. J.* **2018**, *73*, 37–42. [CrossRef]

26. Mateos-Angulo, S.; San-Miguel-Montesdeoca, M.; Mayor-Duarte, D.; Khemchandani, S.L.; del Pino, J. SET analysis and radiation hardening techniques for CMOS LNA topologies. *Semicond. Sci. Technol.* **2018**, *33*, 085010. [CrossRef]
27. Lalchand Khemchandani, S.; del Pino Suarez, J.; Diaz Ortega, R.; Hernandez, A. A fully integrated single core VCO with a wide tuning range for DVB-H. *Microw. Opt. Technol. Lett.* **2009**, *51*, 1338–1343. [CrossRef]
28. Calin, T.; Nicolaidis, M.; Velazco, R. Upset hardened memory design for submicron CMOS technology. *IEEE Trans. Nucl. Sci.* **1996**, *43*, 2874–2878. [CrossRef]
29. Krithivasan, R.; Marshall, P.W.; Nayeem, M.; Sutton, A.K.; Kuo, W.M.; Haugerud, B.M.; Najafizadeh, L.; Cressler, J.D.; Carts, M.A.; Marshall, C.J.; et al. Application of RHBD Techniques to SEU Hardening of Third-Generation SiGe HBT Logic Circuits. *IEEE Trans. Nucl. Sci.* **2006**, *53*, 3400–3407. [CrossRef]

© 2019 by the authors. Licensee MDPI, Basel, Switzerland. This article is an open access article distributed under the terms and conditions of the Creative Commons Attribution (CC BY) license (http://creativecommons.org/licenses/by/4.0/).

Article

Fine-Grain Circuit Hardening Through VHDL Datatype Substitution

Maria Muñoz-Quijada, Samuel Sanchez-Barea, Daniel Vela-Calderon and Hipolito Guzman-Miranda *

Department of Electronic Engineering, Universidad de Sevilla, Camino de los Descubrimientos s/n, 41092 Sevilla, Spain; mamuki92@gmail.com (M.M.-Q.); samuelsanchezbarea@gmail.com (S.S.-B.); danvelcal96@gmail.com (D.V.-C.)
* Correspondence: hguzman@us.es; Tel.: +34-954-481-298

Received: 30 November 2018; Accepted: 23 December 2018; Published: 25 December 2018

Abstract: Radiation effects can induce, amongst other phenomena, logic errors in digital circuits and systems. These logic errors corrupt the states of the internal memory elements of the circuits and can propagate to the primary outputs, affecting other onboard systems. In order to avoid this, Triple Modular Redundancy is typically used when full robustness against these phenomena is needed. When full triplication of the complete design is not required, selective hardening can be applied to the elements in which a radiation-induced upset is more likely to propagate to the main outputs of the circuit. The present paper describes a new approach for selectively hardening digital electronic circuits by design, which can be applied to digital designs described in the VHDL Hardware Description Language. When the designer changes the datatype of a signal or port to a hardened type, the necessary redundancy is automatically inserted. The automatically hardening features have been compiled into a VHDL package, and have been validated both in simulation and by means of fault injection.

Keywords: radiation hardening; hardening by design; TMR; selective hardening; VHDL

1. Introduction

1.1. Background

Ionizing radiation affects the normal operation of electronic circuits. Different kind of effects may produce both physical degradation of the components, like TID (Total Ionizing Dose) or DD (Displacement Damage), or corruption of the logic values stored in the circuit, such as SEU (Single Event Upset), SET (Single Event Transient) or MBU (Multi-Bit Upset) [1]. The former category of effects, known as hard errors, are destructive in nature and must be protected against by using specific technology approaches. Soft errors, on the other hand, induce modifications in the internal states of the circuits, which may or may not then propagate both inside the circuit architectures and to their primary outputs. Errors propagating to the primary outputs of a circuit may escalate to external systems and produce device errors, subsystem failures or even catastrophic mission failures. These soft errors can be mitigated by inserting logic protections in the designs [2,3].

1.2. Problem of Interest

These logic protections can be inserted at different steps during the design flow. Typically, these protections are inserted either during the synthesis process or just after the synthesis process has completed, but before the placement and routing steps. These approaches require design teams to implement changes to their design flows, either by including specific proprietary synthesis tools or extra post-synthesis netlist manipulation software, both of which have to be adapted and configured for the mission requirements, which demands extra effort from the designers.

When developing hardware modules cores that are expected to need some selective protections, but not full redundancy, it would be desirable to include the information on which elements should be hardened in the module code itself, in a non-synthesizer-specific way, since different developers and projects may choose or require different synthesis tools. An ideal situation would allow the designer to easily specify in the VHDL (Very High Speed Integrated Circuit Hardware Description Language) source code which elements should be hardened, with minimal code modifications.

1.3. Literature Survey

There are multiple types of protections that can be inserted in a digital circuit [4], from which the most common one is the full triplication of single memory elements, which is known as Triple Modular Redundancy or TMR. TMR is typically preferred to DMR (Dual Modular Redundancy) since the former can detect and correct single errors, but the latter has only detection, but no correction capabilities. The tradeoff for this is that TMR uses more area and power (around a 3.2× factor, instead of a ~2.1× factor for DMR, compared with the unhardened design [5]). TMR can be applied at both flip-flop level or module level, but DMR is more typically applied at module level.

Selective hardening is a more recent technique that involves identifying the most sensitive modules of a design (for example, by means of fault injection), and then applying TMR only to those modules. This way, a better tradeoff between area/power increase and error mitigation is achieved, since modules that do not contribute much to the Architectural Vulnerability Factor (AVF) of the design [6] are left unmodified and their power/area will not be affected by the aforementioned ~3.2× factor [7,8].

Hardening techniques can be applied during the synthesis process. An example of this are the protections inserted by some proprietary synthesizers that allow hardening of full modules, or even applying local TMR attributes to the specific signals that need to be hardened. The Synopsys Synplify pro [9] and Mentor Precision Hi-rel [10] synthesizers are examples of this.

Another way of inserting mitigation schemas into the designs is to perform post-synthesis netlist manipulation, for example using software such as the Xilinx XTMRtool [11] and the BYU (Brigham Young University) EDIF (Electronic Design Interchange Format) tools [12]. The former allows full module hardening in a Xilinx-specific design flow, and the latter is a software suite that can insert both TMR and DWC (Duplicate With Compare) for the user-selected elements. Mitigation elements may also be manually inserted in the post-synthesis netlist, but this process is error-prone and thus not recommended.

Approaches that insert protections during the synthesis process, or just after it, work at the RTL (Register-Transfer Level) netlist abstraction level and thus do not consider physical implementation aspects that may affect the robustness of the implemented design. Depending on whether the target technology on which the digital design will be implemented is an FPGA (Field Programmable Gate Array) or an ASIC (Application-Specific Integrated Circuit), other complementary approaches can be used at the place and route level to improve the robustness of the implemented design, for example physically separating the redundant copies of a hardened element, which improves tolerance to Domain Crossing Errors (DCE) [13]. For the ASIC design of the hardened microprocessor HERMES [14], both DMR and TMR techniques were implemented, depending on which processor block was to be hardened, and the replicated redundancy domains were physically separated during the circuit layout design phase. Another approach in fine-grain techniques is the one proposed on [15], in which design flip-flops are replaced by self-correcting rad-hard by design (RHBD) flip-flops after synthesis, and triplication is performed on spatially separated regions during the placement phase. For FPGA designs, actions can be taken during the placement and routing implementation stages, such as inserting redundant routing connections [16] or using reliability-oriented place and route algorithms to physically separate the redundant copies and avoid single points of failure [17]. Unused FPGA resources may also be employed for error detection: [18] proposes the use of carry propagation chains, which is a common FPGA resource, as a way to create fine-grained comparators to detect bit upsets, which is complemented by the use of coarse-grain checkers that can determine whether the detected upsets did actually propagate to the main module outputs.

1.4. Scope and Contribution of This Paper

Of the previous approaches that work at the RTL netlist level, there is no single approach that allows for both easy insertion of mitigations by performing minimal modifications in the HDL code, and independence from the synthesis tool. In-code fine-grain selection of which elements should be hardened, that propagates to both arithmetic/logic operations performed, and flip-flops used to store them, would be desirable.

This paper proposes a new technique for performing selective, fine-grain circuit hardening, that allows designers to include the information on which combinatorial and sequential elements should be hardened in the VHDL code. In order to be selective, the technique allows designers to individually choose which elements of the VHDL code to harden. To be useful for designers, the technique only implies minimal code substitution and does not change the functionality of the design in absence of soft errors. The technique is also portable between different VHDL synthesizers and does not require the use of post-synthesis tools to generate the hardened netlist.

The difference between the proposed technique and proprietary approaches such as [9–11] is that the proposed technique can be used across different synthesizers. Also, while [11] must harden complete modules, our technique allows selection of which elements are to be hardened.

Since VHDL allows for both Behavioral and RTL descriptions, the technique can work at both abstraction levels and thus its scope does not include physical layout techniques, but it can be complemented with them.

1.5. Organization of the Paper

The paper is structured as follows: Section 2 describes the proposed approach, with the developed datatypes and operators. Section 3 describes how the approach was verified, both in simulation, to check functional correctness of the hardened designs, and by means of fault injection, to check the correctness of the protection implementations. Finally, the discussion and conclusions are presented in Section 4.

2. The Triple_logic Package

In this article, we propose a new approach to implement fine-grain circuit hardening for digital designs by just changing the datatype of the object to be hardened. By changing the object types, the implementation changes accordingly to introduce the desired redundancy. The designer can then select which nodes of the circuit should be hardened, thus creating redundancy domains for the critical parts of the design. Figure 1 shows a redundant branch of a design, and represents graphically how to pass from a non-hardened domain to a hardened domain, where redundant operations and data storage are performed, and back to the non-hardened domain. It must be noted that, in this context, domain crossing refers to user data passing from the non-hardened to the hardened domain or vice versa, and not to the propagation of errors between redundant copies of the design elements.

We have compiled all the new datatypes and hardening functionality in a VHDL package for ease of use and minimal VHDL code modification. An important feature of the package is avoiding the scenario present in Figure 2, where the robustness of the hardened domain is jeopardized by a Single Point of Failure introduced by premature voting inside the hardened domain. To avoid this situation, transitions between hardened and non-hardened domains are determined by the datatypes of the intervening operands. For example, if an operation receives two hardened operands and must return a non-hardened result, a voter will be inserted, but if the result data type is of a hardened type, no voter will be implemented.

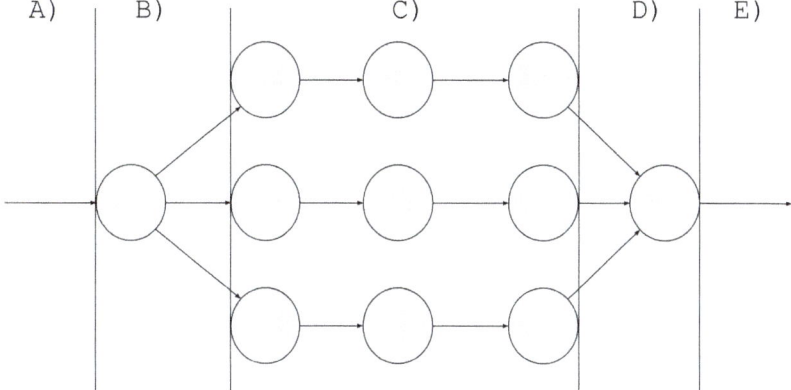

Figure 1. Domain crossing between non-hardened and hardened domains. Each element in the graph may represent either a combinatorial operation or a memory element. (**A**) Non-hardened domain. (**B**) Crossing to hardened domain. (**C**) Hardened domain. (**D**) Crossing to non-hardened domain. (**E**) Non-hardened domain.

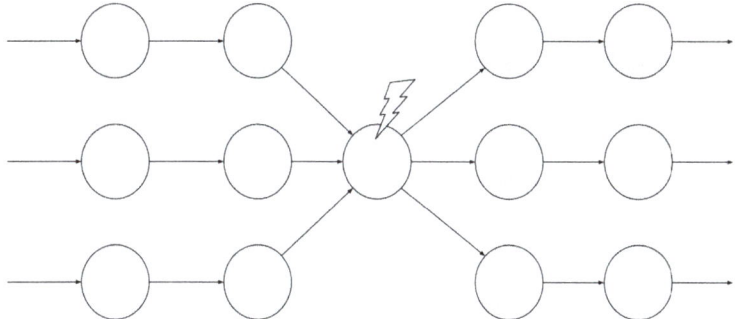

Figure 2. Single Point of Failure introduced inside a redundant domain by premature voting.

2.1. Data Types

Before implementing the automatic hardening functionality mentioned before, the new hardened data types that will compose the hardened domains must be defined. Since the most used standard data types are based on the std_logic data type, defined in the std_logic_1164 package of library IEEE, a triple_logic datatype has been defined that comprises three std_logic values. By defining a vector of triple_logic values, the triple_logic_vector is created. triple_unsigned and triple_signed are hardened vectors with numeric interpretation, just as their non-hardened counterparts. Finally, a triple_integer contains three integers, whose range can be parametrized if using the IEEE Std.1076-2008 revision of the language, more widely known as VHDL-2008 [19]. Table 1 shows the equivalence between hardened and non-hardened data types.

The package defines logic and arithmetic operators for the new datatypes, and for mixed operations between these and the already existing ones. The operator and function overload capability of VHDL will allow an operation (for example, a sum) to receive any combination of datatypes in its input and return operands, and the relevant implementation will be automatically selected depending on the actual data types.

Table 1. Equivalence between non-hardened and hardened data types.

Non-Hardened	Hardened
std_logic	triple_logic
std_logic_vector	triple_logic_vector
unsigned	triple_unsigned
signed	triple_signed
integer	triple_integer

2.2. Hardened to Non-Hardened Domain Crossing

Once all data types and operations have been defined, special consideration must be taken into how to pass data between the non-hardened and hardened domains. The function/operator overload capability of VHDL allows for this domain crossing to be performed automatically for all operator results, but when making a single assignment without any operations this cannot be automatically done, as VHDL is strongly typed and thus the assignment operator cannot be overloaded. We have developed two functions for these cases: a vote() function to pass from the hardened domain to the non-hardened domain (Figure 3), and a triple() function to perform the opposite operation (Figure 4).

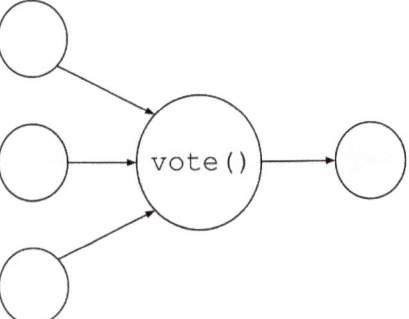

Figure 3. Graphic illustration of vote() function.

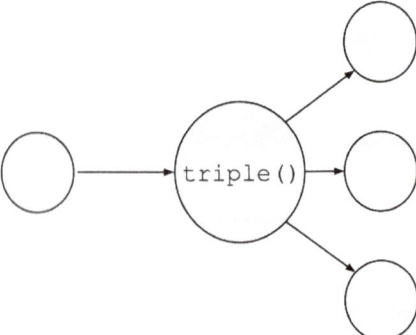

Figure 4. Graphic illustration of triple() function.

Both functions, vote() and triple(), are overloaded so that the user can pass every equivalent data type from the non-hardened domain to the hardened domain, and vice versa, with the same two functions.

2.3. Developed Functionality

After the development of the datatypes and the `vote()` and `triple()` functions, logic, arithmetic and comparison operators were developed for these datatypes.

2.3.1. Operator List

The operators developed for the hardened datatypes are logic (AND, NAND, OR, NOR, XOR, XNOR), comparison (= [is equal], /= [is not equal], > [greater than], >= [greater or equal], < [lower than], <= [lower or equal] and arithmetic operators (+ [addition], - [substraction], * [multiplication], / [division]). Since not every operator is available for every non-hardened datatype (for example, `std_logic_vector` does not have numerical interpretation, and integers do not support bitwise operations), not all operators have been implemented for all datatypes. The list of implemented operators is shown in Table 2.

The assignment operator (<= for signals, := for variables) may not be overloaded since VHDL is strongly typed.

Table 2. List of implemented operators.

Datatype	Logic	Equality/Inequality	Rest of Comparison Operators	Arithmetic
triple_logic	yes	yes	yes	no
triple_logic_vector	yes	yes	no	no
triple_unsigned	yes	yes	yes	yes
triple_signed	yes	yes	yes	yes
triple_integer	no	yes	yes	yes

2.3.2. Operator Variants

Due to operator overload, for each of the operators, we have developed a number of variants. This way, domain crossing is performed by automatically choosing the appropriate operator variant, which is done by the synthesis tools and simulators. For example, the statement A <= B + C will assign a hardened or non-hardened value to A depending on its data type. For unary operators, there are four combinations according to whether the operand and result are hardened or not. For binary operators, there are eight possibilities. All these possibilities are shown in Table 3. Of course, the possibilities that correspond to all values in the non-hardened domain are already defined in the `std_logic_1164` or `numeric_std` packages so they do not need to be defined again.

Table 3. Operator Variants.

Unary Operators	
Operand	Result
unhardened	unhardened
unhardened	hardened
hardened	unhardened
hardened	hardened

Binary Operators		
Left Operand	Right Operand	Result
unhardened	unhardened	unhardened
unhardened	unhardened	hardened
unhardened	hardened	unhardened
unhardened	hardened	hardened
hardened	unhardened	unhardened
hardened	unhardened	hardened
hardened	hardened	unhardened
hardened	hardened	hardened

The current implementation of the hardening functionality includes all operator variants in the same VHDL file, but those operator variants could also be separated into different files, in case the designer wants to automate domain crossing in one direction but not on the other. In that case, the functions that automatically cross from the unhardened to the hardened domain, the functions that automatically cross from the hardened domain to the unhardened one, and the functions that operate only on the hardened domain would be defined in different files. This way, the user could choose one of these four possibilities, depending on which files are included:

1. Automatically cross domains from the unhardened to the hardened one, but manually use the `vote()` function to go back to the unhardened domain.
2. Automatically cross domains from the hardened to the unhardened one, but manually use the `triple()` function to go back to the hardened domain.
3. Automatically perform all domain crossing operations. In this case, qualified expressions of VHDL may be needed to solve ambiguity in some cases. For example, the statement `B <= not (not A);` becomes ambiguous, because even if A and B are known types, the innermost not operator does not know whether it should return a hardened or unhardened result. This is resolved by specifying the desired return type for the intermediate operations, for example: `B <= not std_logic'(not A);`.
4. Manually perform all domain crossing operations.

2.3.3. Hardening Finite State Machines

Hardening Finite State Machines (FSMs) is not trivial when FSMs use an enumerated data type, which is a common practice. A custom solution can be implemented for each FSM, by defining a `decode()` and `triple()` function for the hardened version of their state datatype. Both functions are used for domain crossing: when decoding the state of the FSM, the first function returns the correct state, after correcting errors, and when assigning a new state, the second function converts the enumerated constant to a hardened value. This is a needed tradeoff in order to have fine-grain hardening with minimal code modifications, since on every possible state many signals may be assigned, and the designer may not want to harden all of them.

These functions can be made generic for every enumerated datatype if using VHDL-2008, and can be used with the rest of the package when using a VHDL-2008 capable synthesizer. When full TMR is not needed in the state registers, the technique allows the user to implement his own EDAC (Error Detection and Correction) functions to encode and decode the FSM state instead of triplicating it, for example by defining the functions `encode()` and `decode()` to add Hamming codes to the state registers.

2.4. Usage Examples

A couple of usage examples follow. Figures 5 and 6 show a hardened multiplexer and a hardened generic-width counter, with minimal code modifications, which are underscored. For the designer, it is clear from the signal and port datatypes which objects belong to the hardened domain.

To prevent the synthesizer from removing the redundancy, attributes can be applied to the tripled registers. The name of the specific attribute depends on the chosen synthesis tool, for example, when using Synopsys Synplify the attribute `syn_preserve` can be used, whereas in Xilinx XST (Xilinx Synthesis Technology) the relevant attributes are called `keep` and `equivalent_register_removal`.

```
entity mux2to1 is
    port ( input_l : in   triple_logic;
           input_r : in   triple_logic;
           sel     : in   triple_logic;
           output  : out  triple_logic);
end mux2to1;

architecture arch of mux2to1 is
begin
  comb: process (input_l, input_r, sel)
  begin
    if (sel = '0') then
       output <= input_l;
    else
       output <= input_r;
    end if;
  end process;
end arch;
```

Figure 5. Hardened 2-to-1 multiplexer. Note that the equality comparison operator is overloaded, so sel can be compared to '0'.

```
architecture arch of contparam is
    signal reg_i, p_reg_i: triple_unsigned (N-1 downto 0);
begin

  comb: process (reg_i, enable, updown)
  begin
     if (enable = '1') then
           if (updown = '1') then
                  p_reg_i <= reg_i + 1;
           else
                  p_reg_i <= reg_i - 1;
           end if;
     else
           p_reg_i <=reg_i;
     end if;
  end process;

  sinc: process (clk,rst)
  begin
     if (rst = '1') then
              reg_i <= (others => (others => '0'));
     elsif (rising_edge(clk)) then
              reg_i <= p_reg_i;
     end if;
  end process;

     data_out <= std_logic_vector(vote(reg_i));

end arch;
```

Figure 6. Hardening an N-bit counter architecture. Note that the only modifications are the change in the datatype of the internal count, its reset value, and the voting for the primary output, which belongs to the non-hardened domain.

3. Package Verification

To check the correct behaviour of the package, a number of test cases have been generated. Both basic functionality and designs of increasing levels of complexity have been tested. Synthesis, simulation and fault injection results have been obtained to verify that not only the inserted protections mitigate effectively against SEU, but also that the added functionality does not change the expected circuit functionality in the absence of SEU.

Synthesis has been performed with Xilinx ISE (Integrated Synthesis Environment) 14.7 and Synopsys Synplify v4.2. The simulations have been performed with Xilinx ISim (ISE Simulator) version 14.7. The fault injection campaigns have been performed with the FT-Unshades2 (Fault Tolerance—Universidad de Sevilla Hardware Debugging System) fault injection platform [20], version 3.10, working in ASIC mode, which means injections are performed in the user flip-flops.

The Yosys Open SYnthesis Suite [21] has been used to formally verify design equivalence between the hardened and non-hardened versions of the smaller designs, described below, such as counter and shiftreg. The formal equivalence checker tries to solve a boolean satisfiability problem (abbreviated SAT). In this case, the solver must check if there is any input combination that would make the outputs of the hardened and unhardened design differ, and prove by induction that the design outputs will not differ at any time in the future, for any possible set of input vectors. For some of the other designs, even if full formal equivalence cannot be demonstrated because of their complexity, hundreds of induction steps have been performed without any equivalence error being encountered. The simulations also show that the output of the hardened and unhardened versions of all designs are the same, when no SEU are being injected.

3.1. Primitive Verification

To validate the smallest package functionality, a number of test cases have been generated, which have been checked both in simulation, checking correct behaviour against transient errors, and by reviewing the generated netlist topologies. To check the primitives, synthesis has been performed with the XST synthesizer, but results are expected to be reproducible with any other VHDL synthesizer. No optimization of the inserted protections has been detected when synthesizing with XST, but in the case of these optimizations happening with other synthesizers, VHDL attributes can be added to the hardened signals to avoid removal of the hardening elements. Figures 7 and 8 show the synthesized netlist and a short simulation of one of the developed primitives.

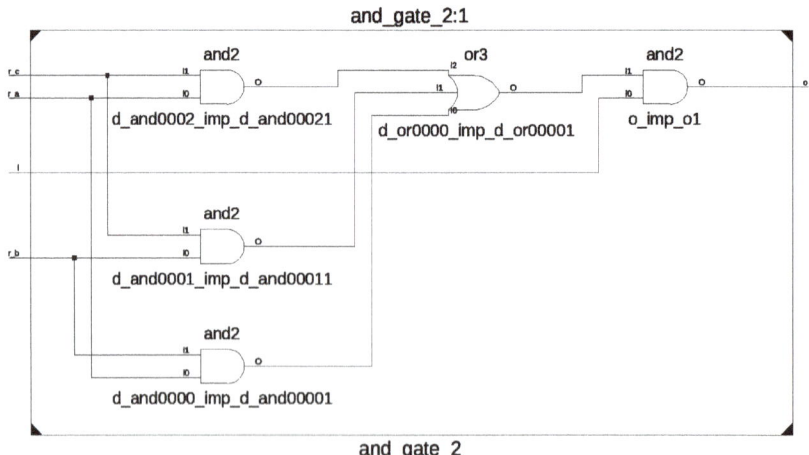

Figure 7. Internal logic structure of AND gate with right port hardened.

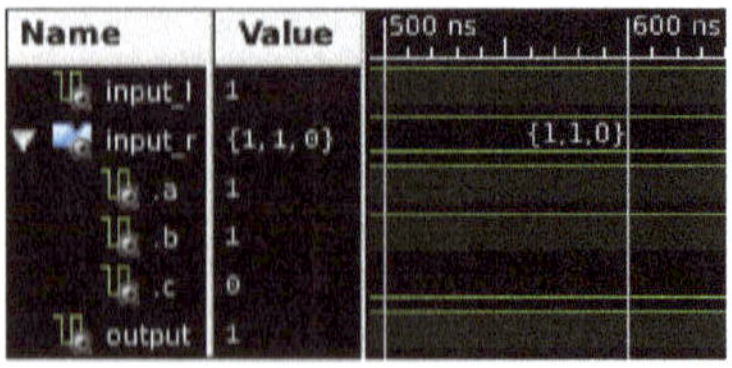

Figure 8. Simulation results of AND gate with right port hardened, with a transient error in its right input.

3.2. Designs Under Test

A number of VHDL designs with increasing levels of complexity have been chosen to validate the hardening capabilities of the package. For each of these designs, an SEU fault injection campaign has been performed with FT-Unshades2, in order to identify the most critical registers, which will be hardened by using the methodology proposed in this work. The hardened versions of the designs have also been subjected to fault injection campaigns, to check the effectiveness of the inserted protections. In order to validate that the technique can be used with different synthesizers, synthesis of both hardened and unhardened designs has been performed with Xilinx ISE and Synopsys Synplify. Finally, synthesis results have also been obtained with the NanoXplore NanoXmap synthesizer version 2.9.1, but the results of this synthesis cannot be tested in the current version of FT-Unshades2, since this synthesizer targets the NanoXplore NG-MEDIUM FPGA and the current version of FT-Unshades2 uses a Virtex-5 FPGA.

- counter

  An 8-bit up counter with an enable signal.

- shiftreg

  An 8-bit shift register. In this example, flip-flops turn into shift registers when synthesis is made with XST so they are optimized even if "keep" attribute is set. To avoid this, "Shift Registers Extraction" and "Equivalent Register Removal" synthesis options have been unselected for this design.

- simple_fsm

  A 4-state simple state machine design, described specifically for this work. In the unhardened version, when using the default synthesis options, XST uses binary codification for synthesis. However, keeping the default synthesis options, in the hardened version, one-hot codification is used for synthesis, so the number of FF (flip-flops) increases from 2 to 12 (4 bits, triplicated). This is the worst case in terms of area overhead, but it can be controlled by the user, by specifying the desired FSM encoding during synthesis. For example, the user can change the FSM encoding from binary to one-hot when hardening the design, in order to reduce the area overhead of hardening the FSM state register, if the timing constraints allow for a slower state decoding. Table 4 shows flip-flop usage for this design in all its possible variants.

  Table 4. Simple_fsm state flip-flops.

FSM Encoding	Unhardened	Hardened
one-hot	4	12
binary	2	6

  For the fault injection experiments both versions (hardened and unhardened) of the simple_fsm design have been synthesized using one-hot encoding, when synthesizing with XST, and binary encoding, when synthesizing with Synplify, to show that the FSM hardening can be performed independently of the encoding.

- adder_acum

  A simple adder-accumulator design that accumulates the sum of 8-bit input values into a 20-bit register.

- fifo

    A generic 256-bit depth and 32-bit width FIFO (First in, First Out) memory buffer with Empty and Full flags [22].

- fft

    Fast Fourier Transform module for usage on FPGA devices [23].

- fir_ri

    A low pass FIR (Finite Impulse Response) filter [24].

- pcm3168

    An I^{2}S interface designed for the PCM3168 audio interface from Texas Instruments [25].

- 8051

    A VHDL model of The Intel 8-bit 8051 micro-controller [26]. This design, which has more complexity than the others, has been tested with a simple program written in C.

3.3. Experimental Results

Injection campaigns have been performed for all the test designs, and their results have been analyzed by comparing the number of flip-flops, AVF and lines of code changed between hardened and unhardened designs. AVF has been estimated by making N injections in a set of FF and dividing the number of injections that produce output errors by the number of total injections (N). For each injection, the complete test vectors are executed by the design.

Designs with a low percentage of total FFs and less number of clock cycles like counter, shiftreg, simple_fsm, fir_ri or adder_acum have been tested with exhaustive campaigns. However, designs with a higher occupancy and more clock cycles, like pcm3168, fft, fifo or 8051 have been tested with random campaigns checking that the number of injections performed on these campaigns is enough to assure less than 5% of error, with a confidence level of 99%, according to [27].

For each design, Table 5 shows the name of registers with damages due to the injections performed in every campaign. The results of these campaigns have been analyzed to determine the AVF of each register with damage, as it can be seen in the fourth column of the table.

To determine which registers will be hardened, the percentage of FFs in the register by FFs in the design has been calculated and those that have more percentage of FF with a higher AVF have been selected to be hardened.

Tables 6 and 7 show results for hardened designs synthesized with XST and Synplify respectively. A comparison between hardened and unhardened designs versions has been done to check the effectiveness of the package. The first column contains the name of the hardened version of the design (in bold) followed by the name of the hardened registers. In the second one, the number of different code lines between the hardened and unhardened versions and the resulting percentage against the total lines in the design are shown. Third and fourth columns present the AVF both for the complete design and each register and the number of FFs obtained in each version.

Table 8 compares synthesis results with three synthesis tools (XST, Synplify and NanoXmap), showing that the proposed technique can be used with different synthesizers, avoiding vendor lockdown.

An estimation of the power consumption for each design is also shown in Tables 9 and 10 for XST and Synplify synthesis respectively. As power consumption depends on which target technology is going to be used, we have made this estimation using the XPower analyzer tool from Xilinx [28], assuming that designs will be implemented for an FPGA, specifically the Virtex-5

XC5VFX70T. According to this, results presented show logic (flip-flops and lookup tables) and signal (interconnections) power consumption estimation both for the unhardened and the hardened versions, and the increase incurred by using our approach, in absolute value and percentage.

Table 5. Designs under test, synthesized with Synplify, with register signals classified by percentage of total number of flip-flops and Architectural Vulnerability Factor.

Signals	Flip-flops	% FFs	AVF
Design: counter *Total Flip-flops: 8*			
reg	8	100	98.60
Design: shiftreg *Total Flip-flops: 8*			
reg	8	100	97.54
Design: simple_fsm *Total Flip-flops: 2*			
state_FSM	2	100	68.75
Design: adder_acum *Total Flip-flops: 8*			
acc_value	20	100	97.50
Design: pcm3168 *Total Flip-flops: 95*			
s_bit_clk1	1	1.05	91.84
s_counter_bit	2	2.11	95.83
s_counter_lr	5	5.26	74.31
s_lr_clk	2	2.11	91.38
v_lr_clk_enable	1	1.05	58.18
DATA_L	24	25.26	40.04
DATA_R	24	25.26	60.63
s_current_lr	1	1.05	52.38
shift_reg	24	25.26	16.54
s_parallel_load	1	1.05	51.22
counter	5	5.26	3.77
DOUT	2	2.11	83.67
Design: fifo *Total Flip-flops: 34*			
looped	1	0.03	100.00
Tail	8	23.52	45.00
Head	8	23.52	60.00
Design: fft *Total Flip-flops: 929*			
o_im	9	0.97	100.00
o_re	9	0.97	100.00

Table 5. Cont.

Signals	Flip-flops	% FFs	AVF
Design: fir_ri Total Flip-flops: 86			
N_bit_reg/Q	9	0.97	95.97
shift_reg	9	0.97	93.40
Design: 8051 Total Flip-flops: 1396			
/alu_op_code/	4	0.29	75.00
/alu_src/	8	0.59	15.79
/p0_out_c/	8	0.59	100.00
/p1_out_c/	8	0.59	100.00
/p2_out_c/	8	0.59	100.00
/p3_out_c/	8	0.59	100.00
/ram_wr	1	0.07	100.00
/U_CTR/exe_state/	3	0.22	100.00
/U_CTR/reg_pc_7/	8	0.59	75.00
/U_RAM/iram/	1024	75.13	5.34
/U_RAM/sfr_acc/	8	0.59	50.00
/U_RAM/sfr_psw/	8	0.59	50.00
/U_RAM/sfr_sp	8	0.59	20.00
/U_RAM/sfr_tmod/	8	0.59	20.00

Table 6. Hardened versions of the designs under test, synthesized with XST.

Design	Code Modif. (Lines)	Code Modif. (%)	AVF Unhardened	AVF Hardened	AVF Decrease (%)	FF Unhardened	FF Hardened	FF Increase (%)
counter_v2	4	10.53	99.09	0.00	100.00	8	24	200.00
reg			99.09	0.00	100.00			
shiftreg_v2	5	11.90	86.89	0.00	100.00	8	24	200.00
reg			86.89	0.00	100.00			
simple_fsm_v2	49	73.13	53.13	0.00	100.00	4	12	200.00
state_FSM			53.13	0.00	100.00			
adder_acum_v2	8	21.05	97.50	0.00	100.00	20	60	200.00
i_acc_value			97.50	0.00	100.00			
pcm3168_v2	44	8.40	41.16	4.14	89.94	90	234	160.00
DATA_L			40.57	0.00	100.00			
DATA_R			56.00	0.00	100.00			
shiftreg			17.43	0.00	100.00			
fifo_v2	13	13.00	18.38	18.27	0.61	8243	8275	0.39
Tail			100.00	0.00	100.00			
Head			100.00	0.00	100.00			
fft_v2	42	8.73	1.53	1.02	33.33	387	723	86.82
o_im			100.00	0.00	100.00			
o_re			100.00	0.00	100.00			
fir_ri_v2	18	16.82	72.66	0.00	100.00	80	240	200.00
N_bit_reg/Q			72.70	0.00	100.00			
8051_v2	31	0.47	99.71	1.30	98.72	1327	1365	2.86
/U_CTR/reg_pc_7/			100	0.00	100.00			
/U_RAM/sfr_acc/			100	0.00	100.00			
/U_RAM/sfr_psw			100	0.00	100.00			

Table 7. Hardened versions of the designs under test, synthesized with Synplify.

Design	Code Modif. (Lines)	(%)	AVF Unhardened	Hardened	Decrease (%)	FF Unhardened	Hardened	Increase (%)
counter_v2 reg	6	15.79	98.60 98.60	0.00 0.00	100.00 100.00	8	24	200.00
shiftreg_v2 reg	5	11.90	85.57 97.54	0.00 0.00	100.00 100.00	8	24	200.00
simple_fsm_v2 state_FSM	53	79.10	68.75 68.75	0.00 0.00	100.00 100.00	2	6	200.00
adder_acum_v2 i_acc_value	11	28.95	97.50 97.50	0.00 0.00	100.00 100.00	20	60	200.00
pcm3168_v2 DATA_L DATA_R shift_ref	42	8.02	38.51 40.04 60.63 17.19	3.93 0.00 0.00 0.00	89.79 100.00 100.00 100.00	95	244	156.84
fifo_v2 Tail Head	13	13.00	17.55 45.00 60.00	17.48 0.00 0.00	0.40 100.00 100.00	34	87	155.88
fft_v2 rot2bf_im rot2bf_re	26	5.41	2.42 100.00 100.00	2.15 0.00 0.00	11.16 100.00 100.00	929	1061	14.21
fir_ri_v2 N_bit_reg/Q	18	16.82	93.61 95.97	0.00 0.00	100.00 100.00	86	240	179.07
8051_v2 /U_CTR/reg_pc_7/ /U_RAM/sfr_acc/ /U_RAM/sfr_psw	34	0.51	8.37 75 50 50	1.35 0.00 0.00 0.00	83.87 100.00 100.00 100.00	1396	1445	3.51

Table 8. Comparison of synthesis results. Data marked with an asterisk (*) corresponds to the synthesizer implementing an internal memory with Flip-flops instead of inferring a Block RAM.

Design	FF XST	FF Synplify	FF NanoXmap
counter	8	8	8
counter_v2	24	24	24
shiftreg	8	8	8
shiftreg_v2	24	24	24
simple_fsm	4	2	4
simple_fsm_v2	12	6	12
adder_acum	20	20	20
adder_acum_v2	60	60	60
pcm3168	90	95	91
pcm3168_v2	234	244	187
fifo	8243 *	34	23
fifo_v2	8275 *	87	55
fft	387	929	447
fft_v2	723	1061	783
fir_ri	80	86	80
fir_ri_v2	240	240	224
8051	1327	1396	1339
8051_v2	1365	1445	1359

Table 9. Power consumption estimation of hardened and unhardened versions of the designs under test, synthesized with XST.

Design	Unhardened Power (mW)		Hardened Power (mW)		Increase	
	(Logic)	(Signal)	(Logic)	(Signal)	Total	Percentage (%)
counter	0.10	0.30	0.33	0.97	0.90	225.00
shiftreg	0.00	0.00	0.00	0.00	0.00	N/A
simple_fsm	0.00	0.02	0.02	0.04	0.04	200.00
adder_acum	0.00	0.01	0.00	0.01	0.00	0.00
pcm3168	0.19	0.33	0.22	0.42	0.12	23.08
fifo	0.18	11.83	0.27	12.79	1.05	8.74
fft	1.55	4.13	2.56	5.68	2.56	45.07
fir_ri	0.00	0.19	0.03	0.27	0.11	57.89
8051	1.35	9.70	1.32	10.90	1.17	10.59

Table 10. Power consumption estimation of hardened and unhardened versions of the designs under test, synthesized with Synplify.

Design	Unhardened Power (mW)		Hardened Power (mW)		Increase	
	(Logic)	(Signal)	(Logic)	(Signal)	Total	Percentage (%)
counter	0.11	0.20	0.34	0.50	0.53	170.97
shiftreg	0.00	0.00	0.00	0.01	0.01	N/A
simple_fsm	0.00	0.03	0.01	0.04	0.02	66.67
adder_acum	0.00	0.01	0.00	0.02	0.01	100.00
pcm3168	0.13	0.48	0.18	0.61	0.18	29.51
fifo	0.11	0.52	0.20	1.33	0.90	142.86
fft	3.30	9.22	3.52	9.63	0.63	5.03
fir_ri	0.00	0.24	0.03	0.28	0.07	29.17
8051	1.75	16.80	1.97	16.60	0.02	0.11

The experimental results show that the selected registers can be hardened with the proposed approach, and that this protection is effective against SEU. When synthesizing the hardened designs with Synplify, it can be observed that the synthesizer not only triples the hardened flip-flops, but also may insert extra memory elements as a means of compensating the increased fan-out needs by the design, in a process known as timing-driven replication. It is very interesting to note that SEUs introduced in these new flip-flops do not produce errors in the output, which means that the relevant voting logic has also been propagated to these new memory elements, so the timing-driven replication does not negatively impact the effectiveness of the inserted protections. The AVF of the FIFO does not show much improvement, because only the flip-flops have been hardened, while SEU may affect the complete memory, which has many more sensitive elements.

Since the inserted redundancy is hardware redundancy and there are no time redundancy operations, the hardened designs take exactly the same number of clock cycles to perform their workload than their unhardened counterparts. Simulation execution time does not grow significantly: in small designs it varies less than a second, and in large designs it is less than 5%. This is coherent with what would be expected since, in the bigger designs, the entire design is not tripled, but only part of it, so the simulation time should not be tripled. Increments in simulation time for other designs will depend on what percentage of the design was hardened.

Finally, the power consumption increase of the hardened designs is in line with what is expected, according to the area increase of each design and an expected multiplication by a ~3.2 factor for each triplicated element.

4. Conclusions

A new approach to implement fine-grain circuit hardening, using datatype substitution, has been developed and validated. As a result, a VHDL package for selective circuit hardening by design has been developed as a new tool for mitigating soft errors on digital circuits, with minimal code modifications. The designer only has to select which signals or ports should be hardened and change their datatype accordingly. Some use of the `triple()` and `vote()` functions can be needed because of the strongly typedness of VHDL.

An interesting feature of this way of performing hardening by design is that the designer, after identifying the critical elements of his/her design using fault injection or other approaches, can embed in the source code of the module the information of which elements should be protected, thus eliminating the need to configure a second tool (such as a post-synthesis netlist processor) with the results of the vulnerability analysis.

Collaboration with synthesis tool vendors would improve the performance of the package to avoid some unwanted optimizations that may happen when performing multiple passes during the synthesis process, for example, when TMR flip-flops that would not be optimized, because correct signal attributes have been used, get converted to SRL16 primitives (Lookup tables used as Shift Registers) which in turn get optimized away. Another case of this is when hardened ports of internal modules get optimized by the synthesizer, because the attributes to avoid redundancy removal have been applied in the wrong object, since some synthesizers require these attributes to be placed in the ports to preserve, and others require them to be placed in the affected architecture. The ideal situation would be that the attributes that avoid redundancy removal could be applied to the hardened datatypes and inherited by all ports, signals and variables of that datatype.

Future work may also include implementing different hardening schemas by using the datatype substitution technique, such as hamming encoding for FSMs or approximate TMR.

5. Licensing

The `triple_logic` package is licensed under the GNU Lesser General Public License (LGPL) v3.0. The code can be downloaded from the website http://ftu.us.es/triplelogic .

Author Contributions: Conceptualization, H.G.-M.; methodology, H.G.-M., M.M.-Q. and D.V.-C.; validation and bug fixing: M.M.-Q., S.S.-B. and H.G.-M., formal analysis (equivalence checking), H.G.-M.; simulation, M.M.-Q., S.S.-B. and H.G.-M.; fault injection, M.M.-Q., D.V.-C.; analysis, M.M.-Q., D.V.-C.; investigation, M.M.-Q., S.S.-B., D.V.-C. and H.G.-M.; supervision, H.G.-M.; project administration, H.G.-M.

Funding: This work was supported by the Spanish Ministerio de Economía y Competitividad, through the project "Diseño de sistemas digitales robustos frente a radiación mediante componentes y tecnologías comerciales" (RENASER3), project reference ESP2015-68245-C4-2-P. This work was also partly supported by the European Commission, through the project "VEGAS: Validation of European high capacity rad-hard FPGA and software tools", project ID 687220.

Acknowledgments: The authors would like to thank: Clifford Wolf from Symbiotic EDA for kindly providing a VHDL-capable version of the Yosys Open SYnthesis Suite. Edouard Lepape, Hervé Baier and Mohamed Gountaf from NanoXplore for kindly providing a version of NanoXmap and promptly answering our support queries. Xilinx University Program (XUP) for kindly providing the ISE software. European Space Agency for funding the development of the FT-Unshades2 Fault Injection Platform (ESA contract 4200022981/09/NL/JK), which has been used to validate the results of this work. Finally, the authors would like to thank José M. Hinojo from Universidad de Sevilla for his support with multiple tool setup and licensing issues, and Luis Sanz for his advice on automating the generation of the different implementations of the test designs and meaningful conversations on how to better structure the operation overload capabilities of the package.

Conflicts of Interest: The authors declare no conflict of interest. The founding sponsors had no role in the design of the study; in the collection, analyses, or interpretation of data; in the writing of the manuscript, or in the decision to publish the results.

Abbreviations

The following abbreviations are used in this manuscript:

ASIC	Application-Specific Integrated Circuit
AVF	Architectural Vulnerability Factor
DCE	Domain Crossing Errors
DD	Displacement Damage
DMR	Dual Modular Redundancy
DWC	Duplicate With Compare
EDAC	Error Detection And Correction
EDIF	Electronic Design Interchange Format
FF	Flip-flop
FIFO	First In, First Out
FPGA	Field Programmable Gate Array
FSM	Finite State Machine
FT-Unshades	Fault Tolerance-Universidad de Sevilla Hardware Debugging System
HDL	Hardware Description Language
ISE	Integrated Synthesis Environment
ISim	ISE Simulator
LUT	Lookup Table
MBU	Multiple Bit Upset
RHBD	Rad-Hard By Design
RTL	Register-Transfer Level
SET	Single Event Transient
SEU	Single Event Upset
TID	Total Ionizing Dose
TMR	Triple Modular Redundancy
VHDL	Very High Speed Integrated Circuit Hardware Description Language
XST	Xilinx Synthesis Technology
Yosys	Yosys Open SYntesis Suite

References

1. Duzellier, S. Radiation effects on electronic devices in Space. *Aerosp. Sci. Technol.* **2005**, *9*, 93–99. [CrossRef]
2. Dominik, L. System mitigation techniques for single event effects. In Proceedings of the 2008 IEEE/AIAA 27th Digital Avionics Systems Conference, St. Paul, MN, USA, 26–30 October 2008; pp. 5.C.2-1–5.C.2-12.
3. Kuwahara, T.; Tomioka, Y.; Fukuda, K.; Sugimura, N.; Sakamoto, Y. Radiation effect mitigation methods for electronic systems. In Proceedings of the 2012 IEEE/SICE International Symposium on System Integration (SII), Fukuoka, Japan, 16–18 December 2012; pp. 307–312.
4. Lima-Kastensmidt, F.; Reis, R.A. *Fault-Tolerance Techniques for SRAM-Based FPGAs*, 1st ed.; Springer: Heidelberg, Germany, 2006.
5. Gomes, I.A.C.; Kastensmidt, F.G.L. Reducing TMR overhead by combining approximate circuit, transistor topology and input permutation approaches. In Proceedings of the 2013 26th Symposium on Integrated Circuits and Systems Design (SBCCI), Curitiba, Brazil, 2–6 September 2013; pp. 1–6.
6. Nair, A.A.; Eyerman, S.; Eeckhout, L.; John, L.K. A first-order mechanistic model for architectural vulnerability factor. In Proceedings of the 2012 39th Annual International Symposium on Computer Architecture (ISCA), Portland, OR, USA, 9–13 June 2012; pp. 273–284.
7. Martinez-Alvarez, A.; Cuenca-Asensi, S.; Restrepo-Calle, F.; Pinto, F.R.P.; Guzman-Miranda, H.; Aguirre, M.A. Compiler-Directed Soft Error Mitigation for Embedded Systems. *IEEE Trans. Dependable Secur. Comput.* **2012**, *9*, 159–172. [CrossRef]
8. Guzmán-Miranda, H.; Barrientos-Rojas, J.; López-González, P.; Baena-Lecuyer, V.; Aguirre, M.A. On the Structural Robustness Assessment of Wireless Communication Systems for Intra-Satellite Applications. *IEEE Trans. Nucl. Sci.* **2014**, *61*, 3244–3249. [CrossRef]
9. Synopsys. *FPGA Synthesis Attribute Reference Manual*; Synopsys: Mountain View, CA, USA, 2018.

10. Precision Hi-Rel Synthesis Software. Available online: http://www.mentor.com/products/fpga/synthesis (accessed on 20 December 2018).
11. Xilinx. TMRTool User Guide. Available online: https://www.xilinx.com/support/documentation/user_guides/ug156-tmrtool.pdf (accessed on 20 December 2018).
12. BYU EDIF Tools Home Page. Available online: http://reliability.ee.byu.edu/edif/ (accessed on 20 December 2018).
13. Quinn, H.; Morgan, K.; Graham, P.; Krone, J.; Caffrey, M.; Lundgreen, K. Domain Crossing Errors: Limitations on Single Device Triple-Modular Redundancy Circuits in Xilinx FPGAs. *IEEE Trans. Nucl. Sci.* **2007**, *54*, 2037–2043. [CrossRef]
14. Clark, L.T.; Patterson, D.W.; Ramamurthy, C.; Holbert, K.E. An Embedded Microprocessor Radiation Hardened by Microarchitecture and Circuits. *IEEE Trans. Comput.* **2016**, *65*, 382–395. [CrossRef]
15. Hindman, N.D.; Clark, L.T.; Patterson, D.W.; Holbert, K.E. Fully Automated, Testable Design of Fine-Grained Triple Mode Redundant Logic. *IEEE Trans. Nucl. Sci.* **2011**, *58*, 3046–3052. [CrossRef]
16. Kastensmidt, F.L.; Filho, C.K.; Carro, L. Improving Reliability of SRAM-Based FPGAs by Inserting Redundant Routing. *IEEE Trans. Nucl. Sci.* **2006**, *53*, 2060–2068. [CrossRef]
17. Sterpone, L.; Violante, M. A new reliability-oriented place and route algorithm for SRAM-based FPGAs. *IEEE Trans. Comput.* **2006**, *55*, 732–744. [CrossRef]
18. Nazar, G.L.; Carro, L. Fast error detection through efficient use of hardwired resources in FPGAs. In Proceedings of the 2012 17th IEEE European Test Symposium (ETS), Annecy, France, 28–31 May 2012; pp. 1–6.
19. IEEE Standard VHDL Language Reference Manual. In *IEEE Std 1076-2008 (Revision of IEEE Std 1076-2002)*; IEEE Computer Society: New York, NY, USA, 2009; pp. 1–626.
20. Mogollon, J.M.; Guzmán-Miranda, H.; Nápoles, J.; Barrientos, J.; Aguirre, M.A. FTUNSHADES2: A novel platform for early evaluation of robustness against SEE. In Proceedings of the 2011 12th European Conference on Radiation and Its Effects on Components and Systems, Sevilla, Spain, 19–23 September 2011; pp. 169–174.
21. Clifford Wolf. Yosys Open SYnthesis Suite. Available online: http://www.clifford.at/yosys/ (accessed on 20 December 2018).
22. VHDL Standard FIFO. Available online: http://www.deathbylogic.com/2013/07/vhdl-standard-fifo/ (accessed on 20 December 2018).
23. VHDL Implementation of FFT Algorithm(s). Available online: https://github.com/thasti/fft (accessed on 20 December 2018).
24. FPGA4student. A Low Pass FIR Filter for ECG Denoising in VHDL. Available online: https://www.fpga4student.com/2017/01/a-low-pass-fir-filter-in-vhdl.html (accessed on 20 December 2018).
25. I^{2}S Interface Designed for the PCM3168 Audio Interface from Texas Instruments. Available online: https://github.com/wklimann/PCM3168 (accessed on 20 December 2018).
26. Dalton Project, Department of Computer Science, University of California, "Synthesizable VHDL Model of 8051". Available online: http://www.cs.ucr.edu/~dalton/i8051/i8051syn/ (accessed on 20 December 2018).
27. Leveugle, R.; Calvez, A.; Maistri, P.; Vanhauwaert, P. Statistical Fault Injection: Quantified Error and Confidence. In Proceedings of the Design, Automation and Test in Europe Conference, Nice, France, 20–24 April 2009.
28. Xilinx. XPower Analyzer. Available online: https://www.xilinx.com/support/documentation/sw_manuals/xilinx14_7/xpa_c_overview.htm (accessed on 20 December 2018).

© 2018 by the authors. Licensee MDPI, Basel, Switzerland. This article is an open access article distributed under the terms and conditions of the Creative Commons Attribution (CC BY) license (http://creativecommons.org/licenses/by/4.0/).

Article

Optimal Physical Implementation of Radiation Tolerant High-Speed Digital Integrated Circuits in Deep-Submicron Technologies

Jeffrey Prinzie [1,*], Karel Appels [1] and Szymon Kulis [2]

[1] Deptartment of Electrical Engineering (ESAT), KU Leuven, 3000 Leuven, Belgium; karel.appels@student.kuleuven.be
[2] European Center for Nuclear Research (CERN), 1211 Meyrin, Switzerland; szymon.kulis@cern.ch
* Correspondence: jeffrey.prinzie@kuleuven.be; Tel.: +32-14-72-13-38

Received: 14 March 2019; Accepted: 12 April 2019; Published: 14 April 2019

Abstract: This paper presents a novel scalable physical implementation method for high-speed Triple Modular Redundant (TMR) digital integrated circuits in radiation-hard designs. The implementation uses a distributed placement strategy compared to a commonly used bulk 3-bank constraining method. TMR netlist information is used to optimally constrain the placement of both sequential cells and combinational cells. This approach significantly reduces routing complexity, net lengths and dynamic power consumption with more than 60% and 20% respectively. The technique was simulated in a 65 nm Complementary Metal-Oxide Semiconductor (CMOS) technology.

Keywords: single-event upsets (SEUs); digital integrated circuits; triple modular redundancy (TMR); radiation hardening by design

1. Introduction

Digital integrated circuits are important in many of today's complex integrated circuits and systems. A wide range of digital circuits are used as stand-alone digital systems such as microprocessors and digital signal processors which are purely digital systems. A second portion of digital blocks can be found in mixed-signal integrated circuits where finite-state machines or counters assist or interface with their analog counterparts.

The operation frequency of a digital module mainly depends on the system requirements such as data throughput. In mixed-signal circuits such as analog-to-digital converters (ADC), phase locked loops (PLL) and clock and data recovery (CDR), the digital logic is clocked at or a derivative of the mixed-signal sampling time which can be as high as several GHz in recent technologies. The latter digital blocks are intrinsically timing critical and only have little timing overhead for circuit level redundancy in harsh environments.

It has been widely known that ionizing radiation can cause Single Event Effects (SEEs) in CMOS integrated circuits [1], especially in scaled technologies [2]. Single Event Upsets (SEUs) in flip-flops result in corrupted data or logic states [3]. An SEU on a flip-flop can occur at any time and may be unrecoverable if no redundancy is applied. Single-Event Transients (SETs) occur when a particle upsets combinational logic which has no memory [4]. The transient is only temporary. Thus, the digital system is only sensitive to SETs when the SET propagates to the data-inputs of a flip-flop during the setup- and hold-times of that flip-flop [5]. Many aspects can be considered regarding the propagation of the SETs such as logic masking [6,7] and pulse shrinking. However, a common conclusion can be made that digital logic which is clocked at high clock frequencies is more sensitive to SETs since the probability to capture an SET in the clock period dramatically increases in the GHz range [8].

Fortunately, Triple Modular Redundancy (TMR) can be added to digital logic to overcome these effects. TMR triplicates the logic and uses majority voters to correct logic signals [9]. TMR relies on the fact that only one logic signal can be upset at once and would therefore fail if two out of three triplicated nets are upset. This was less of a concern in old CMOS technologies where single particles only affected single digital cells. However, as technologies have scaled down, Multi-Bit Upsets (MBU) have become a serious concern such that one particle can affect multiple gates simultaneously [10–12]. With improper placement, the fault tolerance can dramatically reduce, especially in fast designs.

Many forms of TMR have been presented in recent decades, some of them compromising triplication effectiveness for power consumption or area. The most complete form of TMR, and the one which is addressed in this paper, is full TMR [13] where both the flip-flops, clock-tree and combinational logic are triplicated ([14,15]) as is shown in Figure 1. This method is the most reliable but also uses the highest number of resources and power. A competing method is temporal time-redundancy [16]. This smart approach does not triplicate the combinational logic but only triplicates flip-flops which are clocked with 3 delayed clocks [17]. The delay between the clocks is set to be larger than any possible SET, such that only one flip-flop could possibly capture an SET which is later on corrected. This method has proven its usage in space applications but its major drawback is its limited clock frequency. The intentional clock skew places serious timing constraints on the design typically resulting in sub-GHz designs. As such, many high-speed mixed-signal digital module implementations prefer the original TMR approach. Several other methods were reported in [18–22].

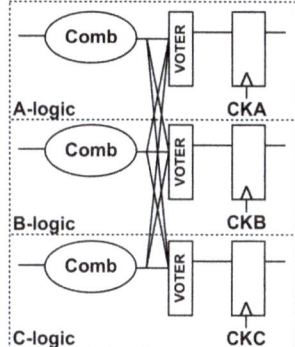
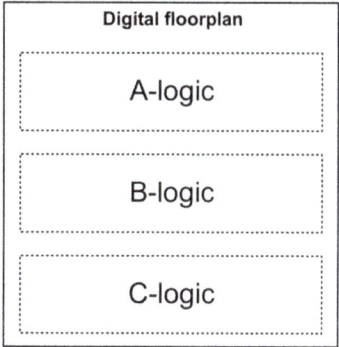

Figure 1. Full TMR design with correcting voters and its associated physical floorplan.

This paper is organized as follows: In Section 2, a novel placement method for high-speed digital TMR designs is presented. In Section 3, a performance analysis is made and compared with conventional methods. In Section 4, a conclusion is drawn.

2. Physical Implementation for TMR Circuits

2.1. Conventional 3-Block Approach

TMR-protected circuits only operate well if only one signal of a TMR signal is upset. Therefore, MBUs should be avoided. Historically, a floorplan was made as is shown in Figure 1. Only flip-flops were constrained to be placed in 3 physical groups respectively A-C and a spacing of 10 µm is ensured between these groups. The idea behind this approach is that the combinational logic and clock trees will follow the constrained flip-flops. This has proven its usage in many high-speed radiation hard designs in the past [23].

The drawback of this approach is the complex routing to interconnect the voters. Each voter has a connection to both A, B and C blocks such that each logic net has 6 cross domain connections. This becomes problematic when the design size increases. Firstly, connections from A to C would require to cross the entire B-logic which results in long nets. Automatic place-and-route tools will always optimize the design to meet the required timing constraints. Therefore, large buffers will be inserted to meet the timing constraints which increases the power consumption of the design. Secondly, these long vertical connections result in a dense metal routing. This may result in a sub-optimal design and may result in worse signal integrity. As the design size increases, so does the problematic physical implementation with this method.

2.2. Novel Interleaved Approach

Ideally, each cell is guided with a spacing constraint to place the standard cells distributed. Such features have become available in the newest place-and-route releases. This can be used to ensure the spacing between flip-flops but not between combinational logic in the data-path. For moderate to complex data-paths such as a fast 8-bit counter, the spacing between logic in a common TMR data-path cannot always be ensured to prevent MBUs in the combinational logic.

The approach proposed in this paper allows a semi-distributed placement to allow maximal freedom to place-and-route tools to optimize the design. It is based on the conventional 3-block approach but uses an interleaved placement constraining method. This is shown in Figure 2a. Multiple repeating small regions allow cells of A-C branches to be placed at different vertical locations in the design. These regions have fixed heights. As the design size increases, only the number of vertical regions increases, not the height of it. As a result, the vertical connections between voters only have to cross a narrow placement region and not 1/3 of the design since the place-and-route tool has more freedom to place the standard cells vertically in the design. The spacing between each region is sufficient to prevent MBUs (10 µm in our case). Grouping the flip-flops to the correct regions (A-C) is straightforward since the naming of these registers includes the TMR group through automatic TMR insertion before synthesis [23]. However, after synthesis, the data-path cells cannot be grouped by name anymore. Therefore, an algorithm was designed to trace-back the combinational logic that drives a flip-flop. As such, if each flip-flop can be segmented by name, the combinational logic can be as well by tracing back the input logic tree. This algorithm finds the fan-in of the flip-flop by searching for the drivers of the input nets of the cells as is shown in Figure 2b. This is done iteratively to find the full fan-in logic tree and stops at a register. A special exception is required when searching the fan-in of a voter block. Only the respective input is used since otherwise, the entire TMR data-path (A-C) is grouped in the same region due to the cross coupled connections before the voters. By means of this method, all cells, including combinational logic can be efficiently constrained to regions. If only the flip-flops were allocated to their respective groups, unconstrained data-path cells might be placed in incorrect regions resulting in MBUs in the combinational logic. This algorithm is executed before placement of the cells and is re-executed after clock tree synthesis and design optimization to.

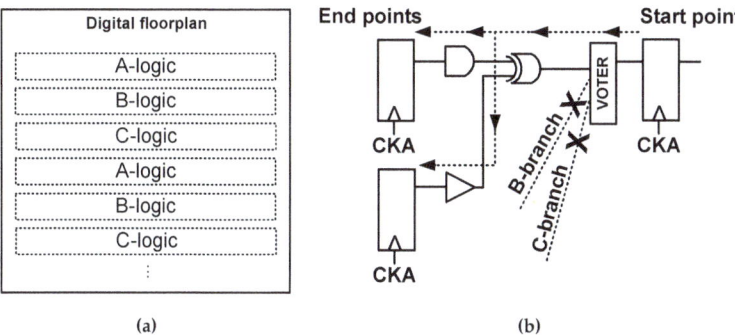

Figure 2. Logic placement of an interleaved design. (**a**) Interleaved physical floorplan; (**b**) Logic fan-in cone traceback.

3. Simulated Performance Analysis

To assess the performance of this new placement and floorplanning method, different comparing tests have been done between the proposed interleaved approach and the conventional 3-block approach. Three digital designs, each with 8 identical independent high-speed counters, were used. The counter sizes of the three designs were varied to implement more complex standardized data-paths. A summary of the designs and associated timing constraints is shown in Table 1. The designs were implemented and benchmarked using Innovus CAD (Computer Aided Design) tools with optimization efforts high. The timing constraints were chosen to be close to the technology limits to ensure a timing critical design. For each design, the power consumption, net length, net capacitance and routing density was analyzed and compared. The designs were implemented in a 9-track standard V_T 65 nm CMOS library. The interleaved method has a slice height and spacing of 7.2 µm while the 3-block method has the same block spacing. The results discussed below are extracted from the timing, power and area reports from place-and-route tools.

Table 1. Benchmarked designs.

	Counter Size	Clock Constraint	Inst. Count	Design Width
Design 1	8 × 8 bit	1.2 ns	2808	175 µm
Design 2	8 × 16 bit	1.7 ns	4560	250 µm
Design 3	8 × 32 bit	1.9 ns	9120	350 µm

The routed designs of "Design 2" are shown in Figure 3a,b, where the 3-block and proposed interleaved placement method are used respectively. It is clear that the proposed approach results in significantly reduced complexity compared to the conventional 3-block approach. More specifically, the vertical routing difficulty is highly reduced since the place-and-route tool has more freedom to optimally place the cells in the design. To quantitatively analyze this, a histogram of the total net length is shown in Figure 4 for both implementations. The histogram shows that a significant portion of the nets has lengths between 1/3 and 1/2 of the design size for the 3-block implementation due to the voter interconnects. This peak is not present anymore in the proposed interleaved implementation method. Most interconnected nets now have a net length of approximately 25 µm. As the design size increases, the difference becomes more significant. However, some long nets are still present in the interleaved placed design. By analyzing pre-Clock Tree Synthesis (CTS) and post-CTS histograms, it becomes clear that these longer nets originate from the clock tree. Clock trees of clocks A-C are now distributed across the entire design while in the 3-block approach, the clock tree was only placed locally in each of the 3 regions. Similar results were obtained from both Design 1 and Design 3.

The standard-cell spacing between cells of the same TMR branch is shown in Figure 5. This plot shows a histogram of the distances between cells of A-branch to respectively B- and C-branch cells. The distances were measured only between cells which implement the same connected logic tree. Again, these results show a significant reduction of the spacing for the proposed method which correlate with the net lengths in Figure 4. With the interleaved method, most cells are spaced within a 45 µm distance which corresponds to roughly 3 small, vertical interleaved banks. From this result, it is clear that the placement engine is given more freedom to place cells closer without compromising radiation hardness. For to the 3-block method, the average cell distance between A-branch and B- or C-cells shows two peaks which correspond to the large A-B-C bock distances in the floorplan. For the proposed interleaved method, the reduced cell spading leads to a reduction of the power consumption and an improved routability of the design.

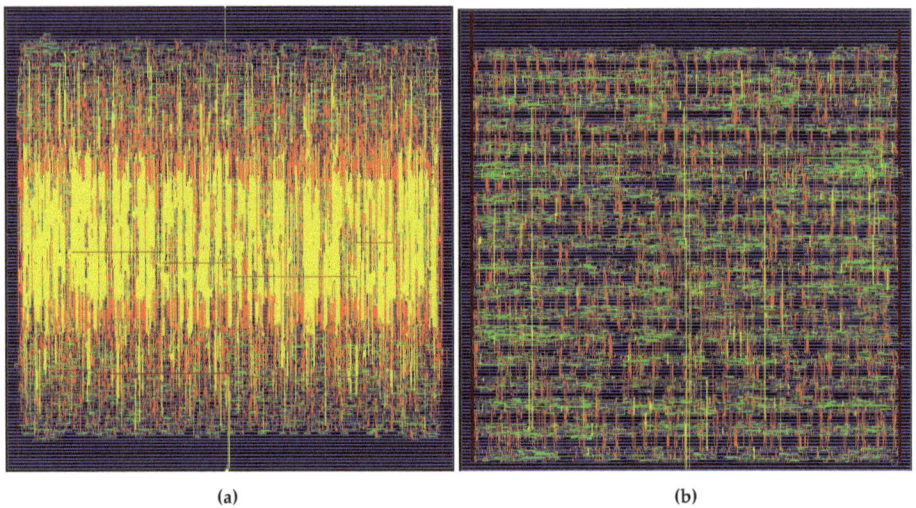

Figure 3. Routed 8 × 16 bit designs with a floorplan size of 250 µm × 250 µm. (**a**) 3-Block approach; (**b**) Proposed interleaved approach.

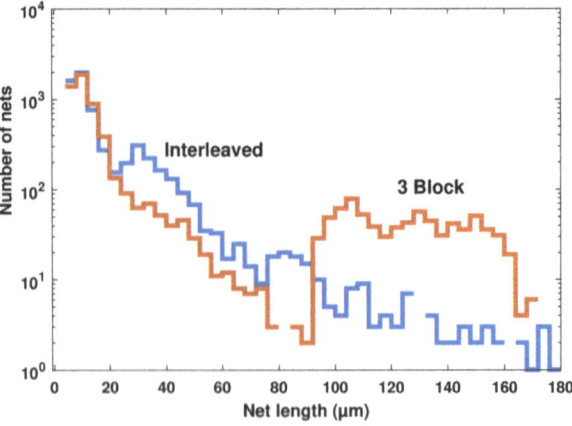

Figure 4. Net length histogram for the 8 × 16 bit design.

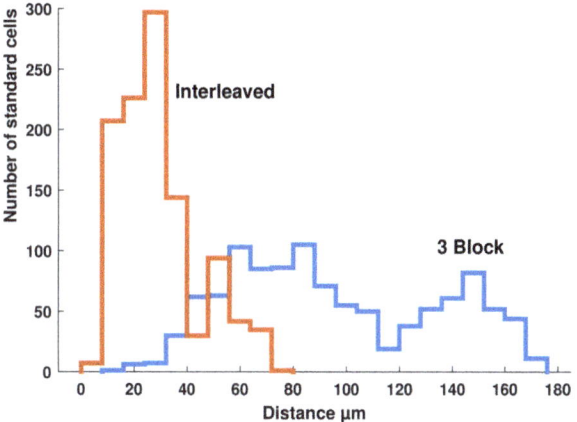

Figure 5. TMR branch logic spacing for the 8 × 16 bit design.

The routing complexity was analyzed by measuring the local metal density for each metal layer on a 100 × 100 square grid across the design. The design examples used a metal stack with 5 routable layers, of which M2 and M4 were vertical routing layers. Figure 6 shows the average metal density across the design floorplan for all different layers. All horizontal layers and M1 (standard cell routing and power rail) show no significant differences between both methods. A significant difference can be found in M2 and M4. Vertical inter-TMR branch connections are routed in these layers such that the density is relatively high for the 3-block implemented designs. However, the interleaved design shows a 50% reduction in M2 and nearly no routing in M4 which is a significant improvement. Near the vertical middle of the design, the peak density of M2 drops from 22% to 8% while that of M4 reduces from 30% to 0.5%. A vertical cross section in the middle of the design is shown in Figure 7. Here, the average metal density as a function of the vertical location in the design is shown. These results clearly indicate that the routing complexity is dramatically reduced which also lead to a reduction of the overall power consumption.

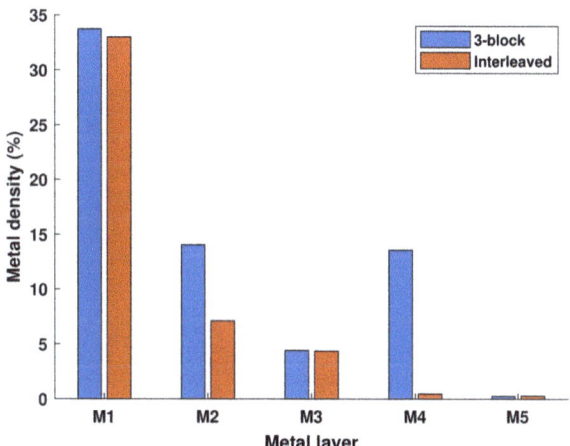

Figure 6. Average design metal density for different metal layers for 3 Block (left) and interleaved (right) methods.

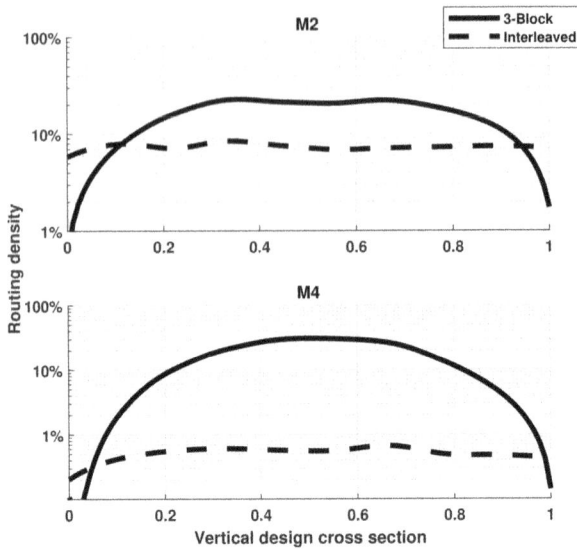

Figure 7. Vertical cross section of metal routing density for M2 and M4 vertical routing layers.

A summary of the power consumption and net lengths is shown in Table 2 for the 3 designs. These figures are extracted after routing and Clock Tree Synthesis (CTS). The internal power is the power consumption of the unloaded standard cells, switching power is the dynamic power consumption due to the switching of the capacitive loads. Total capacitance is the sum of all net and input capacitances of the cells. From these figures, it is clear that the internal power does not significantly change. A small reduction comes from smaller buffers required in the design. However, the switching power changes significantly and scales proportionally with the total capacitance of the nets. A comparison of the results between the 3-block and the proposed interleaved method clearly indicates that a significant reduction of 14% up to 47% can be achieved with our method as the design size increases. These results illustrate the advantages of the proposed placement method. Additionally, it can be seen that the total and average net length of the design reduces by 36% to 65% which is a significant improvement.

Table 2. Performance comparison.

(a) 8 × 8 bit

	3-Block	Interleaved	Difference
Internal power	20.59 mW	19.46 mW	−5.4%
Switching power	10.55 mW	8.712 mW	−17.42%
Total Power	31.14 mW	28.18 mW	−9.51%
Total capacitance	32 pF	26 pF	−14.8%
Total net length	74,425 μm	47,436 μm	−36.26%
Average net length	17 μm	11 μm	−35.46%

(b) 8 × 16 bit

	3-Block	Interleaved	Difference
Internal power	21.95 mW	22.58 mW	+2.8%
Switching power	11.67 mW	9.224 mW	−20.95%
Total Power	33.62 mW	31.8 mW	−5.41%
Total capacitance	45 pF	38 pF	−15.3%
Total net length	142,468 μm	89,096 μm	−37.46%
Average net length	23.9 μm	14.2 μm	−40.47%

(c) 8 × 32 bit

	3-Block	Interleaved	Difference
Internal power	55.17 mW	51.96 mW	−5.82%
Switching power	34.82 mW	18.26 mW	−47.55%
Total Power	89.99 mW	70.22 mW	−21.96%
Total capacitance	108 pF	60 pF	−44.28%
Total net length	368,032 μm	128,320 μm	−65.13%
Average net length	26.8 μm	9.1 μm	−65.89%

4. Conclusions

This paper has presented a novel method for physical implementation of Triple Modular Redundant high-speed digital circuits. The method uses a distributed constraining approach for TMR branches to avoid long interconnects between voters. A TMR logic fan-in search algorithm is used to segment combinational logic in TMR A, B and C groups. The method was tested with increasingly complex digital modules and shows results which improve as the design size increases. For the tested circuits, the total net length reduced up to 65% while the switching power consumption reduced by 44%. Furthermore, the routing complexity was significantly simplified compared to a bulk 3-block physical floorplan.

Author Contributions: Conceptualization, J.P., K.A. and S.K.; methodology, J.P. and K.A. and S.K.; validation, J.P. and K.A.; investigation, J.P. and K.A.; writing—original draft preparation, J.P., K.A. and S.K.

Funding: This research was funded by FWO—Research Foundation Flanders.

Conflicts of Interest: The authors declare no conflict of interest. The founding sponsors had no role in the design of the study; in the collection, analyses, or interpretation of data; in the writing of the manuscript, and in the decision to publish the results.

References

1. Harrington, R.C.; Kauppila, J.S.; Warren, K.M.; Chen, Y.P.; Maharrey, J.A.; Haeffner, T.D.; Loveless, T.D.; Bhuva, B.L.; Bounasser, M.; Lilja, K.; et al. Estimating Single-Event Logic Cross Sections in Advanced Technologie. *IEEE Trans. Nucl. Sci.* **2017**, *64*, 2115–2121.

2. Nsengiyumva, P.; Ball, D.R.; Kauppila, J.S.; Tam, N.; McCurdy, M.; Holman, W.T.; Alles, M.L.; Bhuva, B.L.; Massengill, L.W. A Comparison of the SEU Response of Planar and FinFET D Flip-Flops at Advanced Technology Nodes. *IEEE Trans. Nucl. Sci.* **2016**, *63*, 266–272. [CrossRef]
3. Chen, C.-H.; Knag, P.; Zhang, Z. Characterization of Heavy-Ion-Induced Single-Event Effects in 65 Nm Bulk CMOS ASIC Test Chips. *IEEE Trans. Nucl. Sci.* **2014**, *61*, 2694–2701. [CrossRef]
4. Benedetto, J.; Eaton, P.; Avery, K.; Mavis, D.; Gadlage, M.; Turflinger, T.; Dodd, P.E.; Vizkelethyd, G. Heavy ion-induced digital single-event transients in deep submicron Processes. *IEEE Trans. Nucl. Sci.* **2004**, *51*, 3480–3485. [CrossRef]
5. Benedetto, J.M.; Eaton, P.H.; Mavis, D.G.; Gadlage, M.; Turflinger, T. Digital Single Event Transient Trends With Technology Node Scaling. *IEEE Trans. Nucl. Sci.* **2006**, *53*, 3462–3465. [CrossRef]
6. Dodd, P.E.; Shaneyfelt, M.R.; Felix, J.A.; Schwank, J.R. Production and propagation of single-event transients in high-speed digital logic ICs. *IEEE Trans. Nucl. Sci.* **2004**, *51*, 3278–3284. [CrossRef]
7. Chen, R.M.; Diggins, Z.J.; Mahatme, N.N.; Wang, L.; Zhang, E.X.; Chen, Y.P.; Liu, Y.N.; Narasimham, B.; Witulski, A.F.; Bhuva, B.L. Analysis of temporal masking effect on single-event upset rates for sequential circuits. In Proceedings of the 2016 16th European Conference on Radiation and Its Effects on Components and Systems (RADECS), Bremen, Germany, 19–23 September 2016. [CrossRef]
8. Reed, R.A.; Carts, M.A.; Marshall, P.W.; Marshall, C.J.; Buchner, S.; La Macchia, M.; Mathes, B.; McMorrow, D. Single Event Upset cross sections at various data rates. *IEEE Trans. Nucl. Sci.* **1996**, *43*, 2862–2867. [CrossRef]
9. Ruano, O.; Reviriego, P.; Maestro, J.A. Automatic insertion of selective TMR for SEU mitigation. In Proceedings of the RADECS, Gothenburg, Sweden, 16–21 September 2018; pp. 284–287.
10. Evans, A.; Glorieux, M.; Alexandrescu, D.; Polo, C.B.; Ferlet-Cavrois, V. Single event multiple transient (SEMT) measurements in 65 nm bulk technology. In Proceedings of the 2016 16th European Conference on Radiation and Its Effects on Components and Systems (RADECS), Bremen, Germany, 19–23 September 2016. [CrossRef]
11. Zhang, H.; Jiang, H.; Assis, T.R.; Ball, D.R.; Narasimham, B.; Anvar, A.; Massengill, L.W.; Bhuva, B.L. Angular Effects of Heavy-Ion Strikes on Single-Event Upset Response of Flip-Flop Designs in 16-nm Bulk FinFET Technology. *IEEE Trans. Nucl. Sci.* **2017**, *64*, 491–496. [CrossRef]
12. Bhuva, B.L.; Tam, N.; Massengill, L.W.; Ball, D.; Chatterjee, I.; McCurdy, M.; Alles, M.L. Multi-Cell Soft Errors at Advanced Technology Nodes. *IEEE Trans. Nucl. Sci.* **2015**, *62*, 2585–2591. [CrossRef]
13. Ulloa, G.; Lucena, V.; Meinhardt, C. Comparing 32nm full adder TMR and DTMR architectures. In Proceedings of the 2017 24th IEEE International Conference on Electronics, Circuits and Systems (ICECS), Batumi, Georgia, 5–8 December 2017; pp. 294–297.
14. Balasubramanian, P.; Prasad, K. A Fault Tolerance Improved Majority Voter for TMR System Architecture. *WSEAS Trans. Circuits Syst.* **2016**, *15*, 108–122.
15. Shinohara, K.; Watanabe, M. A double or triple module redundancy model exploiting dynamic reconfigurations. In Proceedings of the 2008 NASA/ESA Conference on Adaptive Hardware and Systems, Noordwijk, The Netherlands, 22–25 June 2008; pp. 114–112.
16. Hasanbegovic, A.; Aunet, S. Heavy Ion Characterization of Temporal-, Dual- and Triple Redundant Flip-Flops Across a Wide Supply Voltage Range in a 65 nm Bulk CMOS Process. *IEEE Trans. Nucl. Sci.* **2016**, *63*, 2962–2970. [CrossRef]
17. Schmidt, R.; García-Ortiz, A.; Fey, G. Temporal redundancy latch-based architecture for soft error mitigation. In Proceedings of the 2017 IEEE 23rd International Symposium on On-Line Testing and Robust System Design (IOLTS), Thessaloniki, Greece, 3–5 July 2017; pp. 240–243.
18. She, X.; McElvain, K.S. Time multiplexed triple modular redundancy for single event upset mitigation. *IEEE Trans. Nucl. Sci.* **2009**, *56*, 2443–2449. [CrossRef]
19. Tooba, A.; Abdus Sami, H.; Hossein, M.; Jeong A.L. Input vulnerability-aware approximate triple modular redundancy: Higher fault coverage, improved search space, and reduced area overhead. *Electron. Lett.* **2018**, *54*, 934–936.
20. She, X.; Li, N. Reducing Critical Configuration Bits via Partial TMR for SEU Mitigation in FPGAs. *IEEE Trans. Nucl. Sci.* **2017**, *64*, 2626–2632. [CrossRef]
21. Matush, B.I.; Mozdzen, T.J.; Clark, L.T.; Knudsen, J.E. Area-Efficient Temporally Hardened by Design Flip-Flop Circuits. *IEEE Trans. Nucl. Sci.* **2010**, *57*, 3588–3595. [CrossRef]

22. Nagpal, C.; Garg, R.; Khatri, S.P. A delay-efficient radiation-hard digital design approach using CSWP elements. In Proceedings of the 2008 Design, Automation and Test in Europe, Munich, Germany, 10–14 March 2008; pp. 345–359.
23. Kulis, S. Single Event Effects Mitigation with TMRG Tool. *J. Instrum.* **2017**, *12*. [CrossRef]

© 2019 by the authors. Licensee MDPI, Basel, Switzerland. This article is an open access article distributed under the terms and conditions of the Creative Commons Attribution (CC BY) license (http://creativecommons.org/licenses/by/4.0/).

Article

Heavy-Ion Induced Single Event Upsets in Advanced 65 nm Radiation Hardened FPGAs

Chang Cai [1,3], Xue Fan [2,*], Jie Liu [1], Dongqing Li [1,3], Tianqi Liu [1,3,*], Lingyun Ke [1,3], Peixiong Zhao [1,3] and Ze He [1,3]

1. Institute of Modern Physics, Chinese Academy of Sciences, Lanzhou 730000, China; caichang@impcas.ac.cn (C.C.); j.liu@impcas.ac.cn (J.L.); lidongqing@impcas.ac.cn (D.L.); kelingyun@impcas.ac.cn (L.K.); zhaopeix16@impcas.ac.cn (P.Z.); heze17@mails.ucas.ac.cn (Z.H.)
2. Chengdu Technological University, Chengdu 611730, China
3. University of Chinese Academy of Sciences, Beijing 100049, China
* Correspondence: x_fan@foxmail.com (X.F.); liutianqi@impcas.ac.cn (T.L.)

Received: 14 January 2019; Accepted: 12 March 2019; Published: 14 March 2019

Abstract: The 65 nm Static Random Access Memory (SRAM) based Field Programmable Gate Array (FPGA) was designed and manufactured, which employed tradeoff radiation hardening techniques in Configuration RAMs (CRAMs), Embedded RAMs (EBRAMs) and flip-flops. This radiation hardened circuits include large-spacing interlock CRAM cells, area saving debugging logics, the redundant flip-flops cells, and error mitigated 6-T EBRAMs. Heavy ion irradiation test result indicates that the hardened CRAMs had a high linear energy transfer threshold of upset ~18 MeV/(mg/cm^2) with an extremely low saturation cross-section of 6.5×10^{-13} cm^2/bit, and 71% of the upsets were single-bit upsets. The combinational use of triple modular redundancy and check code could decline ~86.5% upset errors. Creme tools were used to predict the CRAM upset rate, which was merely 8.46×10^{-15}/bit/day for the worst radiation environment. The effectiveness of radiation tolerance has been verified by the irradiation and prediction results.

Keywords: FPGA; radiation hardening; single event upsets; heavy ions; error rates

1. Introduction

Static Random Access Memory (SRAM) based Field Programmable Gate Array (FPGA) possesses plenty of flexible configuration switches and logics to implement million-gate circuits with a very short development time [1,2], which makes it a valuable Integrated Circuit (IC) for an electronic system. However, the Complementary Metal Oxide Semiconductor (CMOS) based architectures in SRAM-FPGA are very sensitive to radiation effects, which reduces the on-orbit safety and reliability [3–9]. Therefore, it is necessary to evaluate and mitigate the Single Event Effects (SEEs) of SRAM-based FPGAs for potential space applications. Although there have been some radiation hardened SRAM-based FPGAs such as XQVR300 (220 nm) and XQR4VSX55 (90 nm) employed for space mission, a fast system scrubbing is still required to reduce the on-orbit risks [7].

There are several radiation hardening techniques such as Error Correct Code (ECC), Triple Modular Redundancy (TMR) and Dual Interlocked storage Cell (DICE) that can be used in FPGAs, but they are area or power consuming and less effective when the Multiple Bit Upset (MBU) phenomenon occurs frequently [2]. Besides, different circuit blocks in an FPGA have different functions and importance. Hence, hardening technique selection is a tradeoff strategy among function, performance, area and radiation tolerance. For example, errors occurring in Configuration RAMs (CRAMs), which control the routing, the switches and the logic state of an FPGA, have the most critical impact on the FPGA function, and so that the radiation tolerance of CRAMs has to be high, although it costs non-negligible area and power [8–11]. Other resources, e.g. Embedded RAMs (EBRAMs) and registers, can be hardened with a proper combination of several soft error mitigation methods which cost less area and power.

For planar bulk Si CMOS technologies, the operation voltages decrease and frequencies increase with the feature size scaling down, which make CMOS logic circuits, including FPGAs, more sensitive to SEU [12,13] and contribute more complex and diverse upsets phenomenon happening in cosmic rays and solar flare environments [8,10,11]. In most cases, the complexity of SEUs also affect the effectiveness of hardening designs, especially in million-gate deep submicron FPGA. For commercial FPGAs, in [11–13], the significant SEU sensitivities appear, even under low Linear Energy Transfer (LET) heavy-ion radiation. Therefore, more hardening techniques are applied in the deep submicron FPGAs than the submicron hardened FPGAs. As reported in [8], three reasonable radiation hardened 90 nm Xilinx FPGAs using several radiation hardening techniques presented superior performances and radiation tolerance, which are very representative products for space application. Thus, the implementation of reasonable hardening strategies for advanced FPGAs as well as the calculation of FPGAs' convincible error rates in space are extremely necessary and feasible [11,14].

In this paper, a radiation-hardened SRAM-based FPGA was designed and manufactured on a 65 nm CMOS process as the Device Under Test (DUT). This device employed a preferable tradeoff radiation hardening strategy by realizing an effective combination of direct layout reinforcement and error mitigation oriented bitstream configuration design. The basic information about the DUTs is introduced in Section 2. SEE experiments under heavy-ion irradiation were implemented, and experimental details and results are described in Section 3. In Section 4, an effective forecast of the space upset rates is calculated thoroughly. Finally, a deep discussion based on the design purposes and further prediction analysis is explicated.

2. DUT Introduction

2.1. DUT Parameters

The DUT (die area: 1.8 cm × 1.3 cm) was fabricated with a 65 nm bulk silicon epitaxial CMOS process (as the classical 65 nm commercial foundry) with ten metal layers. It was flip-chip packaged in a Ball Grid Array (BGA) type. The ~700 μm silicon substrate was thinned down to 40 μm, which is less than the range of each experimental heavy ion.

The picture of the developed device in the 8-inch wafer is shown in Figure 1a and the whole view of chip's layout is shown in Figure 1b. The DUT contained 600 programmable I/O blocks located in banks. The ~20 Mbit CRAM were used to control the routes and switch boxes. The ~8 Mbit embedded block memories were organized and distributed in DUT. Besides, The ~170 Kbit Debugging Logics (DLs) in circuits were also included to test the actual function of devices by capturing read back signals. Apart from the different architecture and resource distributions, the most essential improvement in the DUT is the combined usage of 8-T and DICE structures in the layout design (programmable modules in Figure 1b).

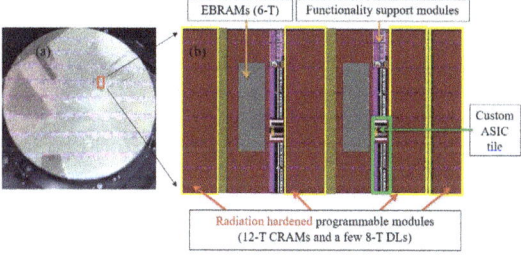

Figure 1. The developed device: (**a**) the 8-inch wafer of designed DUT; and (**b**) the layout and architecture of DUT.

2.2. Radiation Hardening Design

Three kinds of layout hardness techniques were designed in CRAMs, DLs and D flip-flops (DFFs) and the configuration hardness methods were mainly added to EBRAM by Prosice (a self-developed FPGA configuration software). Each DFF was reinforced by double redundant cells. For CRAM, it occupied majority of chip area, and all of them were reinforced by DICE structure (as shown in Figure 2) with large spacing to insulate single-ion charge sharing effect, although the DICE hardened cell had ~2 times area and power cost. Another 8-T structure (as shown in Figure 3) was designed in DL cells to keep the accuracy of read back signals by adding two more anti-disturbance transistors. 8-T cell, which was used to replace the standard 6-T cell, consumed less area.

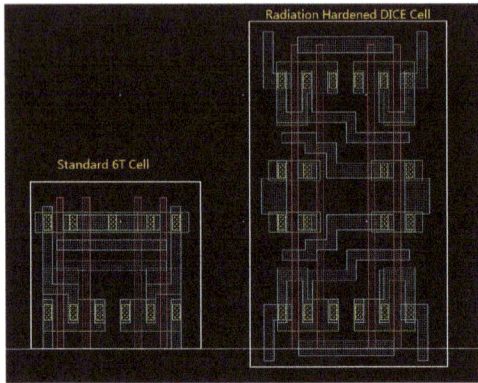

Figure 2. Layout comparison of the DICE hardened cell and standard 6-T cell.

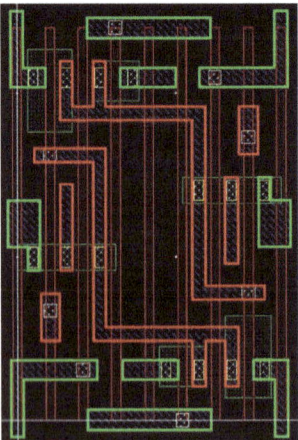

Figure 3. Layout of the 8-T hardened DL.

Configuration hardness techniques such as TMR and ECC created by hardware description language can be employed to verify the effectiveness of error mitigation methods in EBRAM. Thus, four EBRAM modules (Figure 4a) occupying ~100% resources were configured by FPGA bitstreams and tested synchronously. The key method for TMR is the voter design. For each data line, three AND gates and two OR gates, which are generated by FPGA configuration resources, are the TMR circuits (as shown in Figure 4b). For each bit of the data, there is a TMR module to check. To analyze the worst condition in basic 6-T structure, the adjacent RAM resources were included in the TMR test, which means that an event may disturb more than one cell and make the voter invalid. The ECC logics employ hamming codes with five more bits for data check [15]. Hence, the ECC encode and decode circuits shown in Figure 4c,d were created. The 8-bit data became 13-bit data due to the 5-bit redundancy check bits. Eventually, the output data could be reconstructed by the ECC decode design to reveal the error information.

Configuration hardness techniques such as TMR and ECC created by Hardware Description Language (HDL) in Prosice can be employed to verify the effectiveness of error mitigation methods in EBRAM. The special bitstreams, including the initialization and configuration of block memory and logical resources, were generated by Prosice with detailed reports for routing and timing conditions. Both the HDL and the tool command languages are supported in Prosice. The resources of device, the HDL input and the constrained conditions are required for each configuration. A compiled visual control and operation interface can be used to control the SEE testing system and it is compatible for Prosice to achieve a series of operation from the FPGA configuration, bitstream selection, and a real-time debugging. Thus, four EBRAM modules (Figure 4a) occupying ~100% resources were configured by Prosice and tested synchronously. The key method for TMR is the voter design. For each data line, three AND gates and two OR gates, which are generated by FPGA configuration resources, are the TMR circuits (as shown in Figure 4b). For each bit of the data, there is a TMR module to check. To analyze the worst condition in basic 6-T structure, the adjacent RAM resources were included in the TMR test, which means that an event may disturb more than one cell and make the voter invalid. The ECC logics employ hamming codes with five more bits for data check [15]. Hence, the ECC encode and decode circuits shown in Figure 4c,d were created. The 8-bit data became 13-bit data due to the 5-bit redundancy check bits. Eventually, the output data can be reconstructed by the ECC decode design to reveal the error information.

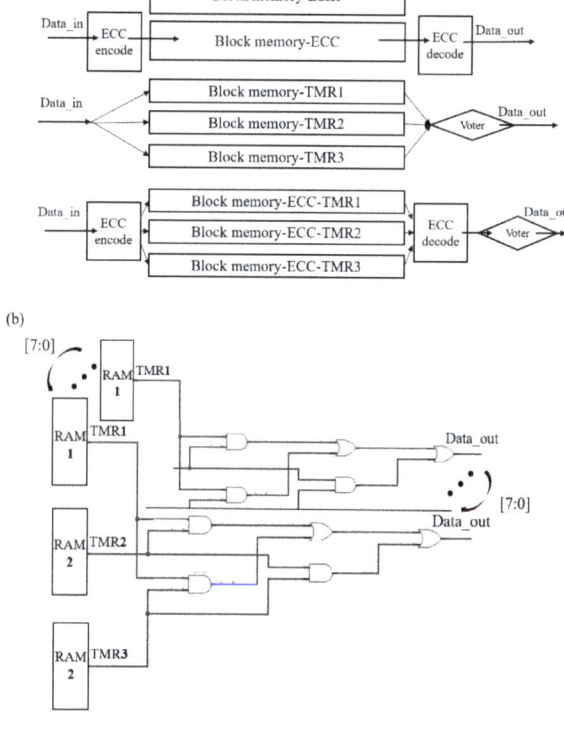

Figure 4. *Cont.*

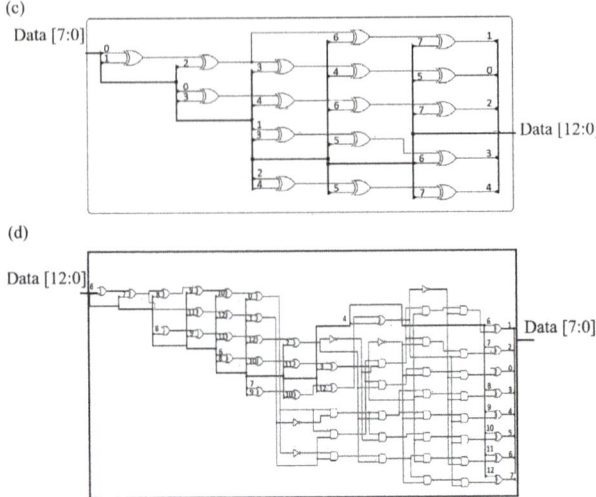

Figure 4. Configured EBRAM: (**a**) the four types of EBRAM; (**b**) the generated 8-channel TMR circuits; (**c**) the generated ECC encode circuits in FPGA; and (**d**) the generated ECC decode circuits in FPGA.

3. Heavy-Ion Test

3.1. Experimental Setup

Heavy ion tests were completed at Heavy Ion Research Facility in Lanzhou (HIRFL) in the Institute of Modern Physics, Chinese Academy of Sciences, and HI-13 Tandem accelerator, China Institute of Atomic Energy. At HIRFL, the tests were carried out in air; and at HI-13 Tandem accelerator, the DUT was located in a vacuum chamber. The type of heavy ions applied for the tests are detailed in Figure 5. To realize a superior accuracy control in LET, for each type of ion, only one energy in high energy region was selected (in the right side of Bragg peak). For each type of heavy ion, its initial energy in DUT surface and the energy in the active layer of DUT (after passing through the thick silicon substrate) are marked in Figure 5. In this study, the LET values in data analysis were all calculated using the ion energy in the active layer.

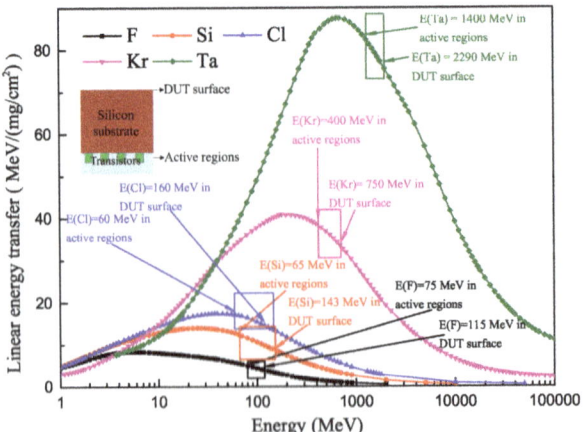

Figure 5. Experimental heavy-ion types and energies.

The DUT was installed on a daughterboard and controlled by a FPGA-based motherboard via digital I/O interface (as shown in Figure 6). The tested resources in DUT, the data configuration, the error data distinction and the communication with Personal Computer (PC) were all completed by the FPGA in the motherboard. During heavy ion irradiation, the upsets (errors) in EBRAM, CRAM, DL and DFF were analyzed automatically by configured FPGA logic and then the error information including the detail of address were recorded and presented on the screen of PC. Besides, the currents in the testing system were monitored and recorded to judge whether there was a latch-up.

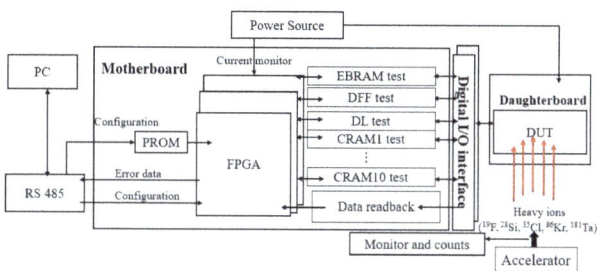

Figure 6. Block diagram of the testing system.

The static mode was employed in tests. The static mode can exclude the influence of the internal address counter SEUs and SET perturbation in peripheral decoding circuit. Considering that the data patterns may influence the SEU cross sections, the checkboard data with "0" and "1" balanced were used in the experiment. For DFFs and EBRAMs, it is easy to design the special bitstream with checkboard data in initialization and then be configured into DUT. However, for configuration modules, plenty of resources have to be used for the whole DUT function, which means the CRAMs cannot be measured at one time. Thus, the CRAMs were divided into ten blocks, with one block for each test to guarantee the function of the DUT.

3.2. Experimental Results

Heavy ion evaluation was focused on the layout hardening and configuration hardening effectiveness. Layout hardening including DICE hardened CRAM, 8-T hardened DL, and double redundant DFF aimed at reducing the SEU sensitivity and system errors, while the configuration hardening was more flexible and applicable to secondarily significant parts such as 6-T EBRAM.

3.2.1. Effectiveness of Layout Hardening

Based on the heavy ion measured upset data, the SEU cross sections could be fitted by Weibull function

$$\sigma_{SEU} = \begin{cases} \sigma_{sat}\left\{1 - exp[-(\frac{LET-LET_{th}}{w})^s]\right\}, & LET \geq LET_{th} \\ 0, & LET \leq LET_{th} \end{cases} \quad (1)$$

where σ_{SEU} (cm^2/bit) indicates the SEU cross-section, σ_{sat} (cm^2/bit) indicates the saturation upset cross-section, LET_th represents the threshold of LET to observe upset errors, s means a dimensionless exponent, and w is a width parameter, as shown in Figure 7. Weibull function is widely used to describe the direct ionization caused by heavy ions, providing great agreement in fitting cross-section thresholds, plateau or limit values. The saturated cross section in EBRAM, DFF, CRAM and DL were 6.2 × 10^{-8} (cm^2/bit), 8.3 × 10^{-8} (cm^2/bit), 6.5 × 10^{-13} (cm^2/bit), and 8.9 × 10^{-11} (cm^2/bit), respectively, for the diverse layout hardening.

DFFs are radiation sensitive resources in FPGA logical blocks. The cross section results of redundant hardened DFF chains are shown in Figure 7. The tested cross sections were still high (8.3 × 10^{-8} (cm^2/bit)). SET may affect sensitive parts in peripheral circuits such as the buffers with error transient creation to disturb the initial data in DFF chains. Besides, the checkboard data were used as the input signals of DFF cells in experiments; therefore, one error of the clock path may cause more than one output error of the DFFs, which can increase the cross sections.

The DL module was distinguished from CRAMs that were used in routing and controlling functions. For the DLs, a superior SEU cross section (8.9 × 10^{-11} (cm^2/bit)) indicates that the 8-T structure with the transient interference excision by two additional transistors (Figure 3) guaranteed its radiation hardening performance. Apparently, the 8-T structure improved its upset threshold and saturated cross section significantly when compared with redundant cell in DFF. More interestingly, the full-DICE protected structure improved CRAM's upset threshold to ~18 MeV/(mg/cm^2). An extremely low SEU saturated cross section at 6.5 × 10^{-13} (cm^2/bit) was tested by ^{181}Ta irradiation because of the successfully isolated sensitive volumes in CRAM.

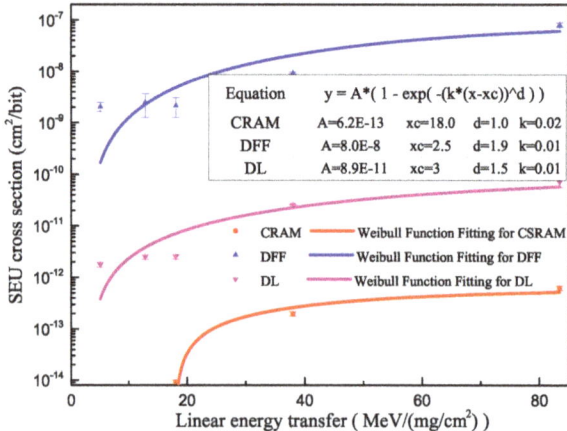

Figure 7. SEU cross section in CRAM, DFF and DL vs. LET.

Additionally, layout hardening declined the Multiple Bits Upset (MBU) significantly. As shown in Figure 8a, Single Bit Upset (SBU) takes high proportion in DICE hardened CRAM. Three-bit or more upsets did not appear during the whole irradiation test. However, in 6-T cell, MBU accounted for a large proportion (54.5%) and four-bit or even five-bit upsets in the unhardened cells were more than 10% (Figure 8b).

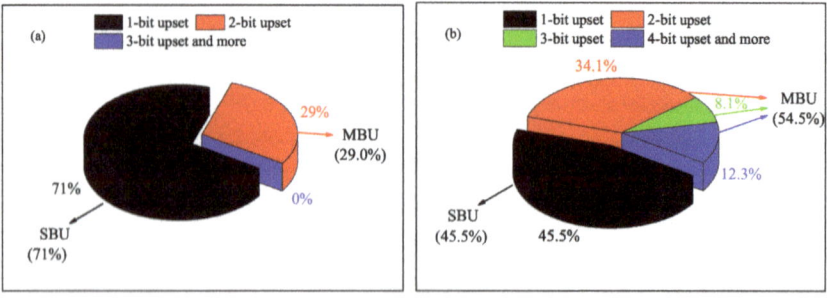

Figure 8. Proportion of SBU and MBU in: (**a**) DICE hardened structure; and (**b**) 6-T structure.

3.2.2. Effectiveness of Configuration Hardening

For the advanced 65 nm CMOS-based EBRAM cells, SEU critical charge declined significantly when compared with sub-micron devices, making the unhardened unit more sensitive to heavy ion irradiation [9,10]. In addition, the carriers created by ionized ions could deposit enough energy in more than one drain regions of off-state MOSFETs in the vicinity, leading to a large MBU rate and high SEU cross section.

Three hardening techniques, namely 8 + 5 ECC codes, TMR and ECC codes plus TMR, were used to evaluate the effectiveness of soft error mitigation. As shown in Figure 9, the ECC codes plus TMR had the lowest upset rates. Even in ^{181}Ta irradiation, the cross section was just 8.5×10^{-9} (cm^2/bit), declined ~86.3% when compared with the unhardened one. Merely ECC or TMR method in EBRAM played a minor role in upset error correction, although they were area and resources consuming. The results matched the high MBU rates in basic 6-T cells presented in Figure 8, leading the separated fault tolerance method (ECC or TMR) becoming invalid.

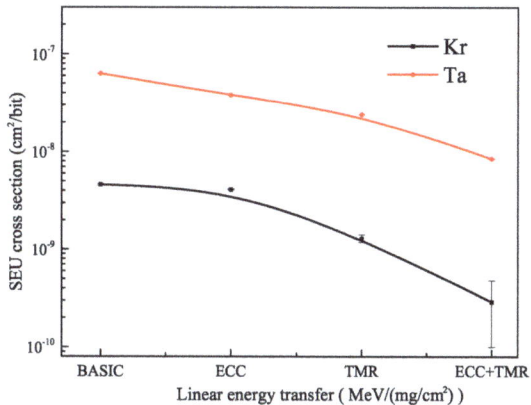

Figure 9. SEU cross section in four kinds of configured EBRAM vs. LET.

4. SEU Rates Prediction

In near-Earth interplanetary or GEO orbit, heavy ions in cosmic ray and solar ray dominate the total SEU rates by direct ionization to produce carriers in the vicinity of transistors' sensitive regions and then lead to SEU phenomenon. The SEU rates induced by heavy-ion direct ionization in space can be calculated by Creme 96 HUP tool [16]. The flux vs. kinetic energy curves of variety of particles under four different radiation environments in GEO were achieved, as shown in Figure 10, through the Creme 96 HUP tool. Then, we obtained the flux distribution vs. LET (Figure 11) with 3 mm aluminum shielding, which is normal for space application. For the calculation of SEU probability (P_{SEU} (/bit/day)), we used P (E_{th}/MeV) to characterize the probability. If the deposited ionization energy of a particle through the track was higher than the energy threshold (E_{th}/MeV) of upset, then an upset was considered to have occurred. For an incident ion, the probability to get an upset is [14,16,17]:

$$P_{SEU} = \frac{1}{\phi_0} \frac{d\phi}{d(LET)}(LET) \times P(E_{th}) \qquad (2)$$

Bradford et al. contributed by expressing the total SEU rates (N) by LET and length, which could also be extended to a statistical result of particle events [18]. Thus, the effective LET spectrum can be integrated with the σ_{SEU} and σ_{sat}. If the surface area S (cm^2) of sensitive volume was determined, a simple formula can be transformed to [14,16,17,19]:

$$N = \frac{S}{4 \times \sigma_{sat}} \int_{LET_{min}}^{LET_{max}} \frac{d\phi}{d(LET)} |_{eff} \sigma_{SEU} \cdot d(LET) \tag{3}$$

where $\frac{d\phi}{d(LET)}$ is a differential *LET* spectrum and the *LET_max* as the maximal *LET* value of the incident particles is 10^5 MeV/(mg/cm^2), as shown in Figure 11.

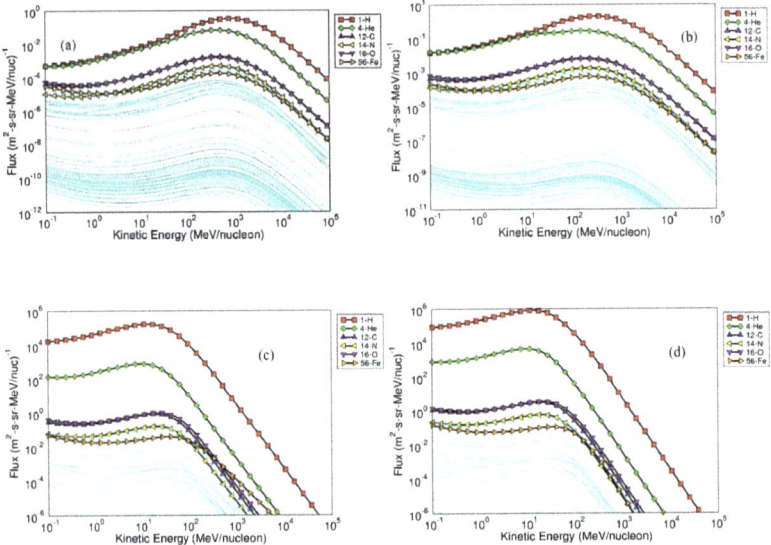

Figure 10. Creme results in GEO: (**a**) cosmic ray min; (**b**) cosmic ray max; (**c**) worst week; and (**d**) worst day.

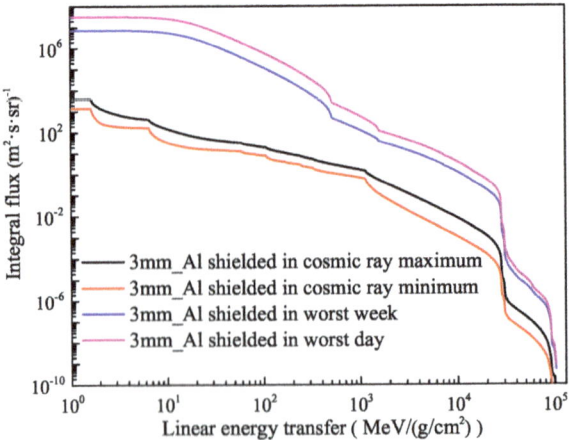

Figure 11. Integral flux behind 3 mm of aluminum in GEO environment.

The SEU rates were calculated and normalized to per bit per day by Creme tools. As shown in Figure 12, CRAM had strong radiation tolerance (8.46×10^{-15}/day/bit) in worst week, even in extremely cruel radiation environments. EBRAM and DFF had upset rates of $\sim 10^{-8}$/bit/day in cosmic ray minimum and maximum periods. Although the rate was many orders of magnitude higher than CRAM, it was still a quite low rate when compared with the results in [7,8]. The calculation results reveal that, even in worst-week radiation condition, only several errors occurred per day per device for EBRAM and lower errors in DFF.

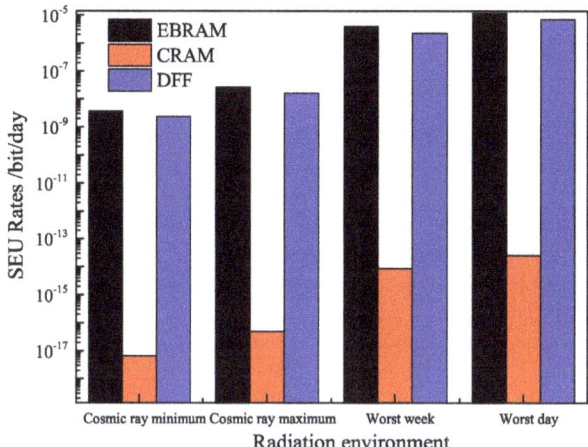

Figure 12. Creme results: SEU rates in isotropic GEO environment behind 3 mm aluminum shielding.

5. Discussion

The special radiation hardening designed circuits in our SRAM-based FPGAs including CRAMs, DLs, EBRAMs and DFFs were manufactured and evaluated. The radiation-hardened DUT and the testing details were all developed to realize the convincible SEU evaluation in large circuit system. All of the hardening information is presented and compared in the previous sections. Next, more details about the purposes of design and the relations between the effectiveness of tradeoff strategy and the actual requirements in space are discussed.

5.1. Radiation Hardening Design

According to the importance on FPGA function, several different hardening techniques are employed in different modules in DUT. The most important part of the FPGA is CRAMs. The upsets in CRAMs contribute the most serious influence to the FPGA system work because routing and logical resources existing in CRAMs are key to FPGA system functions; besides, CRAMs are irradiation sensitive and may cause serious consequent system errors. The errors in CRAMs can be classified into logic block errors and switch box errors. For logic blocks, errors may disturb or change the combinational functions, the multiplexer results, the polarity of reset signal or clock path for DFFs [20]. In switch resources, the open circuit, short circuit or short/open circuit errors may lead to permanent effects with single, multiple or even consequent system errors until the rewritten operation can be carried out. Therefore, the DICE circuit, which costs two times area and power of unhardened CRAM cell, was employed to harden the CRAMs. As shown in Figure 7, the hardened CRAMs had SEU saturation cross section at 6.5×10^{-13} (cm^2/bit). For similar 65 nm space grade SRAM-based FPGA reported in [21], the saturated SEU cross section is $\sim 3 \times 10^{-8}$ (cm^2/bit) for the hardened RAMs. In [7,8], the SEU saturation cross section is only $\sim 6 \times 10^{-8}$ (cm^2/bit) for the configuration cells in 90 nm space grade SRAM-based FPGAs. This means that the SEU data of hardened CRAMs in DUT is 5 orders of magnitude lower than the state-of-the-art space grade SRAM-based FPGAs employing similar CMOS technologies.

Compared with CRAMs, the upset in DFF seems not so critical [8]. For DFFs, not only the heavy ion irradiation but also the CRAM errors or transient pulses can cause upsets. Double redundancy technique was applied for DFF cells and reset control signal ports. The hardened DFFs in DUT had saturation cross section at 8.3×10^{-8} (cm^2/bit), which was almost one order lower than the saturated values at 7.5×10^{-7} (cm^2/bit) and 6.1×10^{-7} (cm^2/bit) of 90 nm radiation hardened FPGAs in [8].

Errors in EBRAM can be mitigated with various methods, which are not suitable for CRAMs and DFFs. EBRAMs are similar to commercial SRAMs, storing the information from users. Thus, some soft error mitigation methods used in general SRAMs such as ECC code and TMR structure with dynamic scrubbing can be replicated in the case of EBRAM. Besides, some soft error mitigation IPs are provided by Electronic Design Automation (EDA) memory generator tools in FPGA configuration procedure having ECC function with Hamming code in a constrained bit width, which can be used easily to decline upsets in EBRAM. These methods are suitable to be used in memory modules to mitigate upset-induced soft errors, making the additional time and area consuming design in layout seems unnecessary.

Besides, the designed 65 nm SRAM-based FPGA had more sufficient resources than other reported devices. The ~8 Mbit EBRAMs in DUT were larger than the same module (<~6.2 Mbi) for FPGAs in [7], which provides more flexible usages and more sufficient data redundancy. The ~20 Mbit CRAMs in DUT had very high radiation resistance (1.69×10^{-7} upset/device/day in worst-week condition in GEO), while the error rate for similar 90 nm hardened FPGA under typical solar conditions in GEO is only ~4 upset/device/day [7]. Although higher radiation tolerances were expected, the overall hardened strategies should depend on the device's importance, upset mechanism, area overhead, performance and the actual radiation tolerant needs in on-orbit missions.

5.2. Convincible Prediction and Perspectives

The drains of the off-state transistors were considered to be the sensitive volumes of memory cells [22]. The spacing of sensitive volume was referred to drain to drain distance in the layout of FPGA in order to get credible results. Based on the accurate curve-fitting results in LET thresholds and saturation cross section, the upset rates we predicted are convincible.

The on-orbit SEU rates can affect the operation security in aircraft and spacecraft systems, although high-speed scrubbing is an essential mitigation technique. However, scrubbing is a time and power consuming method and its frequency must depend on the dynamic changed particle flux and the upset cross-section of device, and the particle flux mainly comprised by the continuous changed cosmic-ray and solar wind was hard to ascertain. After the successful use of radiation hardened techniques in DUT, an extremely low event rate under continuous working conditions was obtained, which was more serious than actual intermittent working modes, indicating an excellent hardening result. Furthermore, the reasonable and effective radiation hardened method used in DUT can be further guided to 28 nm or smaller radiation hardened integrated circuit to enrich the family of urgently needed space-grade devices.

6. Conclusions

In this paper, we present a proper utilization of radiation hardened techniques for SRAM-based FPGA with 65 nm CMOS process. The hardening results are characterized by SEUs of CRAMs, DFFs, DLs and EBRAMs. Both layout hardening techniques including DICE, 8-T, double redundancy, and configuration hardening techniques including ECC and TMR are employed for this FPGA. The heavy-ion results indicate satisfying radiation tolerance, especially for the DICE CRAMs. The convincible low SEU rates for each part of DUT in GEO orbit reported above reveal a good result even without further additional reinforcement. Besides, the heavy ion evaluation results will be also useful for the related CMOS-based integrated circuits.

Author Contributions: Conceptualization, C.C.; methodology, C.C., P.Z. and T.L.; software, D.L. and Z.H.; validation, X.F.; writing—original draft preparation, C.C.; writing—review and editing, X.F., T.L. and L.K.; supervision, J.L.; project administration, and X.F., T.L. and J.L.; funding acquisition, J.L.

Funding: This research was funded by the National Natural Science Foundation of China (No. 11690041, No.11805244 and No.11675233) and the APC was funded by 11805244.

Acknowledgments: The authors concentrate on the research of radiation effects and are devoted to the academic contributions. Thus, they thank the related chip research and development institutes in China for the support of information in DUT design. The DUT and the detail information they provide are very helpful for our research.

Conflicts of Interest: The authors declare no conflict of interest.

References

1. Rose, J.; ElGamal, A.; Sangiovanni-Vincentelli, A. Architecture of Field-Programmable Gate Arrays. *Proc. IEEE* **1993**, *81*, 1013–1029. [CrossRef]
2. Stroud, C.; Nall, J.; Lashinsky, M.; Abramovici, M. Bist-based diagnosis of fpga interconnect. In Proceedings of the International Test Conference, Baltimore, MD, USA, 8–10 October 2002; pp. 618–627. [CrossRef]
3. George, J.; Koga, R.; Swift, G.; Allen, G.; Carmichael, C.; Tseng, C.W. Single Event Upsets in Xilinx Virtex-4 FPGA Devices. In Proceedings of the IEEE Radiation Effects Data Workshop (REDW), Ponte Vedra, FL, USA, 17–21 July 2006; pp. 109–114. [CrossRef]
4. Berg, M.; Poivey, C.; Petrick, D.; Espinosa, D.; Lesea, A.; Label, K.A. Effectiveness of internal versus external SEU scrubbing mitigation strategies in a Xilinx FPGA: Design, test, and analysis. *IEEE Trans. Nucl. Sci.* **2008**, *55*, 2259–2266. [CrossRef]
5. Koga, R.; George, J.; Swift, G.; Yui, C.; Edmonds, L.; Carmichael, C.; Langley, T.; Murray, P.; Lanes, K.; Napier, M. Comparison of Xilinx Virtex-II FPGA SEE sensitivities to protons and heavy ions. *IEEE Trans. Nucl. Sci.* **2004**, *51*, 2825–2833. [CrossRef]
6. Graham, P.; Caffrey, M.; Johnson, D.E.; Rollins, N.; Wirthlin, M. SEU mitigation for half-latches in Xilinx Virtex FPGAs. *IEEE Trans. Nucl. Sci.* **2003**, *50*, 2139–2146, doi:10.1109/TNS.2003.820744. [CrossRef]
7. Gregory, A. Virtex-4QV Static SEU Characterization Summary. In *NASA Electronic Parts and Packaging Program Office of Safety and Mission Assurance*; National Aeronautics and Space Administration: Washington, DC, USA, 2008; pp. 1–15.
8. Swift, G.M.; Allen, G.R.; Tseng, C.W.; Carmichael, C.; Miller, G.; George, J.S. Static Upset Characteristics of the 90 nm Virtex-4QV FPGAs. In Proceedings of the IEEE Radiation Effects Data Workshop (REDW), Tucson, AZ, USA, 14–18 July 2008; pp. 98–105. [CrossRef]
9. Chen, W.; Liu, J.; Ma, X.; Guo, G.; Zhao, Y.; Guo, X.; Luo, Y.; Yao, Z.; Ding, L.; Wang, C.; et al. Research progress of radiation effects mechanisms and experimental techniques in nano-devices. *Chin. Sci. Bull.* **2018**, *63*, 1211–1222. [CrossRef]
10. Liu, T.; Liu, J.; Xi, K.; Zhang, Z.; He, D.; Ye, B.; Yin, Y.; Ji, Q.; Wang, B.; Luo, J.; et al. Heavy Ion Radiation Effects on a 130-nm COTS NVSRAM Under Different Measurement Conditions. *IEEE Trans. Nucl. Sci.* **2018**, *65*, 1119–1126. [CrossRef]
11. Maillard, P.; Hart, M.; Barton, J.; Chang, P.; Welter, M.; Le, R.; Ismail, R.; Crabill, E. Single-event upsets characterization evaluation of Xilinx UltraScale Soft Error Mitigation (SEM IP) Tool. In Proceedings of the IEEE Radiation Effects Data Workshop (REDW), Portland, OR, USA, 11–15 July 2016; pp. 1–4. [CrossRef]
12. Tonfat, J.; Kastensmidt, F.L.; Artola, L.; Hubert, G.; Medina, N.H.; Added, N.; Aguiar, V.A.P.; Aguirre, F.; Macchione, E.L.A.; Silveira, M.A.G. Analyzing the Influence of the Angles of Incidence and Rotation on MBU Events Induced by Low LET Heavy Ions in a 28-nm SRAM-based FPGA. *IEEE Trans. Nucl. Sci.* **2017**, *64*, 2161–2168. [CrossRef]
13. Andrew, M.K.; Timothy, A.W.; Kenneth, B.S.; Michael, J.W. Dynamic SEU Sensitivity of Designs on Two 28-nm SRAM-Based FPGA Architectures. *IEEE Trans. Nucl. Sci.* **2018**, *65*, 280–287. [CrossRef]
14. Inguimbert, C.; Duzellier, S.; Ecoffet, R.; Guibert, L.; Barak, J.; Chabot, M. Using a carbon beam as a probe to extract the thickness of sensitive volumes. *IEEE Trans. Nucl. Sci.* **2000**, *47*, 551–558. [CrossRef]
15. Hamming, R.W. Error detecting and error correcting codes. *Bell Syst. Tech. J.* **1950**, *29*, 147–160. [CrossRef]
16. Sierawski, B.D.; Mendenhall, M.H.; Weller, R.A.; Reed, R.A.; Adams, J.H.; Watts, J.W.; Barghouty, A.F. CREME-MC: A Physics-Based Single Event Effects Tool. In Proceedings of the IEEE Nuclear Science Symposium and Medical Imaging Conference (NSS/MIC), Knoxville, TN, USA, 30 October–6 November 2010; pp. 1258–1261. [CrossRef]

17. Weller, R.A.; Mendenhall, M.H.; Reed, R.A.; Schrimpf, R.D.; Warren, K.M.; Sierawski, B.D.; Massengill, L.W. Monte Carlo Simulation of Single Event Effects. *IEEE Trans. Nucl. Sci.* **2010**, *57*, 1726–1746. [CrossRef]
18. Bradford, J. Geometric Analysis of Soft Errors and Oxide Damage Produced by Heavy Cosmic Rays and Alpha Particles. *IEEE Trans. Nucl. Sci.* **1980**, *27*, 941–947. [CrossRef]
19. Mendenhall, M.; Weller, R. A probability-conserving cross-section biasing mechanism for variance reduction in Monte Carlo particle transport calculations. *Nucl. Instrum. Methods Phys. Res. A* **2011**, *667*, 38–43. [CrossRef]
20. Bellato, M.; Bernardi, P.; Bortolato, D.; Candelori, A.; Ceschia, M.; Paccagnella, A.; Rebaudengo, M.; Reorda, M.S.; Violante, M.; Zambolin, P. Evaluating the effects of SEUs affecting the configuration memory of an SRAM-based FPGA. In Proceedings of the Design, Automation and Test in Europe Conference and Exhibition (DATE), Paris, France, 16–20 February 2004; pp. 584–589. [CrossRef]
21. Gregory, A.; Larry, E.; Chen, W.T.; Gary, S.; Carl, C. Error Detect and Correct Enabled Block Random Access Memory (Block RAM) Within the Xilinx XQR5VFX130. *IEEE Trans. Nucl. Sci.* **2010**, *57*, 3426–3431. [CrossRef]
22. Lei, Z.; Zhang, Z.; En, Y.; Huang, Y. Mechanisms of atmospheric neutron-induced single event upsets in nanometric soi and bulk sram devices based on experiment-verified simulation tool. *Chin. Phys. B* **2018**, *27*, 066105. [CrossRef]

© 2019 by the authors. Licensee MDPI, Basel, Switzerland. This article is an open access article distributed under the terms and conditions of the Creative Commons Attribution (CC BY) license (http://creativecommons.org/licenses/by/4.0/).

Article

A Novel Modular Radiation Hardening Approach Applied to a Synchronous Buck Converter

Solomon Banteywalu [1], Baseem Khan [2,*], Valentijn De Smedt [3] and Paul Leroux [3]

[1] Department of Electrical and Computer Engineering, Addis Ababa Institute of Technology, Addis Ababa University, Addis Ababa P.O. Box 385, Ethiopia; solupa2000@yahoo.com
[2] Department of Electrical and Computer Engineering, Institute of Technology, Hawassa University, Hawassa P.O. Box 05, Ethiopia
[3] Department of Electrical Engineering (ESAT)–ADVISE, KU Leuven, 3000 Leuven, Belgium; valentijn.desmedt@kuleuven.be (V.D.S.); paul.leroux@kuleuven.be (P.L.)
* Correspondence: baseem.khan04@gmail.com

Received: 15 April 2019; Accepted: 4 May 2019; Published: 8 May 2019

Abstract: Radiation and extreme temperature are the main inhibitors for the use of electronic devices in space applications. Radiation challenges the normal and stable operation of DC-DC converters, used as power supply for onboard systems in satellites and spacecrafts. In this situation, special design techniques known as radiation hardening or radiation tolerant designs have to be employed. In this work, a module level design approach for radiation hardening is addressed. A module in this sense is a constituent of a digital controller, which includes an analog to digital converter (ADC), a digital proportional-integral-derivative (PID) controller, and a digital pulse width modulator (DPWM). As a new Radiation Hardening by Design technique (RHBD), a four module redundancy technique is proposed and applied to the digital voltage mode controller driving a synchronous buck converter, which has been implemented as hardware-in-the-loop (HIL) simulation block in MATLAB/Simulink using Xilinx system generator based on the Zynq-7000 development board (ZYBO). The technique is compared, for reliability and hardware resources requirement, with triple modular redundancy (TMR), five modular redundancy (FMR) and the modified triplex–duplex architecture. Furthermore, radiation induced failures are emulated by switching all duplicated modules inputs to different signals, or to ground during simulation. The simulation results show that the proposed technique has 25% and 30%longer expected life compared to TMR and FMR techniques, respectively, and has the lowest hardware resource requirement compared to FMR and the modified triplex–duplex techniques.

Keywords: TMR; FMR; 4MR; triplex–duplex; FPGA-based digital controller; radiation tolerant

1. Introduction

Outer space is full of radiation sources that include solar wind, solar flares, coronal mass ejections, galactic cosmic rays, Van Allen radiation belts, solar particle events, etc. This radiation environment consists of particles such as protons, electrons, neutrons, and heavy ions, [1]. The strike of any of these particles may compromise the normal operation of electronic circuits on board of space systems in this environment. Depending on the type and characteristics of the impinging radiation, different effects, either irreversible or (partially or totally) reversible, may arise. There are two major effects of radiation i.e., total ionizing dose (TID) and single event effect (SEE). TID also called cumulative effect, produce gradual changes in the operational parameters of the devices, which tends to degrade the characteristics of the devices overtime. SEE cause abrupt changes or transient behavior in circuits. Such effects, interfere with space systems' electronics operation, and, in some cases, threaten the survival of such systems. While TID effects reveal themselves gradually often after years of operation before a complete failure, SEEs don't. This work considers alleviating the effects of SEE on electronic circuits used for space applications.

Currently, the study of techniques to keep electronic circuits operational in such hostile environment has increased [2], driven by the increasing number of applications of radiation tolerant circuits, such as space missions, satellites, high-energy physics experiments, etc. [3,4]. This paper considers a module level approach for radiation hardening using fault tolerant method.

Fault tolerant methods use redundancy to mask or get around faults in electronic circuits. Redundancy is one of the most important methods to obtain highly reliable systems. Redundancy techniques have the ability to deliver continuous service in the presence of hardware faults by providing redundant hardware components. Redundancy techniques in general are adopting additional hardware components or additional computation time, which are used for fault detection or for fault masking so that the effect of faults is not reflected on the output signal [5]. The most common radiation mitigation techniques are TMR and FMR methods [6,7]. They are highly-efficient but very costly and are used for situations where high reliability is targeted. Reliability is an important quality measure of a fault tolerant system.

Reliability is defined as the probability of not failing in a particular environment for a particular mission time. Suppose a system consists of N identical components. Let $S(t)$ be the number of surviving components at time t, and $Q(t)$ the number of components that failed up to time t. Then the probability of survival of the components also known as the reliability $R(t)$, which is given by:

$$R(t) = \frac{S(t)}{N} \tag{1}$$

A measure of failure $F(t)$ is defined as the conditional probability that the system fails by time t referred to us unreliability or failure time distribution:

$$F(t) = \frac{Q(t)}{N} \tag{2}$$

Since $S(t) + Q(t) = N$, therefore:

$$R(t) + F(t) = 1 \quad or \quad F(t) = 1 - R(t) \tag{3}$$

Since $F(t)$ is a probability, its derivative is a probability distribution function and defined as,

$$f(t) = \frac{dF(t)}{dt} = \frac{-dR(t)}{dt} \tag{4}$$

where $f(t)$ shows the probability of failures per unit time.

Now, the failure rate λ is defined as the number of failures per unit time, compared with the number of surviving components.

$$Failure\ rate = \frac{The\ number\ of\ failure\ per\ unit\ time}{The\ number\ of\ surviving\ components} \quad or \tag{5}$$

$$\lambda = \frac{1}{R(t)} \times \frac{dF(t)}{dt} \tag{6}$$

Using Equation (3), the failure rate can be written as,

$$\lambda = \frac{-1}{R(t)} \times \frac{dR(t)}{dt} \tag{7}$$

The expression may be integrated from 0 to time t, by considering at time $t = 0$, $R(t) = 1$, and at time t the reliability is $R(t)$, then,

$$\int_0^t \lambda dt = -\int_1^{R(t)} \frac{dR(t)}{R(t)} \tag{8}$$

Often λ is assumed to be constant during the useful life of the system. Thus,

$$\lambda t = -\log R(t) \quad \text{or} \quad -\lambda t = \log R(t) \tag{9}$$

This gives,

$$R(t) = e^{-\lambda t} \tag{10}$$

The mean time to failure (MTTF) for the system is obtained as,

$$MTTF = \int_0^\infty R(t)dt = \frac{1}{\lambda} \tag{11}$$

Assuming independent and identical modules having reliability of R_m and with λ constant failure rate each, and then using the binomial theorem

$$B(r:n, R_m) = \binom{n}{r} R_m^r (1 - R_m)^{n-r} \tag{12}$$

The reliability of TMR is given as,

$$R_{TMR} = \text{Probability of all three modules are functioning} \\ + \text{Probability of any two modlues are functioning} \tag{13}$$

$$R_{TMR} = B(3:3) + B(2:3) = \binom{3}{3} R_m^3 (1-R_m)^0 + \binom{3}{2} R_m^2 (1-R_m)^1 \tag{14}$$

$$R_{TMR} = 3R_m^2 - 2R_m^3 = 3e^{-2\lambda t} - 2e^{-3\lambda t} \tag{15}$$

$$MTTF_{TMR} = \int_0^\infty R(t)dt = \int_0^\infty \left(3e^{-2\lambda t} - 2e^{-3\lambda t}\right)dt = \frac{3}{2\lambda} - \frac{2}{3\lambda} = \frac{5}{6\lambda} \tag{16}$$

For the FMR method:

$$R_{5MR} = B(5:5) + B(4:5) + B(3:5) \tag{17}$$

$$R_{5MR} = \binom{5}{5} R_m^5 (1-R_m)^0 + \binom{5}{4} R_m^4 (1-R_m)^1 + \binom{5}{3} R_m^3 (1-R_m)^2 \tag{18}$$

$$R_{5MR} = 10R_m^3 - 15R_m^4 + 6R_m^5 = 10e^{-3\lambda t} - 15e^{-4\lambda t} + 6e^{-5\lambda t} \tag{19}$$

$$MTTF_{5MR} = \int_0^\infty R(t)dt = \int_0^\infty \left(10e^{-3\lambda t} - 15e^{-4\lambda t} + 6e^{-5\lambda t}\right)dt = \frac{10}{3\lambda} - \frac{15}{4\lambda} + \frac{6}{5\lambda} \\ = \frac{47}{60\lambda} \tag{20}$$

2. Motivation

2.1. The Base Architecture

The proposed method is derived from the architecture presented in [5], which is called triplex–duplex redundancy. In this arrangement there are three primary modules using two duplicate modules each. Thus, a total of six identical modules are computing in parallel, which are grouped in three pairs. The computation result of each pair is compared using a comparator. If the results agree, the output of the comparator participates in the voting. If not, the pair of modules is declared faulty and the switch removes the pair from the system.

The hardware resource requirement is 500% more compared to the simplex system and twice compared to that of TMR technique.

2.2. Modified Triplex–Duplex Architecture

The disadvantage of the triplex–duplex architecture is that it requires two times more hardware resources compared to the TMR method and has the one more module than the FMR method. Both modules in the duplex are removed from the voting as soon as one of the two modules in the duplex fails. This reduces the overall system mean time to failure (MTTF), if no repair is used. Therefore, except for faulty duplex detection, it is similar in operation to TMR.

To increase the reliability of this method, a modified architecture shown in Figure 1 was developed, where the comparator and switch parts are combined and modified in such a way that all duplexes are connected to all disagreement detectors and switch blocks, which allows for any module in the three duplex systems to act as an active spare for any other module in the three duplex systems. Therefore, the overall system will continue to work even if one module in all the three duplexes is failed, or even if only one duplex is left, or two duplexes with one good module each are left. This significantly increases the MTTF of the overall system and helps, if any repair or reconfiguration is used, to reduce the frequency of such repair or reconfiguration compared to TMR or FMR only methods.

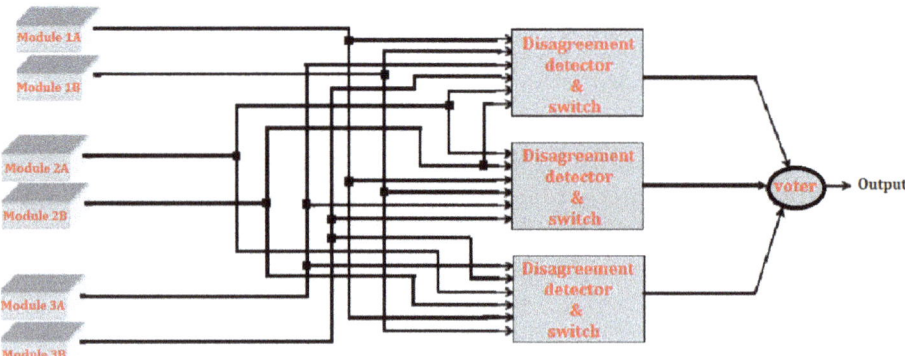

Figure 1. Modified triplex–duplex Redundancy.

This method uses 500% more hardware compared to the simplex system, the same as its base architecture, but with tremendous increase in reliability.

Assuming independent and identical modules having reliability of R_m and with λ constant failure rate each:

$$R_{(triplex-duplex)mod} = B(6:6) + B(5:6) + B(4:6) + B(3:6) + B(2:6) \quad (21)$$

$$R_{(triplex-duplex)mod} = \binom{6}{6} R_m^6 (1-R_m)^0 + \binom{6}{5} R_m^5 (1-R_m)^1 + \binom{6}{4} R_m^4 (1-R_m)^2 + \binom{6}{3} R_m^3 (1-R_m)^3 + \binom{6}{2} R_m^2 (1-R_m)^4 \quad (22)$$

$$R_{(triplex-duplex)mod} = 15R_m^2 - 40R_m^3 + 45R_m^4 - 24R_m^5 + 5R_m^6 \quad (23)$$

$$= 15e^{-2\lambda t} - 40e^{-3\lambda t} + 45e^{-4\lambda t} - 24e^{-5\lambda t} + 5e^{-6\lambda t} \quad (24)$$

$$MTTF_{(triplex-duplex)mod} = \int_0^\infty R(t)dt \quad (25)$$

$$= \int_0^\infty \left(15e^{-2\lambda t} - 40e^{-3\lambda t} + 45e^{-4\lambda t} - 24e^{-5\lambda t} + 5e^{-6\lambda t}\right)dt \quad (26)$$

$$= \frac{15}{2\lambda} - \frac{40}{3\lambda} + \frac{45}{4\lambda} - \frac{24}{5\lambda} + \frac{5}{6\lambda} = \frac{87}{60\lambda} \qquad (27)$$

There is 61% and 66% improvement in MTTF compared to TMR and FMR methods, respectively.

3. Proposed Four Modules Architecture

Besides having the best reliability and consequently MTTF, the disadvantage of the modified triplex–duplex architecture is its high hardware resource utilization. In effort to come up with high reliability and lower resource requirement redundancy, a four module architecture was developed as shown in the Figure 2, which has the highest reliability compared to both TMR and FMR methods and lowest hardware resource requirement compared to FMR and the modified triplex–duplex methods.

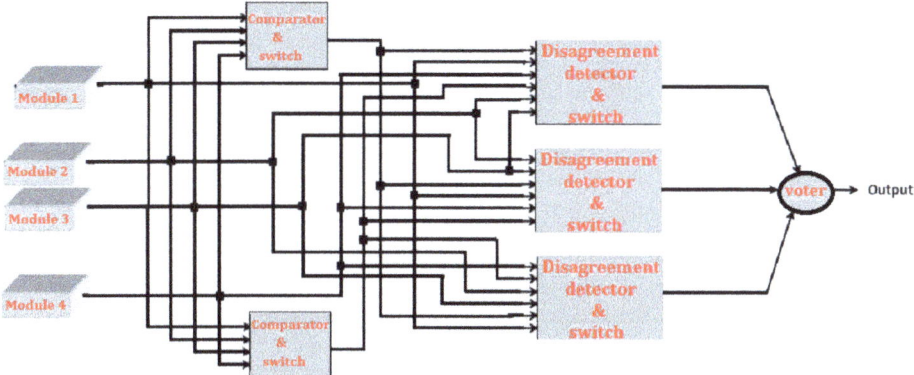

Figure 2. Proposed four modules redundancy.

The operation of this architecture is similar to the modified triplex–duplex architecture above, except that, there are four physical modules and two clone modules reducing the total number of actual duplicated modules to four instead of six. The clone modules were created as long as at least two of the physical modules were fault free, which in effect significantly reduces hardware resource utilization compared to the FMR and the modified triplex–duplex methods. The architecture masks the failure of two physical modules out of four.

The proposed four modules architecture is comparable, in terms of reliability, to the four modules highly reliable self-purging redundancy, [8,9]. Self-purging redundancy uses a threshold voter instead of a majority voter. A threshold voter outputs a 1, if the number of its inputs that are 1 is greater than or equal to the threshold value; otherwise it outputs a 0. The idea of self-purging redundancy is that if only one module fails, then its output will be different from the others. A switch checks if a module's output differs from the output of a threshold voter. If it does differ, then the module is assumed to be faulty and its control flip-flop is reset to 0. This permanently masks the output of the module so that its input to the threshold voter will always be 0.

As pointed out in [8], the self-purging method is not so much popular due to its complex threshold voter architecture. In case of the self-purging technique, faulty module detection is performed by comparing each module's output with the voted output. However, the detection of the faulty module is carried out before voting. In the case of the developed four modules method, it reduces the complexity encountered with a faulty voter especially when using multiple voters in the case of self-purging redundancy. Moreover, the proposed four-module redundancy technique can tolerate the simultaneous failure of two modules, whereas, a four module self-purging redundancy with a threshold of 2 cannot. Self-purging redundancy with a threshold of T can tolerate up-to T-1 simultaneous failures.

Assuming the same conditions as in previous cases for reliability calculation,

$$R_{Four-mod} = B(4:4) + B(3:4) + B(2:4) \tag{28}$$

$$R_{Four-mod} = \binom{4}{4} R_m^4 (1-R_m)^0 + \binom{4}{3} R_m^3 (1-R_m)^1 + \binom{4}{2} R_m^2 (1-R_m)^2 \tag{29}$$

$$R_{Four-mod} = 3R_m^4 - 8R_m^3 + 6R_m^2 = 3e^{-4\lambda t} - 8e^{-3\lambda t} + 6e^{-2\lambda t} \tag{30}$$

$$MTTF_{Four-mod} = \int_0^\infty R(t)dt = \int_0^\infty \left(3e^{-4\lambda t} - 8e^{-3\lambda t} + 6e^{-2\lambda t}\right)dt = \frac{3}{4\lambda} - \frac{8}{3\lambda} + \frac{3}{\lambda} = \frac{13}{12\lambda} \tag{31}$$

There is 25% and 30% improvement in MTTF compared to TMR and FMR methods, respectively. The contributions of the developed methods are as follows:

- Authors proposed a highly reliability redundancy technique called the modified triplex–duplex redundancy, which has 61% and 66% longer expected life than TMR and FMR techniques, respectively, although its hardware utilization is the highest compared to both methods.
- To rectify the hardware consumption drawback of the modified triplex–duplex technique, authors proposed a novel four module redundancy technique derived from the modified triplex–duplex method with the following advantages:

    ○ It is comparable in reliability to the four modules self-purging redundancy with threshold of 2 and to TMR with one spare with the additional advantages of tolerating simultaneous failure of two modules and reducing complexity, which both of the above two techniques lack.
    ○ It gives 30% higher MTTF compared to FMR while utilizing lower hardware resources.
    ○ It gives 25%higher MTTF compared to TMR method.
    ○ Unlike self-purging redundancy that requires a specialized threshold voter, the proposed method is used with both single and triplicated majority voter architectures, since it is based on the modified triplex–duplex architecture.

4. Synchronous Buck Converter Controller Design

4.1. Closed-loop Control System

Figure 3 below shows a synchronous buck converter with its digital control feedback. It consists of four functional blocks: an ADC (analog-to-digital conversion), a compensator (error compensation), a DPWM (digital pulse-width modulation), and a synchronous buck converter power stage.

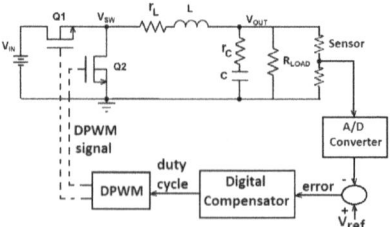

Figure 3. A synchronous buck converters with closed-loop digital control.

In this circuit, the goal is to minimize the difference between V_{ref} and V_o. Therefore, authors need to design a digital PID compensator to track the error and bring it down to as small as possible.

4.2. Digital PID Compensator Design

For control purposes, the block diagram of the buck converter, which is used in this work, is shown in Figure 4.

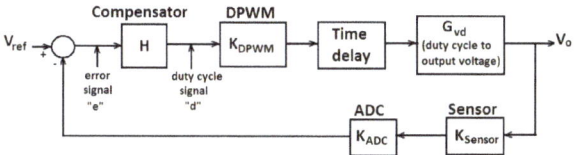

Figure 4. Buck converter control system point view.

The main blocks are the duty cycle-to-output transfer function of the power stage or plant (G_{vd}), the compensator (H), the total time delay of the control loop, the DPWM gain (K_{dpwm}), the ADC gain (K_{adc}) and the output voltage sensor gain (K_{sensor}).

For a buck converter, the small signal control to output transfer function is given by [10].

$$G_{vd}(s) = \frac{V_i(sr_cC + 1)}{s^2LC\left(\frac{R+r_c}{R}\right) + s\left(r_cC\left(\frac{R+r_L}{R}\right) + \frac{L}{R} + r_LC\right) + \left(\frac{R+r_L}{R}\right)} \tag{32}$$

The design parameters considered are shown in Table 1.

Table 1. Design parameters of the converter.

Parameter	Rating Value
Input Voltage (Vi)	12 V, (11–16 V)
Output Voltage (Vo)	5 V
Output Current (Io)	2.5 A, (1.25–5 A)
Inductor (L), ESR	4.75 µH, 10 mΩ
Capacitor (C), ESR	2.466 µF, 5 mΩ
Load (R)	2 Ω, (1–4 Ω)
Switching Frequency (Fsw)	1.5 MHz

With the above design parameters, $G_{vd}(s)$ is given by:

$$G_{vd}(s) = \frac{1.48e^{-07}s + 12}{1.131e^{-11}s^2 + 2.325e^{-06}s + 1.005} \tag{33}$$

The plant transfer function, including the effects of the ADC, DPWM and sensor is given by:

$$G_{vdsys}(s) = K_{sensor}K_{ADC}K_{DPWM}G_{vd}e^{-s(t_{adc} + dT_s + t_{dpwm})} \tag{34}$$

where t_{adc} is the ADC conversion time and t_{dpwm} is the DPWM delay time.

In Equation (34), the exponent term represents the total time delay, which is usually taken equal to the switching period. That is, $T_s = (t_{adc} + dT_s + t_{dpwm})$. Then, the plant transfer function is given by:

$$G_{vdsys}(s) = K_{sensor}K_{ADC}K_{DPWM}G_{vd}e^{-sT_s} \tag{35}$$

The above transfer function presented in Equation (35) is used in the MATLAB control system toolbox to design the compensator in the analog domain. The designed compensator has a gain margin of 12.9 dB and a phase margin of 66.7 degrees. Note that, the phase margin is intentionally made higher to compensate for phase margin loss when converting to the digital form. The compensator so designed is then converted to its equivalent digital form using the bilinear transformation. The final digital PID compensator transfer function is given by:

$$G_c(z) = \frac{1.304e^{-02} - 2.032e^{-02}z^{-1} + 7.916e^{-03}z^{-2}}{1 - z^{-1}} \tag{36}$$

5. FPGA Implementation and Results Obtained

The digital PID compensator, an 8-bit sigma delta ADC and an 8-bit 1.5 MHz DPWM, as well as, all redundancy techniques have been implemented in MATLAB and Xilinx system generator. The overall objective is to properly regulate the output voltage towards the desired output voltage irrespective of the input voltage and any load variations within the given ranges and irrespective of radiation induced failure of any number of the duplicated modules based on the masking ability of the redundancy technique being used.

5.1. Hardware-in-the-Loop Simulation

It is practical to test the embedded controller more efficiently with a powerful method of hardware-in-the-loop (HIL) simulation. By thoroughly testing the controller in a virtual environment before proceeding to real-world tests of the complete system, one can maintain reliability and time requirements in a cost-effective manner. HIL simulation can also allow verifying whether the vendor specific FPGA synthesis tool actually retains the module level design, which is often not the case. Therefore, the HIL block is generated representing the radiation tolerant digital voltage mode controller for the synchronous buck converter.

The manual switches (S1, S2, S3, and S4), shown at the input of the controller HIL block diagram in Figure 5 are used to emulate the radiation faults during simulation; this is accomplished by switching the controller inputs to signals other than expected signals from the feedback system, or switching the inputs to ground (or, switch to zero). The duplicated voter's, Ref [11] error detectors (PIDErr1, PIDErr2, and PIDErr3) and the DPWM signals voter's error detectors (PWMErr1 and PWMErr2), shown at the output of the controller HIL block diagram in the Figure 5 can be used for repair/reconfiguration process initiation [12–14], when radiation faults occur in the respective voters, if such systems are used.

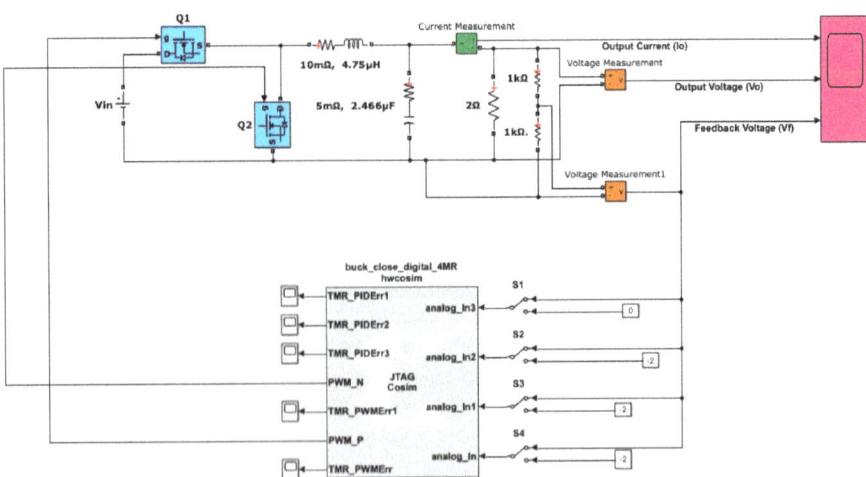

Figure 5. Hardware-in-the-loop (HIL) simulation block including the power stage.

Figure 6 provides the converter output voltage and current without applying radiation fault emulation.

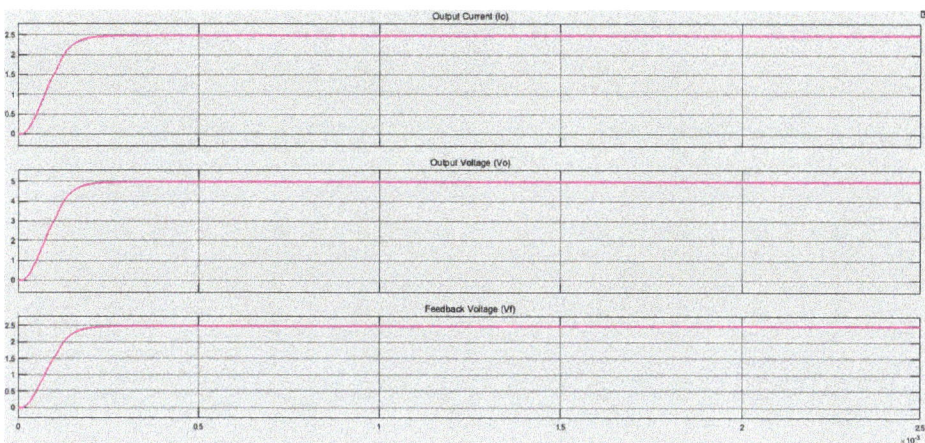

Figure 6. Converter output voltage and current during HIL simulation for the case $V_i = 16$ V and $R_{load} = 2\,\Omega$ without fault emulation.

Figure 7 shows the HIL simulation block during fault emulation of modules 1 and 2.

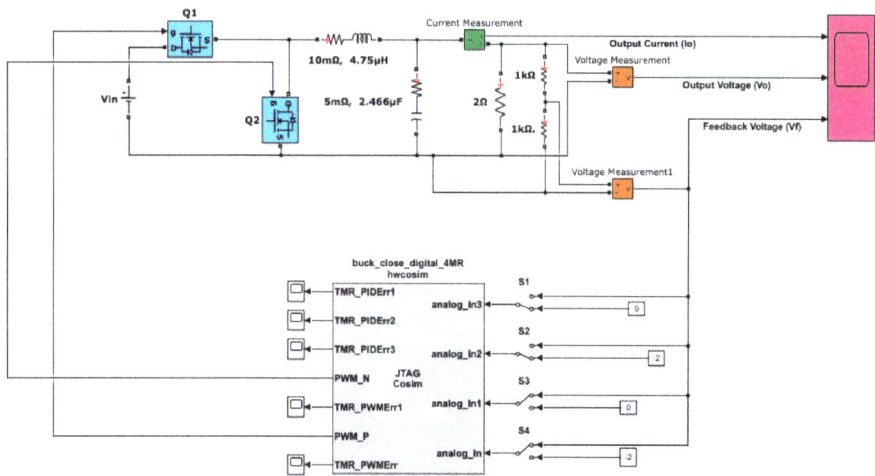

Figure 7. HIL simulation block including the power stage during fault emulation of modules 1 and 2.

Figure 8 presents the output voltage and current of converter under fault emulation of modules 1 and 2. Module 1 is switched to different signal at 0.5 m-second and then module 2 is switched to a different signal at 1 m-second to emulate the radiation fault. As it is clear from the Figure 8, there is a rise in voltage output of converter for short interval when switching the second module. This is due to switching transients.

There are five other different fault emulation cases available. All the other possible fault emulation combinations provided the same output voltages and currents as the case portrayed in the Figure 8.

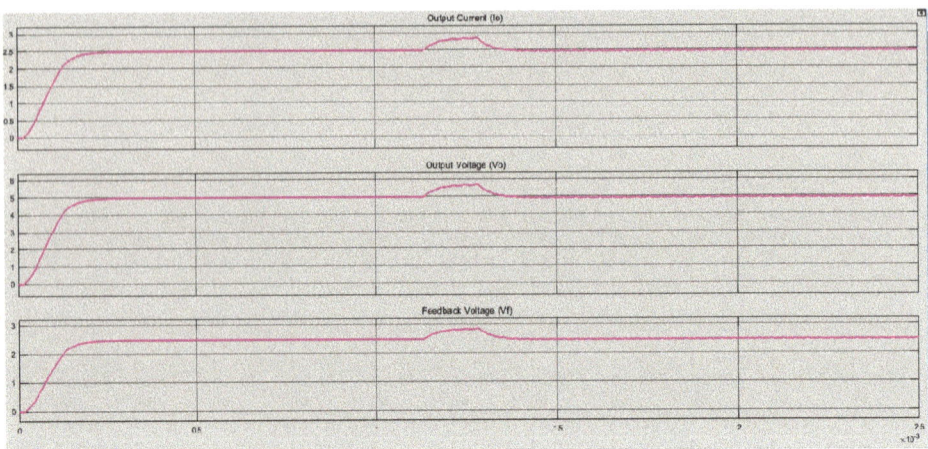

Figure 8. Converter output voltage and current during HIL simulation for the case $V_i = 16$ V and $R_{load} = 2\,\Omega$ with fault emulation of modules 1 and 2.

5.2. Comparison of FPGA Resource Utilization and Reliability

As can be seen from Table 2, the proposed four modules redundancy uses the lowest hardware resources compared to FMR and the modified triplex–duplex redundancies while having the highest reliability compared to TMR and FMR techniques as explained earlier.

Table 2. FPGA resource utilization summary and reliability comparison.

Methods	DSPs			LUTs			Registers			Reliability ($\lambda = 10\%$) Mission Time t	
	Available	Used	Percentage	Available	Used	Percentage	Available	Used	Percentage	t = 1 year	t = 5 Years
Simplex	80	1	1.25	17,600	647	3.68	35,200	301	0.86	0.9	0.6
TMR		3	3.75		1934	10.99		874	2.48	0.9746	0.6574
Proposed Four Module Method		4	5		2710	15.4		1189	3.38	0.9968	0.8282
FMR		5	6.25		3195	18.15		1446	4.11	0.9926	0.6938
Modified Triplex–duplex		6	7.5		4053	23.03		1729	4.91	1.0	0.9620

6. Conclusions

This paper presents a module level design approach to an FPGA based radiation tolerant digital voltage mode controller for a synchronous buck converter. A four-module high-reliability redundancy technique is proposed and implemented on zynq-7000 development board (Zybo). The technique has been compared with three other more common utilized redundancy techniques for reliability and FPGA resource utilization. It is observed that, the developed method has 25% and 30% longer expected life than TMR and FMR techniques, respectively and requires lower FPGA resources compared to FMR and the modified triplex–duplex techniques.

It is shown that the proposed method can be used for radiation tolerant synchronous buck converter design for applications requiring relatively longer mission time, compared to TMR and FMR techniques. The work can be utilized in such applications where fault-masking ability of a system is required. For example space applications, power electronic converters applications, computers, satellites, high-energy physics experiments, etc.

Author Contributions: Conceptualization, S.B.; methodology, S.B.; software, S.B.; validation, S.B., B.K. and P.L.; formal analysis, S.B.; investigation, S.B.; resources, P.L.; writing–original draft preparation, S.B.; writing–review and editing, S.B. and P.L.; visualization, S.B.; supervision, P.L., V.D.S. and B.K.; project administration, P.L. and S.B.

Funding: This research was supported by the Home Grown PhD Program (HGPP) funded by the Ethiopian ministry of Education.

Conflicts of Interest: The authors declare no conflict of interest.

References

1. Maurer, R.H.; Freeman, M.E.; Martin, M.N.; Roth, D.R. *Harsh Environments: Space Radiation Environment, Effects, and Mitigation*; John Hopkins University: Baltimore, MD, USA, 2008.
2. Leonard, C. Challenges for Electronic Circuits in Space Applications. Analog Devices, Inc. Available online: https://www.analog.com/media/en/technical-documentation/tech-articles/Challenges-for-Electronic-Circuits-in-Space-Applications.pdf (accessed on 1 July 2018).
3. Jagannathan, S.; Herbison, D.R.; Holman, W.T.; Massengill, L.W. Behavioral Modeling Technique for TID Degradation of Complex Analog Circuits. *IEEE Trans. Nucl. Sci.* **2010**, *57*, 3708–3715. [CrossRef]
4. Schmidt, F.H., Jr. Fault Tolerant Design Implementation on Radiation Hardened by Design SRAM-Based FPGAs. Master's Thesis, Massachusetts Institute of Technology, Boston, MA, USA, 2013.
5. Jain, M.; Gupta, R. Redundancy Issues in Software and Hardware Systems—An Overview. *Int. J. Reliab. Qual. Saf. Eng.* **2011**, *18*, 61–98. [CrossRef]
6. Kolte, P.P.; Maheshwari, R.; Ajankar, S. Triple modular redundacny using fault tolerant technique—A review. In Proceedings of the International Conference on Emanations in Modern Technology and Engineering ICEMTE-2017, Maharashtra, India, 4–5 March 2017; Volume 5.
7. Istiyanto, J.E.; Harjoko, A.; Putra, A.E. Five modular redundancy with mitigation technique to recover the error module. *Int. J. Adv. Stud. Comput. Sci. Eng.* **2014**, *3*, 1–6.
8. Quintana, J.M.; Avedillo, J.M.; Huertas, J.L. Efficient Realization of a Threshhold voter for Self-purging Redundancy. *J. Electron. Test.* **2001**, *17*, 69–73. [CrossRef]
9. Losq, J. A highly Efficient Redundancy Scheme: Self-purging Redundancy. *IEEE Trans. Comput.* **1976**, *C-25*, 569–578. [CrossRef]
10. Choudhury, S. *Designing the Digital Compensator for a UCD91XX-Based Digital Power Supply*; Texas Instruments Application Notes; Texas Instruments: Dallas, TX, USA, 2007.
11. Carcheri, J.C. *N-Modular Redundancy Techniques for Fault Tolerance in Reconfigurable Logic Devices*; Master's Project Report; UCF: Orlando, FL, USA, 2007.
12. Uznanski, S.; Brugger, M.; Schramm, V.; Thurel, Y.; Todd, B. *Radiation Tolerant Power Converter Design for the LHC*; CERN CH-1211: Genève, Switzerland, 2012.
13. Van Vonno, N.W.; Pearce, L.G. Total dose and single event effect testing of hardened point of load regulator. In Proceedings of the Radiation Effects Workshop (REDW), Denver, CO, USA, 20–23 July 2010; pp. 80–88.
14. Wang, J.J. *Radiation Effects in FPGAs*; Actel Corporation: Mountain View, CA, USA, 2008.

© 2019 by the authors. Licensee MDPI, Basel, Switzerland. This article is an open access article distributed under the terms and conditions of the Creative Commons Attribution (CC BY) license (http://creativecommons.org/licenses/by/4.0/).

Article

A Compact Model to Evaluate the Effects of High Level C++ Code Hardening in Radiation Environments

Leonardo Maria Reyneri [1], Alejandro Serrano-Cases [2], Yolanda Morilla [3], Sergio Cuenca-Asensi [2] and Antonio Martínez-Álvarez [2],*

[1] Department of Electronics, Politecnico di Torino, Corso Duca d. Abruzzi 24, 10129 Turin, Italy; leonardo.reyneri@polito.it
[2] Department of Computer Technology, Ctra. San Vicente del Raspeig s/n, 03690 San Vicente del Raspeig, Spain; aserrano@dtic.ua.es (A.S.-C.); sergio@dtic.ua.es (S.C.-A.)
[3] Centro Nacional de Aceleradores (Universidad de Sevilla, CSIC, JA). Avda. Tomás Alba Edison 7, 41092 Sevilla, Spain; ymorilla@us.es
* Correspondence: amartinez@dtic.ua.es; Tel.: +34 965-903-400

Received: 30 April 2019; Accepted: 6 June 2019; Published: 10 June 2019

Abstract: A high-level C++ hardening library is designed for the protection of critical software against the harmful effects of radiation environments that can damage systems. A mathematical and empirical model to predict system behavior in the presence of radiation induced faults is also presented. This model generates a quick evaluation and adjustment of several reliability vs. performance trade-offs, to optimize radiation hardening based on the proposed C++ hardening library. Several simulations and irradiation campaigns with protons and neutrons are used to build the model and to tune it. Finally, the effects of our hardening approach are compared with other hardened and non-hardened approaches.

Keywords: fault tolerance; single event upset; proton irradiation effects; neutron irradiation effects; soft errors

1. Introduction

Progressive technological down-scaling is reducing the natural resilience of circuits, implying greater susceptibility to radiation faults [1]. In the past, fault-tolerant microprocessors were required for systems working in harsh environments, such as satellites, aircraft, autonomous vehicles, or any kind of autonomous decision-making systems, but today they are increasingly in demand, even at ground level [2] where radiation induced soft errors can frequently occur. Soft-error radiation faults are produced by the effect of incident particles on circuits where, as a consequence, the digital state of a node can be modified (bit-flipping). The developers of critical systems are constantly searching for ways to improve and/or to maximize the reliability of critical applications, due to the presence of soft errors, that can lead to catastrophic failure situations.

Many approaches are shown in the literature to minimize the effect of soft errors. Conventional approaches improve reliability by introducing redundancy at different hardware, software or hardware-software structures, in order to mask the wrong results by majority voting [3] or other redundancy-exploiting methods. For instance, it is common for hardware approaches to apply triple modular redundancy (TMR), to achieve reliability by replicating some physical components (rad-hard processors) [4]. Software-implemented hardware fault tolerance (SIHFT) techniques also introduce redundancy at instruction level by replicating several blocks of code [5] or several critical instructions [6]. Hardware–software hardening techniques, which reduce some weakness of the hardware or software-only techniques, are also possible [7–9]. Other recent approaches represent attempts to gain reliability improvements by introducing no modifications in either the application (code instrumentation) or in the system (specific components). These techniques seek to achieve improvements during the transformation from high level code (source code) to machine code (executable) by altering the code compilation method [10]. Each approach has its own advantages and disadvantages; for example, the disadvantage of producing unwanted overheads in processing time and storage needs can be achieved by applying software hardening techniques.

When comparing the different approaches from the user perspective, there are two that require either high user intervention levels (most TMR-based approaches), or very little intervention (such as the approach proposed in this paper), or no user intervention and the delegation of hardening to some form of Artificial Intelligence (such as MOOGA [10]). The first approaches require lot of human effort, for instance, to change the focus of hardening. The third set requires a lot of CPU time to compare a large number of software versions, while the proposed approach can quickly explore several alternatives simply by changing the type and definition of each variable of interest, in a very fast operation.

In this article we focus on the SIHFT techniques, because they can be implemented in commercial off-the-shelf (COTS) microprocessors, thereby avoiding any internal modification to the microprocessor. More precisely, we are interested in high-level instrumentation techniques capable of deriving the inherent trade-offs, while maintaining flexibility and usability.

In view of the above, the key issues considered during the development of the new SIHFT approach presented in this article are:

1. The approach should be applicable to protect the largest possible amount of software, particularly the intellectual properties (IPs) commonly available on the Internet.
2. Post-compilation interventions must be as limited as possible (possibly none), in order to make software update and optimization fast and reliable.
3. The approach should apply to any COTS processor. It should not rely on any intrinsic radiation hardness of the processor except, obviously, the capability to withstand the total ionizing dose (TID).

The chosen language, C/C++, is compatible with commonly used software development techniques, leaving aside the domain of modern iconic programming. The idea which addressed and solved all the above issues is based on developing a set of C++ classes aimed at protecting program variables and processor registers, mostly by means of TMR.

In the following sections, a new high-level SIHFT technique will be presented, together with a reliability estimation model, to evaluate the impact of the system configuration parameters on program execution and radiation sensitivity. The model was developed from the results of two accelerated radiation campaigns conducted at the National Centre for Accelerators (CNA)—Spain, and Los Alamos Neutron Science Center (LANSCE)– USA.

2. Automatic Hardening Approach Based on C++ Classes

We propose a method that is intended for the protection of software code on COTS processors. In particular, it addresses the following elements of a COTS microcontroller system:

1. Numeric data stored in temporary and long-term storage locations. As it is a C/C++ level approach, there is no explicit distinction between registers and memory, although it provides overall data protection to data stored in the C/C++ variables, regardless of how and where these are allocated by the compiler;
2. Microcontroller configuration registers such as those for interrupts, ports, universal asynchronous receivers-transmitters (UARTs), analog-to-digital converters (ADCs), etc.
3. Program memory, mostly for situations where the program is stored in volatile or radiation-sensitive memory;

4. Protection against single-event functional interrupts (SEFIs).

Under certain circumstances induced by radiation, the microcontroller program may occasionally restart, which is considered acceptable, provided that the results produced at the end of execution are correct. In particular, some aspects of protection rely on inducing an automatic program reset when a SEFI is detected.

It is worth noting that most benefits of the proposed approach may also apply to temporary faults induced by other causes, such as electromagnetic interferences, allowing technological transfer to other ground-based activities, such as functional safety in automotive electric/electronic systems, and detect and correct errors in high performance computing (HPC), among others.

2.1. Using C++ Classes for Data Protection

The proposed methodology is based on a C++ template class called TD<DataType> (standing for "triple data") which can be applied to any numeric DataType (e.g., TD<char>, TD<int>, TD<float>). A TD<DataType> class transparently protects, by means of TMR, a numeric variable of any given DataType. This class has been designed to allow total reuse of existing code, only changing the definition of all the variables to be protected, while maintaining the rest of the code unchanged.

The internal architecture of a TD<DataType> (see Table 1) class contains three private variables (i.e., concealed from the user) of the type DataType, storing as many replicas of the same data (d1, d2, d3).

Table 1. Internal architecture of TD<DataType>.

template <typename DataType> class TD { private: volatile DataType d1, d2, d3; }

A seamless use of the proposed class requires: (i) the appropriate re-definition of all possible numeric, comparison and logical operators; (ii) writing the code to implement each of them in a redundant way.

For instance, for the assignment operator (=), the kernel of the code and its usage are showed in Table 2.

Table 2. Definition and functioning for the = operator.

definition	operator=(TD<DataType>& val) { d1 = val.d1; d2 = val.d2; d3 = val.d3; }
usage	TD<int> a, b; a = b;
equivalent to	int a1, a2, a3, b1, b2, b3; a1 = b1; a2 = b2; a3 = b3;

Which implies that, despite the apparently identical usage of the assignment (a = b) to standard C variables, the usage of operator = of class TD<DataType> implies (in a transparent way) that the value of each replica of b is assigned to the corresponding replica of a. A similar approach applies to all algebraic operators (e.g., +, -, *, /), comparison operators, logical operators, etc.

In our library, the casting operators to/from TMR data and plain data have been overloaded for transparent conversion between data types. Conversion from TMR to plain data implicitly implements majority voting, while conversion from plain to TMR implicitly implements triplication.

The following simple example compares a simple piece of C code which sums up two variables and stores the result in a third variable (See Table 3). The same code is written in fully unprotected and partially protected ways, respectively, together with a possible manual protection.

Table 3. Comparison of how to sum up two integer variables using different protection levels: original code, protection using our technique and manual TMR protection.

Unprotected	Protected		Manual TMR

int a, b;
int c;
c = a + b;
	TD<int> a, b; \\protected
	TD<int> c; \\protected
	c = a + b;
		TD<int> a, b; \\protected
		int c; \\unprotected
		c = a + b;
			int a1, a2, a3, b1, b2, b3;
			int c1, c2, c3;
			c1 = a1 + b1;
			c2 = a2 + b2;
			c3 = a3 + b3;

As a consequence, by writing, c = a + b; the compiler automatically generates the code that will sum up and store each corresponding replica of the DataType in a completely transparent way and will preserve (by construction) the correspondence of each replica. A process that is quite unlike the TMR manual approach, which would be quite prone to coding errors.

The TD<DataType> class is designed to support any operator and constructor (e.g., vectors and structures) commonly used inside C programs. The potential risks of pointers are normally to be avoided and they can be applied with greater safe by using, for instance, a TMR-protected TD<int*>, as redundancy significantly reduces the risk of pointer corruption.

In the case of single-event upsets (SEUs) (or any other transient fault) affecting one of the three replicas, the original value can be recovered by majority voting, again in a transparent way. For instance, the simple piece of code of Table 4

Table 4. Example of triplicating and voting automation for two hardened and non-hardened variables.

TD<int> a; int b;
b = a; \\ majority vote a's replicas into b
a = b; \\ triplicate b into a's replicas
a = (int) a; \\ compact form: vote+TMR

First converts and copies a redundant variable, a, into a non-redundant variable, b, by enforcing majority voting, and it then stores the three replicas of the voted value, b, back into variable a. In other words, it re-synchronizes the replicas by majority voting. The last line is a compact form which does exactly the same thing.

Any existing program can therefore be hardened, by a mere redefinition of the variables used, while the active part of the code requires no single modification. This idea per se is not novel, as TMR is widely used to achieve data protection, but the way it has been implemented and optimized with respect to radiation tolerance is new and easy to use.

2.2. Protecting Other Elements of a Program

A complex program not only relies on data memory, which can be protected by means of the TD<...> class. Other elements have also been considered.

Configuration registers cannot be triplicated in the same way as normal memory location, as they are unique in hardware and their TMR would require redesigning the manufacturing masks. Protecting the configuration registers is therefore supported by another type of class, called TDreg<...>, which automatically stores two other copies of the register in data memory and periodically re-synchronizes the hardware register by majority voting with the other two stored replicas.

Periodical refresh can be implemented in different ways, depending on system and mission requirements. For instance: (i) a timer-driven interrupt routine which refreshes all variables, set at, for instance, every minute; (ii) at the beginning or at the end of each program loop (if any); (iii) by voting whenever a critical variable is used.

Since configuration registers of commercial processors mix read-only and write-only bits, the definition of TDreg<...> supports this feature and synchronization is automatically limited to writable bits.

Program memory should nominally be read-only, as it only contains machine code and numeric constants. We explicitly omit consideration of self-modifying codes, as those are considered too dangerous for critical applications.

As a consequence of a nominally constant program code, its protection is limited to computing a "signature" of the code area, on a periodic basis, and verifying it against a golden sample. As soon as a SEU affects the program area, its signature will no longer match and the program will automatically be reset, downloading the program again from a more rad-tolerant ROM. This is implemented by means of the TDcode class.

Internal control registers (namely internal state machines, program counter and stack pointer) are more difficult to protect and are the most common cause of program hang, therefore causing SEFIs. Our approach offers periodical verification of stack-pointer consistency, but the other control registers (e.g., program counter and status register) can be protected only to a very limited extent.

The only means available to compensate SEFIs is the use of a watchdog timer (or equivalent methods) already commonly used in these situations. Yet Section 4.1 shows that protection of program counter and stack pointer will not usually improve hardness significantly.

Interrupt handlers are normal routines that can be protected with the same techniques described above. In addition, interrupts also rely on interrupt enable bits, which are part of configuration registers; these can be protected by means of the TDreg<...> class described above.

2.3. Performance Issues

The use of TMR, on the one hand, significantly increases the hardness of a program to single-event effects (SEEs) but, on the other hand, it also impacts on aspects of performance, particularly speed and memory size. In theory, execution time should increase by a factor of three at most (the same as redundancy), although the increased flexibility and safety made available by the use of the C++ classes causes an additional overhead by another factor of two, on average, mostly due to the periodic necessity of majority voting. This overhead has been strongly optimized by means of the many features of state-of-the-art optimizing compilers, although it cannot be completely removed for several reasons.

As a consequence, program execution, for a program with variables that are totally triplicated will take six times more time to execute, on average. An appropriate selection of which variables to protect and which ones need no protection significantly reduces the impact on program speed. Section 4.1 gives some hints on both how to select storage blocks and which specific variables to protect and which ones need not be protected, allowing a quick performance trade-off customized for specific mission requirements.

3. A Compact Reliability Estimation Model

During the process of hardening a piece of code (or even a complete HW/SW design), it is of the utmost importance to analyze a number of different configurations and to evaluate the impact of configuration parameters (e.g., data triplication, register refresh, error checking, register optimization, inlining, interrupts, etc.) on program execution and radiation sensitivity.

We developed a mathematical and empirical model, for quick evaluation of several reliability vs. performance trade-offs and for the optimization of radiation hardening without excessively compromising performance. The model offers valuable advance information on system behaviour in the presence of radiation induced faults. Firstly, it predicts the occurrence frequency of faults that affect program execution for any combination of processor, high level language, compilation parameters, hardening techniques and selection of protected variables. Secondly, it estimates the impact of each storage area, variable or data structure on the overall reliability, to concentrate hardening efforts on the block that has the highest impact on radiation sensitivity.

It is worth highlighting other works that either compare different methods with real radiation data (making the approach quite expensive and time consuming, and therefore ruling out the possibility of comparing large numbers of alternatives) and with simulated campaigns (cheaper and faster approach, but of lower reliability). The proposed approach is, instead, based on a compact parametric mathematical model, the parameters of which are first evaluated once and for all on real radiation measurements, then the model is applied in an iterative way, thereby permitting a wider search in the space of hardening alternatives.

3.1. Model Preliminaries

The model is based on cycle-accurate simulations using the OVPsim simulator [11] while randomly corrupting: registers (**R**), data memories (**D**), and program memory (**P**). Each storage block may have a different hardware implementation, so it may therefore have its own cross section per byte α_R (respectively, α_D and α_P); in addition, data storage may be distributed between a number of memories, each one having a different cross section α_{D1}, α_{D2}, $\alpha_{D...}$ (e.g., FLASH, ferroelectric, static and dynamic RAMS).

We assume that the cross section is different for each type of storage, and we relate each one to the basic cross section of main processor RAM (D1), that is $\alpha_{D1} \equiv \alpha$. We therefore state that: $\alpha_R = K_R \cdot \alpha$, $\alpha_{D2} = K_{D2} \cdot \alpha$, $\alpha_P = K_P \cdot \alpha$, ..., where K_X are appropriate coefficients and, by definition, $K_{D1} = 1$. In particular, K_P, is the coefficient of either ROM or RAM, depending on where the program is executed.

For each given processor, algorithm, language, compilation flags, hardening effort, etc., OVPsim simulations are set up to induce one random SEU per run of the compiled program, in either of the aforementioned storage blocks (R, D, P). The fault injection can be performed in each memory block at different abstraction levels. It means, for example, that we can induce an error in an SRAM or a DRAM device on any possible address from their available addressing space or only induce faults on single C/C++ variables of interest (vectors, matrices, ...).

Injected faults are classified according to their effect on program behavior, in a similar way to the first proposals of Mukherjee et al. [12]. Faults which neither hang program execution nor affect expected program output are called unnecessary for architecturally correct execution (unACE). On the other hand, faults which visibly affect program execution are called architecturally correct execution (ACE), which comprise the two categories specifically considered in this paper: (i) faults which allow the program to terminate normally, but produce corrupted results, called silent data corruption (SDC); and, (ii) faults which cause abnormal program termination or infinite execution loops, called HANG.

Each simulation set was configured to inject 1000 faults per register in the register file and 18,000 faults in the memory section allocated by the benchmark. This arrangement implies a total of at least 72,000 faults injected per program version, achieving a statistical error of ±1% at a 99% confidence level, according to the statistical model proposed by Leveugle et al. [13].

3.2. Model Description

Simulations provide, as an output, the number of SDCs (respectively SD_R, SD_{D1}, SD_{D2}, $SD_{D...}$, SD_P) and HANGs (respectively HG_R, HG_{D1}, HG_{D2}, $HG_{D...}$, HG_P) out of R_R program executions (respectively, R_{D1}, R_{D2}, $R_{D...}$, R_P). The size of each storage area being S_R words (respectively, S_{D1}, S_{D2}, $S_{D...}$, S_P).

For every configuration we define, for each storage area (where Z is either R, D1, D2, P, ...) the equivalent block size for SDCs of that area, expressed in bytes, as:

$$\beta_Z \triangleq K_Z \cdot \frac{SD_Z}{R_Z} \cdot S_Z \tag{1}$$

We also define the equivalent block size for HANGs of that area, expressed in bytes, as:

$$\gamma_Z \triangleq K_Z \cdot \frac{HG_Z}{R_Z} \cdot S_Z. \tag{2}$$

The two formulas provide two factors (β_Z, γ_Z) which are proportional to: the increased sensitivity of the specific memory area (K_Z) to radiation; the failure probability as estimated by simulations ($\frac{SD_Z}{R_Z}$, $\frac{HG_Z}{R_Z}$) and the size S_Z of the memory block, which is proportional to the probability of a particle hitting that block.

From these factors, we can find the total equivalent size for SDCs and for HANGs of the whole program, respectively:

$$\beta_{TOT} = \beta_R + \beta_{D1} + \beta_{D2} + \beta_{D...} + \beta_P + \beta_X \tag{3}$$

$$\gamma_{TOT} = \gamma_R + \gamma_{D1} + \gamma_{D2} + \gamma_{D...} + \gamma_P + \gamma_X. \tag{4}$$

The two additional parameters, β_X and γ_X, are the equivalent block size for SDCs and HANGs of the internal control unit and the state machines, which cannot be simulated by the OVPsim simulator and are therefore empirically estimated.

3.3. System Reliability

Given β_{TOT} and γ_{TOT}, our model predicts the probability of SDCs and HANGs per execution:

$$P(SD) = \Phi \cdot \alpha \cdot (T_E \cdot \beta_{TOT}) \tag{5}$$

$$P(HG) = \Phi \cdot \alpha \cdot (T_E \cdot \gamma_{TOT}), \tag{6}$$

where Φ is the radiation flux (particles/s/cm^2), while α is the cross section per byte (cm^2/byte) of storage, and T_E is nominal program execution time (s).

The two expressions between brackets are called the size-time figures for SDCs ($\chi_{SDC} = T_E \cdot \beta_{TOT}$) and HANGs ($\chi_{HANG} = T_E \cdot \gamma_{TOT}$), respectively, of the given configuration.

An innovative aspect of the proposed approach is that the size-time figures, χ_{SDC} and χ_{HANG}, mean that the impact of each data storage, each data structure, and even each individual variable on overall radiation performance can be easily assessed and the hardening efforts may be therefore be concentrated where the effect is highest and to reduce the impact of hardening to a minimum.

Depending on the application, we can estimate, in the first place, the mean work to failure (MWTF) (i.e., the average number of program executions between two failures), in the following way:

$$MWTF = \begin{cases} \frac{1}{P(SD)} = \frac{1}{\Phi \cdot \alpha \cdot \chi_{SDC}} & \text{for SDCs} \\ \frac{1}{P(HG)} = \frac{1}{\Phi \cdot \alpha \cdot \chi_{HANG}} & \text{for HANGs} \\ \frac{1}{P(ERR)} = \frac{1}{\Phi \cdot \alpha \cdot (\chi_{SDC} + \chi_{HANG})} & \text{for any error} \end{cases}, \tag{7}$$

which depends on: (i) radiation flux Φ; (ii) processor's cross section α; and (iii) size-time figures of given program configuration (χ_{SDC} or χ_{HANG}).

In second place, for time-sampled systems, where the program starts every T_S (sample time), executes over a certain time, T_E, then stops until the next sample, we can compute the mean time to failure (MTTF), which is the average time between two failures:

$$MTTF = T_S \cdot MWTF, \tag{8}$$

which also depends on sample time T_S.

3.4. Model Validation under Radiation

The proposed model was evaluated against real radiation measurements. Table 5 shows the relevant model parameters for protons and neutrons measured during the two radiation campaigns described below. The model showed a good accuracy for the estimation of reliability. In fact, the figures shown in the last two columns of Table 6 have an error of −30% + 50% with respect to radiation measurements (not shown in the table).

The device under test (DUT) selected for the irradiation experiments was the ZYBO board. The DUT is equipped with a 28nm CMOS Xilinx ZYNQ XC7Z010 system on chip (SoC). This SoC is divided into two parts, an FPGA area (programmable logic—PL) and a 32-bit ARM cortex A9 microprocessor (processing system—PS). The processor has a 13-stage instruction pipeline that includes a branch prediction block and support for two levels of cache. In addition, the microprocessor has a little built-in memory called on chip memory (OCM), where the bootloader or the program under test can be loaded.

The DUT was controlled by an external computer, the RaspberryPi 3 Model B, the main task of which is to receive and log all the messages sent by the DUT. The DUT was configured to send a state message every five seconds in the absence of errors, otherwise the message is notified instantly and the external computer resets and reprograms the DUT.

Tested programs present a rich variety of flow structures and data. For example, BubbleSort (BB) is a well-known sorting algorithm that achieves its objective by making use of several nested loops. The second algorithm considered here is the Dijkstra algorithm (DK), also known as the shortest path problem, which uses an adjacency matrix that is stored in the memory where the weights of all paths are located.

3.4.1. Proton Irradiation Campaign

The test campaign was carried out in mid-2018 at the National Centre for Accelerators (CNA), in Spain [14]. Irradiation tests were performed using the external beam line, installed in the cyclotron laboratory. Although the proton energy delivered by this cyclotron was fixed to 18 MeV, the beam was extracted to the air up to reach the DUT (device under radiation) position with 15.2 MeV energy. The flux fluctuated within ±5% during each run. Beam uniformity under these experimental conditions was better than 90% in the area of interest.

3.4.2. Neutron Irradiation Campaign

The neutron SEE campaigns were performed at the Los Alamos Neutron Science Center (LANSCE) in September 2018 [15,16]. The neutron beam was provided by a tungsten spallation source at approximately 30 degrees to the left of the main beam. During the campaign, the DUT remained at 23 m from the neutron source, and the beam was collimated, so that a spot was obtained in the order of 30 mm of diameter. This size covers the active area with uniformity better than 90%. A constant neutron flux of $1.7 \cdot 10^5$ n/(s · cm^2), above 10MeV, was obtained.

4. Reliability Issues

In the last step of this activity, our model has been used to identify the most critical storage areas, variables and data structures, that is, those which most affect reliability, in order to concentrate hardening efforts on the most relevant areas. In addition, the performance of the proposed C++ classes against other optimization techniques proposed by the same authors in [10] was compared.

The proposed C++ classes have been used to protect a variety of programs on an ARM Cortex-A9 processor and our model has identified the most critical storage areas which deserve more hardening effort. Some results are shown in Table 6, namely for a BubbleSort sorting algorithm and a Dijkstra shortest path finder algorithm, with both on-chip memory (OCM), an external rad-hard memory (EXT), as well as neutron and proton irradiation. All these results were also verified during the two radiation campaigns briefly described in Section 3.4

Table 5. ARM Cortex-A9 parameters estimated during two radiation campaigns.

Particles	α (cm^2/byte)	K_R	K_P	β_X (byte)	γ_X (byte)
proton	$1.39 \times 10^{-14} \pm 10\%$	35 ± 5	1.6 ± 0.2	40 ± 10	25 ± 25
neutron	$4.85 \times 10^{-14} \pm 10\%$				

4.1. Performance Considerations

We draw a few considerations here, which can be found by analyzing the results shown in Table 6, where a few C++ hardening configurations are compared with other configurations with hardening on specific aims [10]: mean work to failure (MWTF) maximization, fault coverage maximization (Max-ACE), trade-off optimization among execution time, memory size and fault coverage (Pareto), baseline compilation (O0) and code optimization (O3). All C++ versions were compiled using the -O3 optimization flag. We can observe that:

Table 6. Execution time, T_E,(for 666MHz clock) plus equivalent block sizes for SDCs and HANGs and total time-size figures for a BubbleSort and a Dijkstra program, for different compilation flags, use of C++ classes vs. other hardening techniques, for four storage blocks (registers, data memory, stack, and code memory, taken as examples, for an ARM Cortex-A9 processor, using either on-chip memory (OCM) and external rad-hard memory (EXT). Highlighted values are those referenced in the text for the sake of clarity.

	Configuration				T_E	REG		Data+BSS		STACK		CODE		TIME-SIZE	
	Hardening Strategy	ID	MEM	Part	(μs)	β_R (B)	γ_R (B)	β_{D1} (B)	γ_{D1} (B)	β_{D2} (B)	γ_{D2} (B)	β_P (B)	γ_P (B)	χ_{SDC} (B·ms)	χ_{HANG} (B·ms)
BubbleSort	Max-ACE	BB-C1	OCM	prot	67	778	12.9	401	13.8	0	5	420	516	110	37
	O0	BB-C2	OCM	prot	335	53	12.6	256	145	3	3	153	594	169	252
	MWTF	BB-C3	OCM	prot	58	81	16.2	395	24.8	0	4	535	553	61	35
	Pareto	BB-C5	OCM	prot	60	64	15.6	377	14.3	0	6	99	461	35	29.8
	O3	BB-C10	EXT	prot	980	90	15.6	19.9	0.7	0.0	0	4.5	30	151	46
	C++ (O3)	BB-C14	OCM	prot	522	16.8	17.6	4.1	8	0	11	0	807	32	440
	C++ (O3)	BB-C11	EXT	prot	12821	16.8	17.6	0.2	0.4	0	1	0	40	731	755
	C++ (O3)	BB-L4	OCM	neut	517	6.6	9	2.7	9.6	0	18	0	568	25.5	312
Dijk.	O0	DK-L1	OCM	neut	2377	31	4.8	5297	1986	11.8	18.7	205	2052	13,279	9656
	C++ (O3)	DK-L3	OCM	neut	9676	226	18.9	13.1	707	16.8	22.9	742	2890	10,042	35,216

- in the BubbleSort algorithm the influence of stack (β_{D2} and γ_{D2}) is close to zero, therefore negligible with respect to the influence of other storage blocks (β_R, γ_R, β_P and γ_P); in this situation, it is useless to protect the stack. In the Dijkstra algorithm, the influence of stack on SDCs (β_{D2}) is comparable to that of data storage (β_{D1}), at least for one configuration (DK-L3); in this situation, it might also be worth protecting the stack;
- the use of C++ classes (BB-C14, BB-L4, DK-L3) increases execution time by a factor of between 2 and 10 times, depending on configuration (without considering BB-C11 which runs on an external, slower, rad-hard memory), but it reduces the influence of data memory on SDCs (β_{D1}) by a factor of 100 and almost nullifies the influence of data memory on SDCs (β_P); the effects of the C++ classes on HANGs are negligible;
- for configurations not protected by means of the C++ classes, the effect of registers on SDCs and HANGs is negligible despite the register's very high cross section (see K_R in Table 5); when protecting the program by means of the C++ classes, the effect of registers (mostly for SDCs)

is almost the only relevant one, therefore increasing protection requires an additional effort to protect the registers, which are not protectable by means of the C++ classes;
- using an external, slower, rad-hard memory (configuration BB-C11, based on the proposed C++ classes, without cache) offers the lowest equivalent block sizes for all data storage (except obviously registers), despite it increasing execution time, T_E, by a factor of 20 to 25.
- by looking at the total size-time figures (two last columns), which are the most relevant overall parameter directly affecting MWTF and MTTF, the reduction of equivalent program size often counteracts an increase in execution time. The best performance for SDCs was achieved using the proposed C++ classes, while the best performance for HANGs was achieved with configurations BB-C3 and BB-C5.

4.2. Optimization Process

This section shows how an appropriate use of the compact model can rapidly optimize the usage of the C++ classes. We took as an example an optimized BubbleSort algorithm (different from the one used for Table 6) running at 666MHz on a Cortex A9 processor and irradiated by protons. We simulated the few configurations shown in Table 7, both for SDCs and for HANGs.

Each row shows different configurations: first and second configurations are plain C code with no optimization (-O0) and highest optimization (-O3), respectively. Each column shows the equivalent size of : registers (REG); whole data memory (β_D); only the first, the second, and the third C variables of the program ($\beta_{D,V1}$, $\beta_{D,V2}$, $\beta_{D,V3}$, respectively); the other five variables were less relevant, taken together ($\beta_{D,V4}$); program memory (PROG); the other three columns show the equivalent size, the execution time and the size-time figure of the whole program; the last two columns show the expected MWTF and MTTF for a given irradiation level (see caption of Table 7).

From the table, it is, for instance, clear that the variable V2 for SDCs has by far the highest relevance (namely, highest size, 261B/388B) among all the C variables. It would therefore be worth hardening only that variable by means of the C++ classes. The hardening of other variables would add significantly to the execution time while reducing total equivalent size by a negligible amount.

Consequently, one variable, V2, when hardened (by changing the data type to the proposed C++ class), yields the results shown in the third line of the table, which shows the lowest size-time figure χ_{SDC} from among all the configurations. We also evaluated the fourth configuration of the table, for comparative purposes, by applying the C++ classes to all the program variables.

It is clear that the configuration with only one hardened variable, V2, showed the best equivalent size (135 B) and size/figure performance (29 B·ms) from among all of them, despite the higher execution time (215 µs). The same configuration also shows the highest MWTF (about ten times higher than the -O0 and two times higher than the -O3 without the C++ classes; slightly lower for HANG) and MTTF (also about ten times higher than the -O0 and two times higher than the -O3 without C++ classes), proving the effectiveness of the proposed method. Table 8 shows the global MWTF and MTTF metrics (including both SDC and HANG). As can be seen, the configurations hardened by C++ classes provide the best overall reliability.

Table 7. Equivalent sizes (β_{TOT} and γ_{TOT}), size-time figures (χ_{SDC} and χ_{HANG}) and reliability metrics (MWTF and MTTF) of a selected program (optimized BubbleSort) in a few different configuration. The individual impact of Registers (R), total data memory (D), individual memory variables (V1 through V3), other variables (V4), code area (P) and total (TOT), for an ARM Cortex A9 processor, with on-chip memory (OCM) tuning at 666MHz clock frequency. The last two columns refer to the estimated proton irradiation results with radiation flux of 5.45×10^5 particles/cm²/s and sample time $T_S = 20$ ms. Highlighted values are those referenced in the text for the sake of clarity.

	Configuration	REG	MEM					PROG	TOTAL			MWTF	MTTF
		β_R (B)	β_D (B)	$\beta_{D,V1}$ (B)	$\beta_{D,V2}$ (B)	$\beta_{D,V3}$ (B)	$\beta_{D,Vx}$ (B)	β_P (B)	β_{TOT} (B)	T_E (µs)	χ_{SDC} (B·ms)	runs ($\times 10^3$)	hrs
SDC	-O0	56.0	261	1.67	261	1.92	2.45	210	527	562	296	446	2.5
	-O3	102	401	0.04	388	0.00	0.00	105	608	87.1	53.0	2493	13.8
	C++ -O3 (V2)	81.2	0.00	0.01	0.00	0.00	0.00	53.5	**135**	215	**29.0**	4552	25.3
	C++ -O3 (all)	82.6	0.00	0.04	0.00	0.00	0.00	68.7	151	208	31.5	4184	23.2
		γ_R (B)	γ_D (B)	$\gamma_{D,V1}$ (B)	$\gamma_{D,V2}$ (B)	$\gamma_{D,V3}$ (B)	$\gamma_{D,Vx}$ (B)	γ_P (B)	γ_{TOT} (B)	T_E (µs)	χ_{HANG} (B·ms)	runs ($\times 10^3$)	hrs
HANG	-O0	455	152	0.3	**133**	0.1	1.2	714	1321	562	742	178	1.0
	-O3	679	17.8	0.0	0.0	0.0	0.0	434	1130	87.1	98.4	1341	7.5
	C++ -O3 (V2)	386	24.1	0.0	0.0	0.0	0.0	281	691	100	69.1	1910	10.6
	C++ -O3 (all)	274	14.6	0.0	0.0	0.0	0.0	371	660	100	66.0	2000	11.1

Table 8. Total MWTF and MTTF for different configurations. Highlighted values are those referenced in the text for the sake of clarity.

Configuration	-O0	-O3	C++ -O3 (V2)	C++ -O3 (All)
MWTF runs ($\times 10^3$)	127	872	1345	1353
MTTF hours	0.7	4.8	7.5	7.5

4.3. Further Improvements

It is clear from Table 7 that the proposed C++ classes significantly reduced the influence of data storage for SDCs and slightly reduced the influence of data storage for HANGs, although they significantly increased the execution time.

The reason is that all the variables were protected for the configurations shown in Table 6. Nevertheless, the proposed approach can be individually used to address the effect of each variable, by splitting data storage into smaller blocks (D1, D2, D...), namely one per variable or group of variables, and to evaluate the effect of each of them on execution time and equivalent program sizes. From this analysis, the best trade-off between what to protect and what not to protect can be assessed.

Another parameter that can be addressed is the rate of data recovery; each data verification in the C++ classes takes time and data recovery takes even longer. Frequent verifications and recovery increase execution times, while less frequent verifications can increase the risk of double faults. A trade-off may also be established in this case, by means of the proposed approach.

5. Conclusions

A new hardening approach has been proposed on the basis of a set of C++ classes, to ease the protection of existing and new software programs.

A simple though accurate reliability model has also been proposed, to support the optimization of the usage of C++ classes, and to compare the performance of those classes with the performance of other hardening methodologies.

A relevant feature of this model is that it provides two compact figures (namely the size-time figure χ_{SDC} and χ_{HANG}) that directly relate to the reliability figures (MWTF and MTTF for SDCs and HANGs, respectively), by taking into account both the increased computation time-typical of SIHFT-and the improvement in robustness-typical of TMR.

The basic results showed that programs protected with the C++ classes were slower, but less subject to radiation-induced effects. Yet the two effects partially canceled out when considering the mean time or mean work between consecutive program HANGs, while the lower sensitivity to radiation was more relevant than the increase in execution time when considering the mean time or mean work between consecutive SDCs. It has been shown that a straightforward usage of C++ classes improved the reliability of a software system against corrupted results, but had less effect on program HANGs. A targeted application of the proposed C++ classes to specific variables significantly improved both effects.

In conclusion, the use of appropriate C++ classes shown in this paper has greatly facilitated the use of TMR. Also, the availability of an easy-to-use performance estimation model could be used for quick and effective radiation tolerance optimization of the COTS microcontroller systems.

Author Contributions: Conceptualization, L.M.R. and A.M.-Á.; Methodology, L.M.R., A.S.-C. and A.M.-Á.; Validation, L.M.R., A.S.-C., Y.M., S.C.-A. and A.M.-Á.; Writing—Original Draft Preparation, L.M.R., A.S.-C., S.C.-A. and A.M.-Á.

Funding: This work was funded by the Spanish Ministry of Economy and Competitiveness and the European Regional Development Fund through the following projects: *'Evaluación temprana de los efectos de radiación mediante simulación y virtualización. Estrategias de mitigación en arquitecturas de microprocesadores avanzados'* and *'Centro de Ensayos Combinados de Irradiación'*, (Refs: ESP2015-68245-C4-3-P and ESP2015-68245-C4-4-P, MINECO/FEDER, UE).

Acknowledgments: Thanks to the the National Centre for Accelerators (CNA)—Spain, and Los Alamos Neutron Science Center (LANSCE)– USA for all the support in the the irradiation campaigns.

Conflicts of Interest: The authors declare no conflict of interest.

References

1. Baumann, R.C. Radiation-induced soft errors in advanced semiconductor technologies. *IEEE Trans. Device Mater. Reliab.* **2005**, *5*, 305–315. [CrossRef]
2. Michalak, S.E.; Harris, K.W.; Hengartner, N.W.; Takala, B.E.; Wender, S.A. Predicting the number of fatal soft errors in Los Alamos national laboratory's ASC Q supercomputer. *IEEE Trans. Device Mater. Reliab.* **2005**, *5*, 329–335. [CrossRef]
3. Nicolaidis, M. Design for soft error mitigation. *IEEE Trans. Device Mater. Reliab.* **2005**, *5*, 405–418. [CrossRef]
4. Ruano, O.; Maestro, J.A.; Reviriego, P. A Methodology for Automatic Insertion of Selective TMR in Digital Circuits Affected by SEUs. *IEEE Trans. Nucl. Sci.* **2009**, *56*, 2091–2102. [CrossRef]
5. Quinn, H.; Baker, Z.; Fairbanks, T.; Tripp, J.L.; Duran, G. Software Resilience and the Effectiveness of Software Mitigation in Microcontrollers. *IEEE Trans. Nucl. Sci.* **2015**, *62*, 2532–2538. [CrossRef]
6. Martínez-Álvarez, A.; Cuenca-Asensi, S.; Restrepo-Calle, F.; Palomo, F.R.; Guzmán-Miranda, H.; Aguirre, M.A. Compiler-Directed Soft Error Mitigation for Embedded Systems. *IEEE Trans. Dependable Secur. Comput.* **2012**, *9*, 159–172. [CrossRef]
7. Martínez-Álvarez, A.; Restrepo-Calle, F.; Cuenca-Asensi, S.; Reyneri, L.M.; Lindoso, A.; Entrena, L. A Hardware-Software Approach for On-Line Soft Error Mitigation in Interrupt-Driven Applications. *IEEE Trans. Dependable Secur. Comput.* **2016**, *13*, 502–508. [CrossRef]
8. Reyneri, L.M.; Roascio, D.; Passerone, C.; Iannone, S.; de los Rios, J.C.; Capovilla, G.; Martínez-Álvarez, A.; Hurtado, J.A. Modularity and Reliability in Low Cost AOCSs. In *Advances in Spacecraft Systems and Orbit Determination*; Ghadawala, R., Ed.; IntechOpen: Rijeka, Croatia, 2012; Chapter 5.
9. Peña-Fernandez, M.; Lindoso, A.; Entrena, L.; Garcia-Valderas, M.; Philippe, S.; Morilla, Y.; Martin-Holgado, P. PTM-based hybrid error-detection architecture for ARM microprocessors. *Microelectron. Reliabil.* **2018**, *88–90*, 925–930. [CrossRef]

10. Serrano-Cases, A.; Morilla, Y.; Martín-Holgado, P.; Cuenca-Asensi, S.; Martínez-Álvarez, A. Non-intrusive automatic compiler-guided reliability improvement of embedded applications under proton irradiation. *IEEE Trans. Nucl. Sci.* **2019**. [CrossRef]
11. Open Virtual Platforms. OVPsim Simulator. Available online: www.ovpworld.org/technology_ovpsim (accessed on 1 April 2019).
12. Mukherjee, S.S.; Weaver, C.; Emer, J.; Reinhardt, S.K.; Austin, T. A systematic methodology to compute the architectural vulnerability factors for a high-performance microprocessor. In Proceedings of the 36th Annual IEEE/ACM International Symposium on Microarchitecture, San Diego, CA, USA, 5 December 2003; pp. 29–40. [CrossRef]
13. Leveugle, R.; Calvez, A.; Maistri, P.; Vanhauwaert, P. Statistical fault injection: Quantified error and confidence. In Proceedings of the 2009 Design, Automation & Test in Europe Conference & Exhibition Nice, France, 20–24 April 2009; pp. 502–506. [CrossRef]
14. Centro Nacional de Aceleradores. Seville, Spain. Available online: http://www.cna.us.es (accessed on 1 April 2019).
15. Wender, S.A.; Lisowski, P.W. A white neutron source from 1 to 400 MeV. *Nucl. Inst. Methods Phys. Res. B* **1987**, *24–25*, 897–900. [CrossRef]
16. Lisowski, P.W.; Schoenberg, K.F. The Los Alamos Neutron Science Center. *Nucl. Instrum. Methods Phys. Res. Sect. A Accel. Spectrometers Detect. Assoc. Equip.* **2006**, *562*, 910–914. [CrossRef]

© 2019 by the authors. Licensee MDPI, Basel, Switzerland. This article is an open access article distributed under the terms and conditions of the Creative Commons Attribution (CC BY) license (http://creativecommons.org/licenses/by/4.0/).

Article

Total Ionizing Dose Effects on a Delay-Based Physical Unclonable Function Implemented in FPGAs

Honorio Martin [1,*], Pedro Martin-Holgado [2], Yolanda Morilla [2], Luis Entrena [1] and Enrique San-Millan [1]

1. Departement of Electronics Technology, Universidad Carlos III de Madrid, 28911 Leganés, Spain; entrena@ing.uc3m.es (L.E.); enrique.sanmillan@uc3m.es (E.S.-M.)
2. Centro Nacional de Aceleradores (CNA, Universidad de Sevilla, CSIC), 41092 Sevilla, Spain; pmartinholgado@us.es (P.M.-H.); ymorilla@us.es (Y.M.)
* Correspondence: hmartin@ing.uc3m.es; Tel.: +34-91-624-8390

Received: 6 August 2018; Accepted: 23 August 2018; Published: 24 August 2018

Abstract: Physical Unclonable Functions (PUFs) are hardware security primitives that are increasingly being used for authentication and key generation in ICs and FPGAs. For space systems, they are a promising approach to meet the needs for secure communications at low cost. To this purpose, it is essential to determine if they are reliable in the space radiation environment. In this work we evaluate the Total Ionizing Dose effects on a delay-based PUF implemented in SRAM-FPGA, namely a Ring Oscillator PUF. Several major quality metrics have been used to analyze the evolution of the PUF response with the total ionizing dose. Experimental results demonstrate that total ionizing dose has a perceptible effect on the quality of the PUF response, but it could still be used for space applications by making some appropriate corrections.

Keywords: physical unclonable function; FPGA; total ionizing dose; Co-60 gamma radiation; ring-oscillator

1. Introduction

Securing sensitive information on low-cost satellite applications has become a major challenge for the space industry. Typical approaches that include very expensive cryptographic primitives, non-volatile memory and analogous blocks cannot be afforded in these small space systems. In this context, commercial Field Programmable Gate Arrays (FPGAs) have turned out to be a good solution due to their flexibility and cost. Among their many uses, FPGAs can be dedicated to ensuring secure satellite data.

A popular solution to provide security in resource constrained applications, such as those using FPGAs, is on-chip Physical Unclonable Functions (PUFs). PUFs are a very promising security primitive used for authentication and key generation in IC and FPGAs. These security primitives are based on the impossibility of creating two physically exactly identical ICs due to the influence of random and uncontrollable effects during the manufacturing process. These uncontrollable influences leave measurable random marks on some features which possess the potential to generate encryption keys directly associated to a device [1]. Thus, PUFs work as an unclonable specific feature that can identify a circuit, just as a fingerprint can identify a human being. Among the various device properties that can be used for this purpose, delay-based PUFs deserve special attention due to their straightforward implementation. A delay-based PUF exploits the delay dependency on the random process variations [2]. Well-known examples of this type of PUF are the Arbiter PUF [3] and the Ring Oscillator PUF [2].

PUFs can be used to solve an important issue related to the generation of secure encryption keys in satellite communications, removing the necessity for key storage. Nonetheless, as PUF response depends on some circuit features that may be affected by the operational conditions; it is important to assert the suitability of these primitives in harsh environments subjected to ionizing radiation. Ionizing radiation induces charges in the semiconductor material that can be trapped in the oxide, altering the electrical characteristics of electronic devices. This effect, known as Total Ionizing Dose (TID), is cumulative and produces a gradual degradation of major electrical parameters, such as threshold voltage and leakage current, that can eventually result in device failure at a certain dose. To the best of our knowledge, the effects of TID in delay-based PUFs have not been studied before. There is one work that studies the TID effects in a CMOS silicon PUF based on transistor breakdown [4].

In this work, we evaluate the effects of ionizing radiation on a well-known delay-based PUF [2] implemented in a SRAM-FPGA. To that end, we have performed an extensive test with two different devices exposed to a radiation source, periodically collecting the PUF response as TID increased. All the external influences that can affect the PUF response (temperature, humidity, voltage, etc.) were controlled. Several major quality metrics have been used in order to assess the impact of radiation in the PUF response.

The remainder of this paper is organized as follows. Section 2 introduces the RO-PUF under study and the typical effects of ionizing radiation on SRAM-FPGAs. Section 3 describes the implementation of the RO-PUF and the TID test setup. Section 4 reports the impact of the ionizing dose on the RO-PUF. Several metrics are presented and analyzed in this section. Finally, Section 5 summarizes the conclusions of this work.

2. Background

2.1. Ring Oscillator Based PUF

A Ring Oscillator (RO) is a delay loop that oscillates at a particular frequency. Thanks to its straightforward implementation in FPGAs, ROs have been widely used in the implementation of secure primitives such as true random number generator (TRNG). RO-PUFs are delay-based PUFs that use Challenge-Response scheme as a chip authentication mechanism. A traditional RO-PUF [2] makes use of many identically laid out ROs to quantify the manufacturing variability. RO oscillation frequencies depend on (i) fixed conditions established at the design phase (i.e., number of stages, place&route, etc.); (ii) random process variations (that once manufactured are fixed for each single device); (iii) dynamic conditions derived from the operation environment (i.e., supply voltage, temperature, surrounding logic, etc.). Figure 1 depicts the traditional RO-PUF scheme [2] that consists of many identical ROs, counters and comparators.

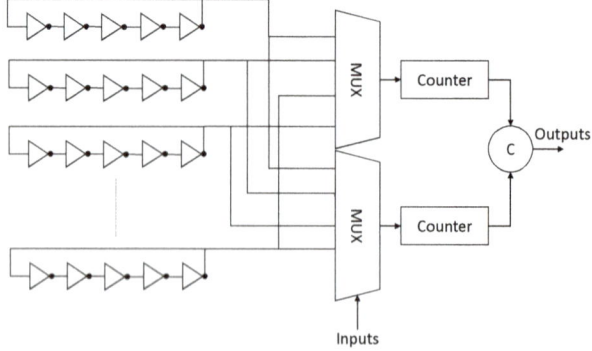

Figure 1. Ring Oscillator Physical Unclonable Function (RO-PUF) general scheme.

During the authentication process, a pair of ROs (selected by a user challenge) are quantized by measuring and comparing the RO frequencies (f_a, f_b) and generating a response bit r:

$$r = \begin{cases} 1 & \text{if } f_a > f_b \\ 0 & \text{otherwise} \end{cases} \quad (1)$$

An n-bit signature of the chip is computed from n different comparisons between RO frequencies. The quality of the PUF is evaluated by analysing the signatures. In most of the literature, major security metrics like uniformity, reliability and uniqueness are used to assess the PUF quality [5].

- **Uniformity** is a metric related to the entropy of the system. This metric estimates the ratio of '1' vs. '0' in all the response bits generated by a PUF. The uniformity is computed as follows for an n-bit PUF response:

$$Uniformity = \frac{1}{n} \sum_{l=1}^{n} r_{i,l} \times 100\% \quad (2)$$

where $r_{i,l}$ is the l-th bit of an n-bit PUF response. An uniform distribution of 0's and 1's (50%) is expected in PUFs that have full entropy.

- **Reliability** is a metric that quantifies how stable the PUF response is over varying operating conditions (voltage, temperature, aging, etc.). To that end, a specific challenge is applied to the RO-PUF in order to obtain an n-bit reference response (r_i) for normal conditions (room temperature, ideal power supply voltage). The same n-bit response is collected at different operating conditions (r_i'). Finally, the reliability is obtained using the Hamming distance (HD) analysis of responses.

$$Reliability = \frac{1}{x} \sum_{y=1}^{x} \frac{HD(r_i, r_{i,y}')}{n} \times 100\% \quad (3)$$

where x is the number of samples for each condition and $r_{i,y}'$ is the y-th sample of (r_i'). A lower value of intra-die HD leads to a higher reliability.

- **Uniqueness** is a measure of how different the PUF responses of different chips among a set of chips are. The uniqueness of a population of k-chips is obtained by computing the inter-HD of the n-bit responses for the same challenge:

$$Uniqueness = \frac{2}{k(k-1)} \sum_{i=1}^{k-1} \sum_{j=i+1}^{k} \frac{HD(r_i, r_j)}{n} \times 100\% \quad (4)$$

where r_i and r_j are the PUF responses of chips i and j ($i \neq j$). An ideal uniqueness of 50% is desired for the complete set of chips.

2.2. TID on SRAM FPGAs

TID effects on SRAM-based FPGAs, including non-radiation hardened, have been widely studied in recent years [6–8]. TID causes a degradation of the transistors as ionizing radiation accumulates on the component. This degradation leads to creating trapped charges that will slowly affect the electrical parameters of the device (threshold voltage (V_{th}) and leakage current) [9]. In this context, NMOS and PMOS transistors behave differently. The trapped charges will negatively affect the threshold voltage increasing the leakage current in NMOS transistors. Conversely, in PMOS transistors the threshold voltage will be increased and the leakage reduced. In addition, a deterioration of noise parameters can be observed [10]. All these effects are dependent on many factors such as dose rate, the type of radiation applied, temperature, etc.

At device level, an increase of the propagation delay of the circuits instantiated in the FPGA is the main aspect to consider. Faults can appear when the timing constraints are violated due to this increment [11].

3. Experimental Setup

3.1. Device Under Test

The radiation experiments have been carried out on two Xilinx XC3S500E FPGA, manufactured on 90 nm CMOS technology. In the remainder of the paper, these Devices Under Test (DUTs) are referred to as FPGA1 and FPGA2. The clock is set using the on-board 50MHz crystal oscillator. Two high precision voltage sources (programmable HP 66103A DC power modules) have been used in order to set and monitor the core voltage (1.2 V) and the I/O voltage (3.3 V).

A conventional RO-PUF consisting of 512 identically laid-out ROs has been implemented. Each of the ROs consists of four inverters and a NAND gate, the latter being able to enable/disable the oscillation. A hard macro that occupies one Configurable Logic Block (CLB) has been created in order to guarantee the same placement and routing for all the ROs. This hard macro has been replicated in the middle of the FPGA creating a 16 × 32 array. During the experiment, each RO is activated at a time during 13,000 clock cycles using its enable signal. The rest of the RO-PUF logic (multiplexers, counters, decoders, etc.) have been implemented in other FPGA zones in order to limit the impact of the surrounding logic on RO frequencies. An RS232 communication protocol has been used in order to transfer the measured RO frequencies to the host computer. A CRC has been implemented in order to ensure the integrity of the communication.

The operating conditions of the room (temperature, pressure and humidity) have been controlled in order to guarantee that these conditions do not affect the RO frequencies.

3.2. TID Setup

The TID tests have been performed at the RADLAB facility, the Gamma Radiation Laboratory installed in the Centro Nacional de Aceleradores (CNA), Spain. The RADLAB [12,13] is based on a Co-60 radioactive source, placed into a Gamma beam X200 irradiator. The average value for the photons energy is 1.25 MeV, which is usually established for testing purposes.

The irradiator has a conical opening which contains a variable collimator, providing different square irradiation fields. During a first irradiation setup, the maximum irradiation field was used and no shielding was applied on the board, so all the components of the PCB were exposed to radiation, not only the DUT (FPGA). As a consequence, some issues were observed before detecting any effect in the DUT. For the subsequent campaigns, the irradiation field was reduced to focus the main gamma beam on the FPGA under study. Moreover, the setup was additionally improved with a custom partial shielding on the board, significantly decreasing the dose rate on the most sensitive components of the PCB (Figures 2 and 3).

Since the DUT is located in one specific position of the PCB submitted to radiation, a dummy board was placed for each setup preparation in the same position as the SAMPLE in order to carry out the dosimetry (Figure 4), that is, to measure the dose rate on the DUT, before starting the irradiation test.

The dosimetry system is composed by a Farmer ionization chamber connected to the UNIDOS Webline electrometer, both of them by PTW. First air kerma rate is obtained and the dose rate in silicon (Si) is calculated taking into account the conversion factors. The dose rate uniformity in the radiation field was better than 95% and the expanded uncertainty associated with the measurement was ±4.2%.

Figure 2. Shielding of the PCB attached to the cover of the filter box.

Figure 3. Profile view of the shielding between the PCB and the cover of the filter box.

Figure 4. PCB customized to carry out the dosimetry measurements with the reference point of the ionization chamber placed exactly in the position of the device under test (DUT).

The dosimetry and the irradiation run were performed using a filter box, with the DUT inside, according to the TID standard from European Space Agency [12]. This container has 1.5 mm Pb (lead) with an inner lining of 2 mm Al (aluminium). The front cover is made of Al, except in the region close to the DUT, where the build-up material is polymethilmetacrilate (PMMA). In Figure 5 is depicted the final setup.

The FPGA1 and FPGA2 were exposed to a total dose of 500 krad(Si), with the dose rates of 5.2 krad(Si)/h and 5.3 krad(Si)/h, respectively.

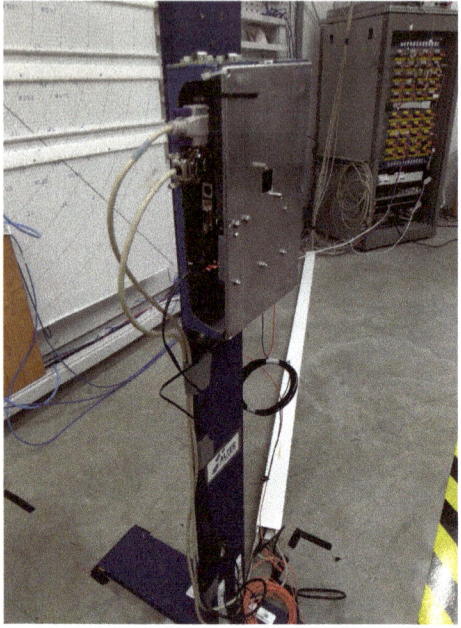

Figure 5. Filter Box with the PCB (and DUT) inside and the connections to the different ports and power supply.

4. Experimental Results and Discussion

This section describes the results of the irradiation experiments on the two DUTs.

4.1. FPGA Parameters and Functionality

The core currents and voltages were continuously measured during the entire experiment. The pre-irradiation operation currents were measured to be 23.9 mA for FPGA1 and 29.09 mA for FPGA2. At the end of the experiment [500 krad(Si)], the currents reached 95.97 mA and 83.35 mA, respectively. These internal core currents are within the limits of the vendor recommendations for this FPGA (typical: 25 mA; maximum: 106 mA). Nevertheless, the first failure was registered at 410 krad(Si) in FPGA1. This failure was related to the RS232 communication protocol and a reprogramming of the FPGA was necessary in order to recover normal functionality. This kind of error was reproduced until the end of the experiment. In FPGA2, no failures were registered. Both DUTs could not be reconfigured any more after the deposited dose reached 440 krad(Si). The faulty behaviour that appeared only in FPGA1 can be explained by the batch difference on the DUT or the lower dose rate (the results typically show marginally higher dose degradation threshold at lower dose rates [14]).

Figure 6 depicts the core currents of FPGA1 and FPGA2 through the irradiation experiment. It is noteworthy that for both DUTs the core current increased linearly with the dose. This increase of leakage current is fully accounted for in the creation of electron-hole pairs due to the ionization of SiO_2 [15].

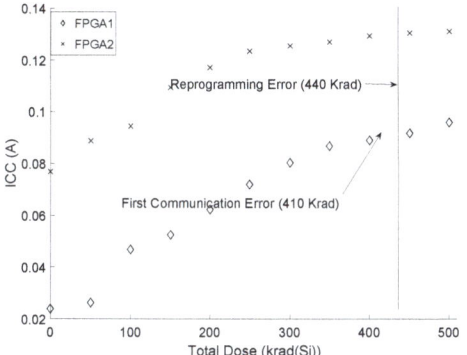

Figure 6. Internal Core Current vs. radiation.

4.2. Delay Variation of the RO Loop

RO frequencies play a key role in the RO-PUF authentication scheme, therefore they have been subjected to exhaustive analysis during the whole experiment. For each RO, we have carried out 100 frequency acquisitions in order to improve the measurement error by averaging the values. At pre-irradiation conditions, the average RO frequencies of FPGA1 and FPGA2 are 196.12 MHz and 199.16 MHz respectively. At the end of the experiment [500 krad(Si)], the average RO frequencies of FPGA1 and FPGA2 have decreased to 195.1 MHz (0.5%) and 197.92 MHz (0.6%) respectively. Figure 7 shows how the 512 RO average frequencies of FPGA2 changed during the experiment. It can be appreciated that after a first sharp decrease at 10 krad(Si), all the ROs follow the same tendency. It is also worthy of note that there are not many intersections between the different lines, which indicates a good frequency stability. These results are very similar to those reported in [16], where the effects of ageing in a RO-PUF were studied.

Figure 8 shows the distribution of the average frequencies at pre-irradiation conditions and at the end of the experiment. Once again, the frequency stability can be highlighted. The results for FPGA1 are analogous to those depicted in Figures 7 and 8.

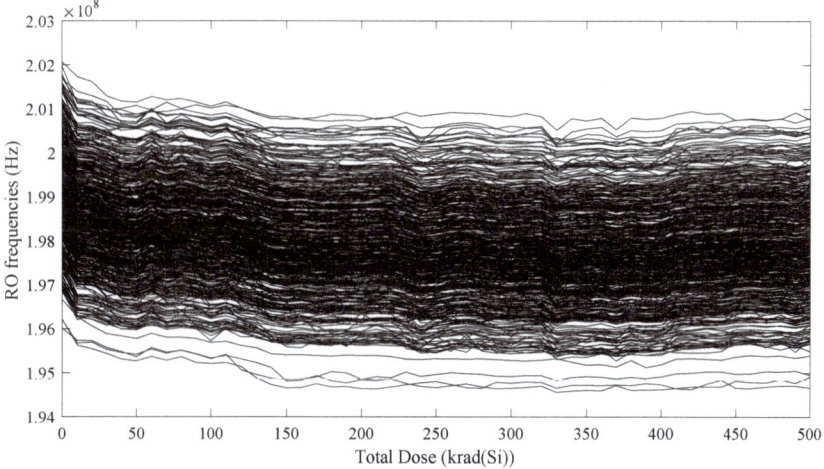

Figure 7. RO frequencies vs. radiation.

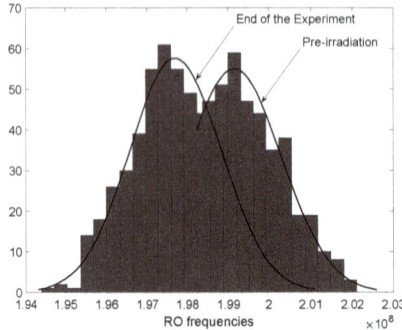 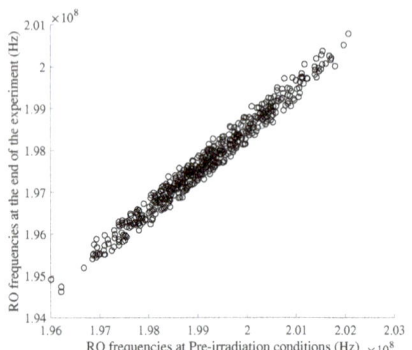

Figure 8. FPGA2 frequency distribution and scatter plot at pre-irradiation conditions and at 500 krad(Si).

4.3. RO-PUF Quality Factors

We have calculated the main metrics related to PUF quality in order to evaluate the suitability of the RO-PUF for space applications. To that end, a 511-bit response has been generated for each accumulated doses. This 511-bit response is extracted by comparing the average frequencies of adjacent pairs of ROs in the array.

4.3.1. Uniformity

Figure 9 depicts the uniformity of the PUF response during the experiment. For both DUTs, the response bits are fairly evenly distributed among '0' and '1', showing almost an ideal distribution throughout the entire experiment. The average number of 1's in the PUF response for FPGA1 and FPGA2 are 51.43% and 49.61%, respectively.

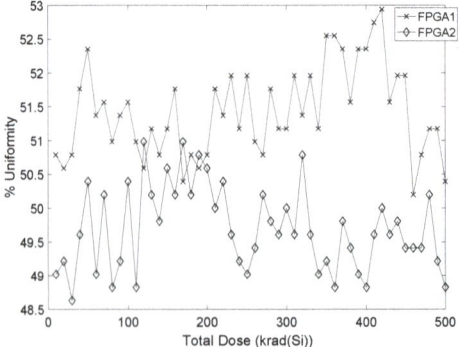

Figure 9. Uniformity vs. radiation.

4.3.2. Reliability

Figure 10 shows the intra-die Hamming distance calculated using Equation (2). The pre-irradiated 511-bit response has been set as the reference response. In both FPGAs, the initial intra-die HD is low (>3%) and it increases with the accumulated dose. This means that there is an increasing degradation of the RO performance that is proportional to the radiation dose.

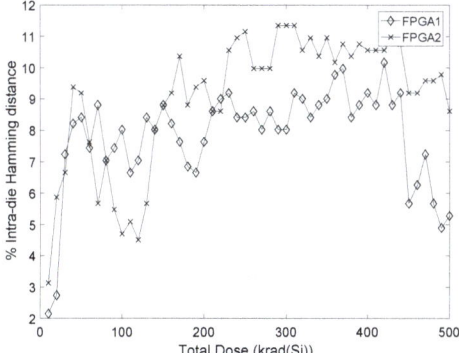

Figure 10. Reliability vs. radiation.

4.3.3. Uniqueness

Figure 11 presents the inter-die Hamming distance obtained using Equation (3). Once again, a loss on the uniqueness can be observed and is proportional to the accumulated dose.

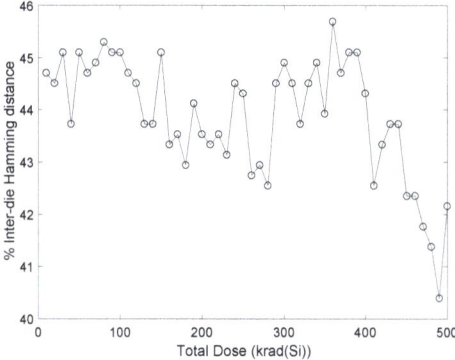

Figure 11. Uniqueness vs. radiation.

4.4. Result Analysis

The increase in the core currents are consistent with other results reported in the literature for the same FPGA [17]. In the case of FPGA1, the first failure occurred at 410 krad(Si), which is also similar to the first failure reported in [17], where the DUT worked properly until 345 krad(Si). Regarding the RO-frequencies, the experiments have shown that there is a good stability on the frequencies and the changes due to the accumulated dose are negligible.

On the other hand, the quality metrics show that the accumulated dose makes the responses produced by the PUF unreliable. Nonetheless, as the uniformity metric shows, the randomness of the response remains unaffected by the total dose. This may be due to the decrease of the noise parameters that have a direct influence on the randomness. Reliability is the key metric to evaluate after the deployment of PUFs in space. If at pre-irradiation conditions all the metrics have acceptable values, only a decrease in the reliability due to ionizing radiation can affect the rest of the metrics. In this case, the reliability metric shows a little degradation that can be corrected using some typical countermeasures such as using only RO pairs with maximal frequency difference [2] or using quantizers with reliability guarantees [18]. Regarding uniqueness, as a collateral effect of unreliability, the results also show a degradation of the metric.

All in all, we can conclude that total ionizing dose has a perceptible effect on the quality of the RO-PUF response. However, with some post-processing, this RO-PUF could be used for space applications.

5. Conclusions

RO-PUFs leverage minor delay variations that exist between devices to support security functions such as authentication and key generation. As TID significantly affects propagation delays, the response of an RO-PUF may be jeopardized in space. In this work, we have performed a comprehensive analysis of the effects of ionizing radiation on the quality of RO-PUFs implemented in FPGAs. RO-PUFs responses were collected as dose accumulated in order to evaluate uniformity, reliability, uniqueness and frequency stability. The external environment was controlled and the core currents and voltages were continuously measured to ensure the results were not biased.

Experimental results showed that RO frequencies show good stability and that the randomness of the response is not affected by TID. However, the reliability and uniqueness of the response shows a little degradation. Nevertheless, this degradation can be corrected by using some typical countermeasures. With these corrections, we can conclude that RO-PUFs implemented in FPGAs can be used in space applications.

Author Contributions: Conceptualization, H.M. and L.E.; Methodology, H.M., P.M.-H. and Y.M.; Writing—Original Draft Preparation, H.M., P.M.-H., Y.M., L.E. and E.S.-M.

Funding: This work was partially supported by the Spanish Ministry of Economy and Competitiveness under contracts ESP2015-68245-C4-1-P and ESP-2015-68245-C4-4-P.

Acknowledgments: Thanks to the ALTER TECHNOLOGY company for making possible the monitoring and recording of the biasing and environmental test conditions during the irradiation test runs.

Conflicts of Interest: The authors declare no conflict of interest. The founding sponsors had no role in the design of the study; in the collection, analyses, or interpretation of data; in the writing of the manuscript, and in the decision to publish the results.

References

1. Maes, R. *Physically Unclonable Functions: Constructions, Properties and Applications*, 1st ed.; Springer: Berlin, Germany, 2016.
2. Suh, G.E.; Devadas, S. Physical Unclonable Functions for Device Authentication and Secret Key Generation. In Proceedings of the 44th ACM/IEEE Design Automation Conference, San Diego, CA, USA, 4–8 June 2007; pp. 9–14.
3. Lee, J.W.; Lim, D.; Gassend, B.; Suh, G.E.; van Dijk, M.; Devadas, S. A technique to build a secret key in integrated circuits for identification and authentication applications. In Proceedings of the 2004 Symposium on VLSI Circuits. Digest of Technical Papers, Honolulu, HI, USA, 17–19 June 2004; pp. 176–179, doi:10.1109/VLSIC.2004.1346548.
4. Wang, P.F.; Zhang, E.X.; Chuang, K.H.; Liao, W.; Gong, H.; Wang, P.; Arutt, C.N.; Ni, K.; McCurdy, M.W.; Verbauwhede, I.; et al. X-ray and Proton Radiation Effects on 40 nm CMOS Physically Unclonable Function Devices. *IEEE Trans. Nucl. Sci.* **2018**, *65*, 1, doi:10.1109/TNS.2017.2789160.
5. Maiti, A.; Gunreddy, V.; Schaumont, P. A Systematic Method to Evaluate and Compare the Performance of Physical Unclonable Functions. In *Embedded Systems Design with FPGAs*; Springer: New York, NY, USA, 2013; pp. 245–267, doi:10.1007/978-1-4614-1362-2_11.
6. Barnaby, H.J. Total-Ionizing-Dose Effects in Modern CMOS Technologies. *IEEE Trans. Nucl. Sci.* **2006**, *53*, 3103–3121, doi:10.1109/TNS.2006.885952.
7. Yao, X.; Hindman, N.; Clark, L.T.; Holbert, K.E.; Alexander, D.R.; Shedd, W.M. The Impact of Total Ionizing Dose on Unhardened SRAM Cell Margins. *IEEE Trans. Nucl. Sci.* **2008**, *55*, 3280–3287, doi:10.1109/TNS.2008.2007122.
8. Zebrev, G.I.; Orlov, V.V.; Gorbunov, M.S.; Drosdetsky, M.G. Physics-based modeling of TID induced global static leakage in different CMOS circuits. *Microelectron. Reliab.* **2018**, *84*, 181–186, doi:10.1016/j.microrel.2018.03.014.
9. Quinn, H. Radiation effects in reconfigurable FPGAs. *Semicond. Sci. Technol.* **2017**, *32*, 044001.

10. MacQueen, D.M. Total Ionizing Dose Effects on Xilinx Field-Programmable Gate Arrays. Master's Thesis, University of Alberta, Edmonton, AB, Canada, 2000.
11. Ma, N.; Wang, S.; Liu, D.; Peng, Y. A run-time built-in approach of TID test in SRAM based FPGAs. *Microelectron. Reliab.* **2016**, *64*, 42–47, doi:10.1016/j.microrel.2016.07.128.
12. Morilla, Y.; Muniz, G.; Dominguez, M.; Martin, P.; Jimenez, J.; Praena, J.; Munoz, E.; Sanchez-Angulo, C.I.; Fernandez, G. New Gamma-Radiation Facility for Device Testing in Spain. In Proceedings of the 2014 IEEE Radiation Effects Data Workshop (REDW), Paris, France, 14–18 July 2014; pp. 1–5, doi:10.1109/REDW.2014.7004580.
13. Costantino, A.; Muschitiello, M.; Zadeh, A.; Romero, G.F.; Holgado, P.M.; Morilla, Y.; Muniz, G.; Standaert, L.; Vanhees, J. Dosimetry Inter-Laboratory Comparison between ESTEC, CNA-ALTER/RADLAB, and UCL. In Proceedings of the 15th European Conference on Radiation and Its Effects on Components and Systems (RADECS), Moscow, Russia, 14–18 September 2015; pp. 1–8, doi:10.1109/RADECS.2015.7365606.
14. Fabula, J.; Bogrow, H. Total Ionizing Dose Performance of SRAM-based FPGAs and supporting PROMs. In Proceedings of the 2001 MAPLD International Conference, Laurel, MD, USA, 11–13 September 2001.
15. Schwank, J.R.; Shaneyfelt, M.R.; Fleetwood, D.M.; Felix, J.A.; Dodd, P.E.; Paillet, P.; Ferlet-Cavrois, V. Radiation Effects in MOS Oxides. *IEEE Trans. Nucl. Sci.* **2008**, *55*, 1833–1853, doi:10.1109/TNS.2008.2001040.
16. Maiti, A.; McDougall, L.; Schaumont, P. The Impact of Aging on an FPGA-Based Physical Unclonable Function. In Proceedings of the 21st International Conference on Field Programmable Logic and Applications, Chania, Greece, 5–7 September 2011; pp. 151–156, doi:10.1109/FPL.2011.35.
17. Da Silveira, M.A.G.; Santos, R.B.B.; Leite, F.G.H.; Araujo, N.E.; Medina, N.H.; Porcher, B.C.; Aguiar, V.A.P.; Added, N.; Vargas, F. X-Ray-Induced Upsets in a Xilinx Spartan 3E FPGA. In Proceedings of the 15th European Conference on Radiation and Its Effects on Components and Systems (RADECS), Moscow, Russia, 14–18 September 2015; pp. 1–4, doi:10.1109/RADECS.2015.7365696.
18. Günlü, O.; Kernetzky, T.; İşcan, O.; Sidorenko, V.; Kramer, G.; Schaefer, R.F. Secure and Reliable Key Agreement with Physical Unclonable Functions. *Entropy* **2018**, *20*, 340, doi:10.3390/e20050340.

 © 2018 by the authors. Licensee MDPI, Basel, Switzerland. This article is an open access article distributed under the terms and conditions of the Creative Commons Attribution (CC BY) license (http://creativecommons.org/licenses/by/4.0/).

Article

Protecting Image Processing Pipelines against Configuration Memory Errors in SRAM-Based FPGAs

Luis Alberto Aranda [1,*], Pedro Reviriego [2] and Juan Antonio Maestro [1]

1. ARIES Research Center, Universidad Antonio de Nebrija, Madrid 28040, Spain; jmaestro@nebrija.es
2. Departamento de Ingeniería Telemática, Universidad Carlos III de Madrid, Leganés 28911, Spain; revirieg@it.uc3m.es
* Correspondence: laranda@nebrija.es

Received: 24 October 2018; Accepted: 13 November 2018; Published: 15 November 2018

Abstract: Image processing systems are widely used in space applications, so different radiation-induced malfunctions may occur in the system depending on the device that is implementing the algorithm. SRAM-based FPGAs are commonly used to speed up the image processing algorithm, but then the system could be vulnerable to configuration memory errors caused by single event upsets (SEUs). In those systems, the captured image is streamed pixel by pixel from the camera to the FPGA. Certain local operations such as median or rank filters need to process the image locally instead of pixel by pixel, so some particular pixel caching structures such as line-buffer-based pipelines can be used to accelerate the filtering process. However, an SRAM-based FPGA implementation of these pipelines may have malfunctions due to the mentioned configuration memory errors, so an error mitigation technique is required. In this paper, a novel method to protect line-buffer-based pipelines against SRAM-based FPGA configuration memory errors is presented. Experimental results show that, using our protection technique, considerable savings in terms of FPGA resources can be achieved while maintaining the SEU protection coverage provided by other classic pipeline protection schemes.

Keywords: Image processing; line buffer; SRAM-based FPGA; single event upset (SEU); configuration memory; soft error

1. Introduction

Image processing has an important role in space applications enhancing the images captured by spacecrafts and robotic vehicles [1]. However, space radiation can affect electronic devices and image sensors causing different malfunctions in the image processing system. These malfunctions can be produced by energetic particles that collide with vulnerable parts in the device leading to, for example, single event upsets (SEUs), a type of soft error that changes the value of a flip-flop or memory cell [2].

The effects of the soft errors depend, amongst other things, on the device that is implementing the image processing algorithm. Microprocessors, for example, are widely used in the image processing field, so soft errors in some critical parts of the processor such as the program counter register can cause unexpected crashes or hangs [3]. Likewise, field-programmable gate arrays (FPGAs) are also used to accelerate image processing algorithms since their logic blocks can be configured to exploit parallelism or specific data features [4]. Depending on the technology used to manufacture the FPGA, they can be more or less susceptible to the mentioned radiation effects. In particular, SRAM-based FPGAs consist of two-dimensional arrays of logic cells and programmable blocks that can be configured by loading a bitstream into the SRAM cells of their configuration memory. Consequently, if an energetic particle strikes an SRAM cell, the loaded design functionalities can change permanently until the device is partial or completely reconfigured [5].

Certain image processing operations based on local filters such as median or rank filters, need to process the image locally instead of pixel by pixel, so the entire image has to be stored in memory for further pixel reading/writing operations. As an alternative, some particular pixel caching pipelines composed of registers and first-in first-out (FIFO) line buffers can be used to process the pixel stream as it arrives from the camera. These line-buffer-based pipelines allow the local filter to process several image rows in parallel. However, local filters and pixel caching designs implemented in SRAM-based FPGAs may have malfunctions due to the mentioned configuration memory errors [6], so an error detection or correction technique is required depending on the criticality of the application.

In this paper, a novel method to protect line-buffer-based image processing pipelines against SRAM-based FPGA configuration memory errors is presented. The technique uses two additional 8-bit registers to store pixels temporarily for later output comparisons. There are other techniques used to protect pipelines or shift registers based on modular redundancy [7], cyclic redundancy check (CRC) [8], or duplication and encoding [9], but an XOR-signature scheme has been chosen for comparison with our proposed scheme due to its overhead-detection tradeoff, and its similar store-and-compare procedure with the proposed technique [10]. The proposed and the XOR-based techniques have been compared in terms of FPGA resource usage and error detection capabilities through exhaustive fault-injection campaigns. Experimental results show that the error detection capabilities of the proposed technique are similar to the XOR-based technique, but our design uses considerably less FPGA resources. In addition to this, the proposed technique is designed in a manner that the identification of the damaged part of the pipeline can be easily inferred, so once the error is detected, a partial reconfiguration can be performed to remove the error.

The rest of the paper is organized as follows. In Section 2, a brief introduction to local filtering and pixel caching is presented. Section 3 explains the proposed error detection techniques for line-buffer-based image processing pipelines. These techniques are evaluated in Section 4. Finally, Section 5 concludes the paper.

2. Pixel Caching

In digital images, local filters are defined as operations in which the output pixel value is a function of the pixel values within a window centred on the currently analyzed input pixel [11,12]. In Figure 1, the local filtering procedure is illustrated for a 3 × 3 square window.

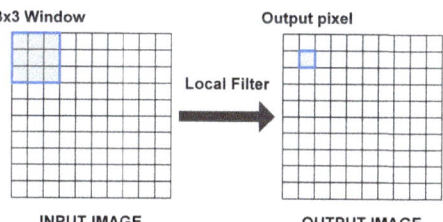

Figure 1. Local filtering process. Grey-shaded pixels on the input image represent the square window that feeds the local filter. The resulting filtered pixel replaces the center pixel of the window. The window is moved along the image to generate all the output pixel values.

As mentioned before, the two-dimensional window is applied to an image region and replaces its original center pixel value with the resulting filtered value of the pixels contained within the window. In order to generate the output filtered image, the window must be moved along the entire image to process each input pixel. In FPGAs, moving the window along the image means storing the whole image frame in memory for subsequent pixel readings. However, the sliding window procedure can also be achieved using a N-by-N pixel stream that sequentially passes through the local filter. In other words, moving the window along the image is equivalent to streaming the image through the window.

This procedure is commonly known as pixel caching, and it is based on the fact that each pixel is reused in multiple window positions. There are several pixel caching structures [13], but they all consist of line buffers and registers. The main difference between those structures is how the connections of the mentioned line buffers are performed. FIFO line buffers can be connected in series or in parallel with the window as shown in Figures 2 and 3.

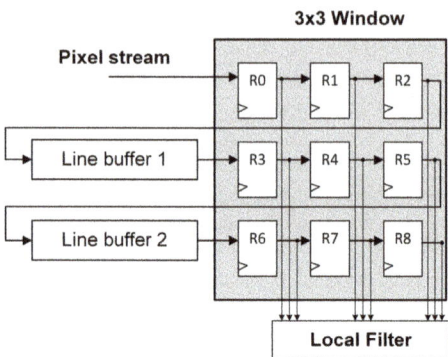

Figure 2. Line buffer pipeline in series with window.

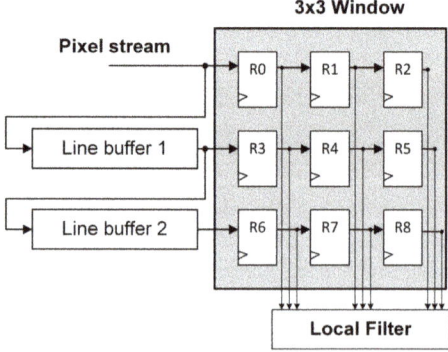

Figure 3. Line buffer pipeline in parallel with window.

The structures presented in Figures 2 and 3 are functionally equivalent. They temporarily store the pixel stream in order to enable a 3 × 3 local filter operation. As can be noticed, these structures does not deal with the image borders. For instance, those pixels at the end of a row are followed by the pixels at the beginning of the next row. Therefore, the border pixels will usually be invalid. There are several approaches to deal with border pixels such as duplication, mirroring with duplication, or mirroring without duplication [13]. However, the original structures presented in Figures 2 and 3 have been studied since, in most applications, the useful information in the image is typically located in the center of the image.

In order to test both line-buffer-based pipelines, a standard 8-bit 128 × 128 pixels grayscale image has been chosen (see Figure 4). As will be explained later, the presented protection techniques are independent of the image size, so the relative resource overhead added to protect the original design decreases as the image size increases. Therefore, a reasonably small image size has been chosen to evaluate the pipelines in a realistic but unfavourable case scenario.

The standard Lena image in Figure 4 has been selected to test the pipelines because its grayscale properties make it widely used in the image processing field [14]. The size of the line buffers and registers is dependent on the image color depth, so for an 8-bit grayscale image, the size of both line buffers and registers must be 8-bits. If, for example, a three-channel RGB color image was used, the pipelines would have to be replicated in parallel for each channel. Then, the proposed error detection scheme can be used to individually protect each of them. This protection technique is presented below.

Figure 4. Standard 8-bit grayscale Lena image.

3. Proposed Techniques

As mentioned in the Introduction, configuration memory errors in SRAM-based FPGAs modify the design functionalities permanently until the original bitstream is reloaded. For this reason, it is more practical to detect the error and then perform a partial or complete reconfiguration of the device to rewrite the affected configuration bits. In contrast to complete reconfiguration, partial reconfiguration avoids reloading the whole bitstream on the device by changing parts of the configuration memory frames while the FPGA is working, so the application does not have to be stopped [15]. Therefore, protection techniques that provide enough information about the damaged part of the design are helpful in facilitating the partial reconfiguration of the device.

Figure 5 illustrates the proposed protection technique for the series structure shown in Figure 2, while the protected parallel structure is presented in Figure 6 (output register connections to "Local Filter" block have been omitted for a better visualization). As can be observed, both techniques are based on including a couple of 8-bit detection registers (*DR1* and *DR2*), a counter, and a three-input comparator to the original unprotected design.

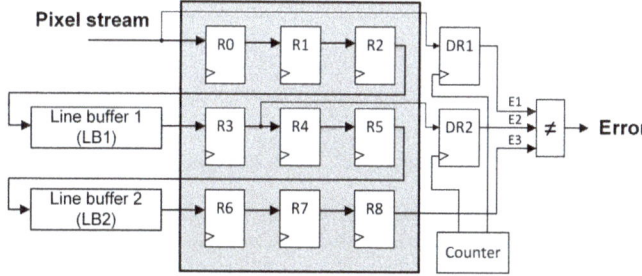

Figure 5. Protected series pipeline.

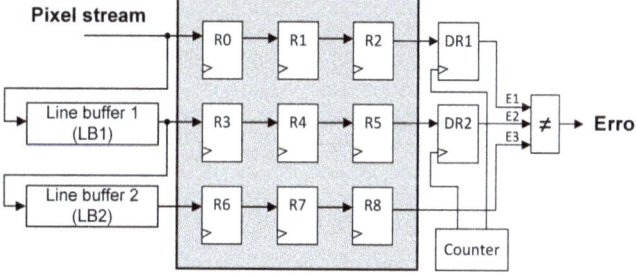

Figure 6. Protected parallel pipeline.

It should be noticed that both error detection registers are different from the window registers. Every register in Figures 5 and 6 is a synchronous 8-bit register, but *DR1* and *DR2* are also controlled by a pixel counter that enables the reading/writing operations. This mechanism is needed to perform the comparison of the three output signals (*E1*, *E2*, and *E3*) at some particular time periods. For example, in the protected parallel pipeline, the pixel at time t is stored in *R0* and *LB1*, then, at time $(t + 3)$ this pixel is stored in *DR1*. The same pixel will be stored in *DR2* at $(t + 3 + line\ buffer\ length)$, and will be outputted by *R8* $(t + 3 + 2 \cdot line\ buffer\ length)$ clock cycles later. At this precise moment, the error signals comparison has to be performed to detect if a configuration memory error has modified the pixel value during its travel through the pipeline. Therefore, for the selected 128×128 pixels grayscale image, one pixel every $128 + 128 + 3 = 259$ clock cycles is checked. The proposed error detection technique is based on the fact that a configuration memory error modifies permanently the configuration of the design, so once the error has damaged the pipeline, most of the subsequent pixel values will be corrupted.

As can be observed in Figures 5 and 6, the error detection registers have been connected to different parts of the pipelines. The series structure is a long unique delay line, so the connections are at the beginning, the middle, and the end of the pipeline in order to check if the pixel has been corrupted during its trip through the pipeline. In the parallel structure, the connections have been made at the end of the branches to detect errors in each of them. These connections also enable the identification of the damaged part of the pipeline. Using the three error detection output signals *E1*, *E2*, and *E3*, a table can be created to evaluate the different possible scenarios and the part of the design that has to be reconfigured to remove the configuration memory error. These errors affect equally both logic elements and routing, so the partial reconfiguration has to be performed in the damaged components shown in Table 1 and their route-related configuration bits. Moreover, the counter always has to be reconfigured along with the possible damaged parts of the pipeline since all the erroneous output scenarios can be caused by an error on it.

Table 1. Possible Damaged Part of the Pipeline.

Output Scenario	Series	Parallel
E1 = E2 = E3	None	None
(E1 = E2) but $\neq$ E3	R4 to R8/LB2	R6 to R8/LB2
(E1 = E3) but $\neq$ E2	DR2	R3 to R5/DR2
(E2 = E3) but $\neq$ E1	R0 to R3/LB1/DR1	R0 to R2/LB1/DR1
E1 $\neq$ E2 $\neq$ E3	Comparator	Comparator

Table 1 summarizes the possible damaged parts of the studied pipelines depending on the values of the output error detection signals. For example, in the series pipeline, if a configuration memory error affects the window register *R4*, then the pixels stored in *DR1* and *DR2* will be equal, but the pixel could be corrupted as it passes through *R4*, so the pixel that outcomes from *R8* may not be equal to them. This means that *E1* and *E2* would have the same values but *E3* would be different from them. Conversely, the same scenario in the parallel pipeline means that only *DR2* would store a different value since the output of *R5* is not connected to the rest of the pipeline, so *E2* would have a different value from *E1* and *E3*. The rest of the scenarios presented in Table 1 can be similarly deduced. With the information provided in Table 1, a partial reconfiguration of the damaged part of the pipeline can be performed to remove the error, instead of reconfiguring the entire FPGA. The quantitative benefits of partial reconfiguration against complete reconfiguration are not discussed in this paper, but they are considered as future work.

4. Technique Evaluation

In order to compare performance in terms of resource utilization and error detection rate, an XOR-based signature technique has been chosen. This technique creates a signature of the input image using an XOR gate and a couple of 8-bit registers. The input image pixels are XORed and the result is stored for a later comparison to the similarly calculated output pixels signature. If an error alters the pixel values that pass through the pipeline, a signature mismatch occurs and the error is detected. The described XOR-based signature calculation module is shown in Figure 7. In this figure, *Reg 1* is a simple 8-bit synchronous register, however, *Reg 2* also uses a read/write enable signal controlled by a counter to temporarily store the signature, as happened with the error detection registers *DR1* and *DR2* from the proposed techniques.

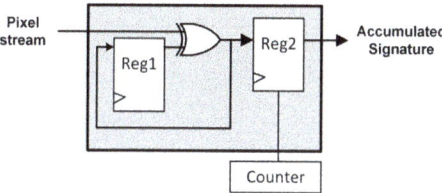

Figure 7. XOR-based signature generation module.

This module (named "XOR" in Figures 8 and 9) has to be placed before and after the series pipeline to calculate the input and the output image signatures for later comparison. However, in the parallel pipeline, three XOR modules have to be used to be able to detect errors in the three pipeline branches. This connection is similar to the one performed in Figure 6 for the proposed technique.

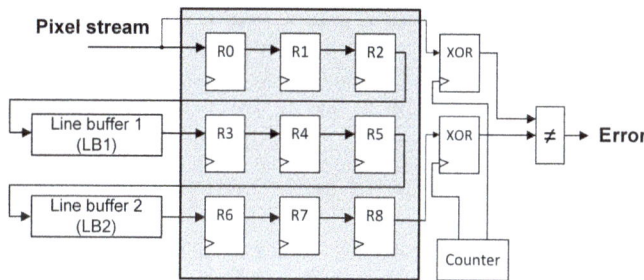

Figure 8. XOR-based protected series pipeline.

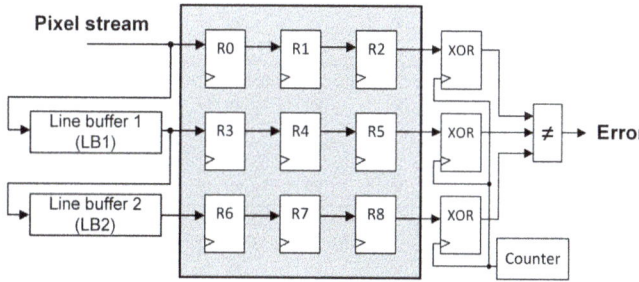

Figure 9. XOR-based protected parallel pipeline.

The proposed error detection structures presented in Section 3 and the XOR-signature technique have been implemented in the SRAM-based FPGA part of a Xilinx Zynq-7000 All Programmable System on a Chip (SoC) obtaining the utilization reports presented in Tables 2 and 3.

Table 2. Utilization Report (Series).

	Unprotected	XOR Signature	Proposed
LUTs	91 (0%)	112 (23.1%)	114 (25.3%)
FFs	99 (0%)	149 (50.5%)	127 (28.3%)

Table 3. Utilization Report (Parallel).

	Unprotected	XOR Signature	Proposed
LUTs	73 (0%)	119 (63%)	92 (26%)
FFs	107 (0%)	183 (71%)	117 (9.3%)

In Table 2, the number of look-up tables (LUTs) and flip-flops (FFs) used by each series structure is presented, while the parallel pipeline resources are shown in Table 3. The percentages in these tables show the overhead of LUTs or FFs added to the unprotected design to implement each protection technique. In the parallel structure, the XOR-based scheme needs three XOR signature generation modules so, as can be observed in Table 3, significantly more resources are required. In the series pipeline, the number of LUTs is slightly higher in the proposed technique. This is due to the use of a two-input comparator in the XOR-based scheme instead of a three-input comparator but, in general, the total FPGA resource usage of the proposed technique is lower.

The error detection rate of the techniques has been measured through fault-injection. First, the techniques are validated against the standard 8-bit grayscale Lena image presented in Figure 4 to obtain the "golden" output pixel values that should outcome from the pipeline if no error is injected. These golden outputs are then stored for later golden comparisons. Once the golden outputs are obtained, an exhaustive fault-injection campaign is executed using the Xilinx Soft Error Mitigation (SEM) IP Controller [16]. The SEM IP is an independent circuit that has to be loaded along with the design under test (DUT) in order to perform read/write operations in the DUT-related configuration bits through the internal configuration access port (ICAP). The fault-injection campaign is sequentially performed in an injection-correction loop. In each iteration, a configuration bit is flipped, the test image is processed, and the golden comparison results are stored. Finally, the bit flip is corrected by the SEM IP and the loop is repeated until all the DUT-related configuration bits are covered. It is worth mentioning that SEUs have been injected since they are the worst case scenario for the error detection techniques considered in this paper. This is because more errors imply more opportunities to detect a malfunction in the pipeline.

For a better understanding of the experimental set-up, the fault-injection framework is presented in Figure 10. This figure illustrates the different modules that are loaded in the SRAM-based FPGA. Those modules that are grouped together inside the grey-shaded "Testbench" box are not affected by the fault injector.

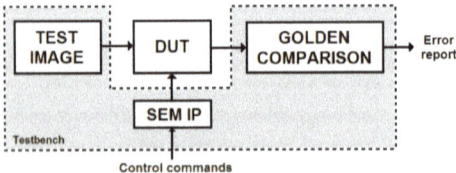

Figure 10. Soft Error Mitigation (SEM) IP-based fault-injection framework.

The fault-injection campaign has been performed for the unprotected original design, the XOR-based protection scheme, and the proposed protection technique, obtaining the results summarized in Tables 4 and 5. Table 4 presents the error detection capabilities for the series structure of the line-buffer-based pipeline, while Table 5 shows the results for the parallel pipeline.

Table 4. Error Detection Report (Series).

	Unprotected	XOR Signature	Proposed
Corrupted images	4393 (100%)	4810 (100%)	4559 (100%)
- Detected	N/A	4430 (92.1%)	4202 (92.2%)
- Undetected	4393 (100%)	380 (7.9%)	357 (7.8%)
Uncorrupted images	20,588 (100%)	25,825 (100%)	23,943 (100%)
- Normal operation	20,588 (100%)	24,345 (94.3%)	22,775 (95.1%)
- False positives	N/A	1480 (5.7%)	1168 (4.9%)
MSE of Undetected	310	28	26

Table 5. Error Detection Report (Parallel).

	Unprotected	XOR Signature	Proposed
Corrupted images	3972 (100%)	4768 (100%)	4533 (100%)
- Detected	N/A	4404 (92.4%)	4186 (92.3%)
- Undetected	3972 (100%)	364 (7.6%)	347 (7.7%)
Uncorrupted images	21,269 (100%)	24,590 (100%)	28,634 (100%)
- Normal operation	21,269 (100%)	22,461 (91.3%)	27,290 (95.9%)
- False positives	N/A	2129 (8.7%)	1344 (4.1%)
MSE of Undetected	287	27	26

It can be observed in these tables that the detection rate is approximately the same in both pipelines and between both studied techniques. However, the proposed technique seems to be more effective than the XOR-based scheme due to its fewer false positives. This is particularly noticeable in the parallel pipeline, in which the percentage of false positive detections is more than twice the proposed technique. False positives are strongly related to the number of FPGA resources used for the error detection part, so an error affecting these configuration bits can trigger the "error detected" signal while the pipeline outputs are still correct. By comparing Tables 2 and 3, it can be noticed that the XOR-based scheme requires more LUTs and FFs for the parallel pipeline than for the series pipeline since it needs a third XOR module to perform the error detection. As can be observed in Table 5, these additional resources imply more false positive detections that will lead to more FPGA reconfigurations. Consequently, the availability of the FPGA will be increased when our proposed technique is implemented.

In order to measure the quality of the outputted image when the error is not detected, the averaged Mean Square Error (MSE) of the undetected corrupted images has also been calculated for each pipeline scheme and included in Tables 4 and 5. It can be observed that there is a significant reduction of the MSE in both protection schemes compared to the unprotected pipelines and that the averaged MSE obtained for the proposed protection schemes is slightly lower than the XOR-based technique. This means that, when an error is not detected by our proposed technique, the outputted corrupted image will still have (on average) better quality than the image outputted by the XOR-based scheme and, therefore, it should have a lower impact on subsequent image processing algorithms.

5. Conclusions and Future Work

In this paper, a novel method to protect line-buffer-based image processing pipelines against SRAM-based FPGA configuration memory errors is presented. Our proposed method is used to protect two types of line-buffer pipelines (series and parallel). They store the image pixels temporarily using two additional 8-bit registers. These pixels are later compared and, if a pixel mismatch is found, the error is detected and then a partial or complete reconfiguration can be performed.

The proposed technique has been compared in terms of FPGA resource usage and error detection rate with an XOR-based signature scheme commonly used to protect pipelines. Both detection techniques have been implemented in a Xilinx SRAM-based FPGA and fault-injection campaigns have been performed using a Xilinx injection IP core.

Experimental results show that the proposed technique presents lower FPGA resource usage, similar error detection rate, and fewer false positive detections in comparison with the XOR-based scheme. In addition to this, the proposed technique has also been designed in a manner that the identification of the damaged part of the pipeline can be easily inferred, so the partial reconfiguration of the damaged module is facilitated. This means that, using our proposed error detection method, the image processing system does not have to be stopped and rebooted, as usually happens when a classic complete reconfiguration is performed. Implementing and measuring the effectiveness of the partial reconfiguration in terms of time and power consumption is considered as future work.

Author Contributions: L.A.A. conceptualized the main idea of this research project; L.A.A. designed and conducted the experiments with the help of P.R. and J.A.M.; P.R. and J.A.M. checked the results. L.A.A. wrote the whole paper; P.R. and J.A.M. reviewed and edited the paper.

Conflicts of Interest: The authors declare no conflict of interest.

References

1. Mourikis, A.I.; Trawny, N.; Roumeliotis, S.I.; Johnson, A.E.; Ansar, A.; Matthies, L. Vision-Aided Inertial Navigation for Spacecraft Entry, Descent, and Landing. *IEEE Trans. Robot.* **2009**, *25*, 264–280. [CrossRef]
2. Baumann, R.C. Radiation-induced soft errors in advanced semiconductor technologies. *IEEE Trans. Device Mater. Reliab.* **2005**, *5*, 305–316. [CrossRef]
3. Irom, F. Radiation Induced Effects in Microprocessors. In *Guideline for Ground Radiation Testing of Microprocessors in the Space Radiation Environment*; Jet Propulsion Laboratory, National Aeronautics and Space Administration: Pasadena, CA, USA, 2008; Chapter 3, pp. 7–11.
4. Pedre, S.; Krajnik, T.; Todorovich, E.; Borensztejn, P. Accelerating embedded image processing for real time: A case study. *J. Real-Time Image Process.* **2016**, *11*, 349–374. [CrossRef]
5. Asadi, G.; Tahoori, M.B. An analytical approach for soft error rate estimation of SRAM-based FPGAs. In Proceedings of the Military and Aerospace Applications of Programmable Logic Devices (MAPLD), Washington, DC, USA, 8–10 September 2004.
6. Aranda, L.A.; Reviriego, P.; Maestro, J.A. A Comparison of Dual Modular Redundancy and Concurrent Error Detection in Finite Impulse Response (FIR) Filters Implemented in SRAM-based FPGAs through Fault Injection. *IEEE Trans. Circuits Syst. II* **2018**, *65*, 376–380. [CrossRef]
7. Tarrillo, J.; Tonfat, J.; Tambara, L.; Kastensmidt, F.L.; Reis, R. Multiple fault injection platform for SRAM-based FPGA based on ground-level radiation experiments. In Proceedings of the 16th Latin-American Test Symposium (LATS), Puerto Vallarta, Mexico, 25–27 March 2015; pp. 1–6.
8. Kim, H.; Choi, I.; Byun, W.; Lee, J.; Kim, J. Distributed CRC Architecture for High-Radix Parallel Turbo Decoding in LTE-Advanced Systems. *IEEE Trans. Circuits Syst. II Express Briefs* **2015**, *62*, 906–910. [CrossRef]
9. Reviriego, P.; Ruano, O.; Flanagan, M.; Pontarelli, S.; Maestro, J.A. An Efficient Technique to Protect Serial Shift Registers against Soft Errors. *IEEE Trans. Circuits Systems II* **2013**, *60*, 512–516. [CrossRef]
10. Werner, M.; Wenger, E.; Mangard, S. Protecting the control flow of embedded processors against fault attacks. In Proceedings of the 14th International Conference on Smart Card Research and Advanced Applications, Bochum, Germany, 4–6 November 2015; pp. 161–176.
11. Szeliski, R. Point operators. In *Computer Vision: Algorithms and Applications*; Springer: Berlin/Heidelberg, Germany, 2010; Chapter 3, Section 1, pp. 101–103.
12. Aranda, L.A.; Reviriego, P.; Maestro, J.A. Error Detection Technique for a Median Filter. *IEEE Trans. Nucl. Sci.* **2017**, *64*, 2219–2226. [CrossRef]
13. Bailey, D.G. Local Filters. In *Design for Embedded Image Processing on FPGAs*, 1st ed.; John Wiley & Sons: Singapore, 2011; Chapter 8, Section 1, pp. 233–239.
14. Hutchison, J. Culture, Communication, and an Information Age Madonna. *IEEE Prof. Commun. Soc. Newslett.* **2001**, *45*, 1–7.

15. Bolchini, C.; Miele, A.; Santambrogio, M.D. TMR and Partial Dynamic Reconfiguration to mitigate SEU faults in FPGAs. In Proceedings of the 22nd IEEE International Symposium on Defect and Fault-Tolerance in VLSI Systems (DFT 2007), Rome, Italy, 26–28 September 2007; pp. 87–95.
16. Xilinx. Soft Error Mitigation Controller; LogiCORE IP Product Guide (PG036). Available online: https://www.xilinx.com/support/documentation/ip_documentation/sem/v4_1/pg036_sem.pdf (accessed on 23 October 2018).

© 2018 by the authors. Licensee MDPI, Basel, Switzerland. This article is an open access article distributed under the terms and conditions of the Creative Commons Attribution (CC BY) license (http://creativecommons.org/licenses/by/4.0/).

MDPI
St. Alban-Anlage 66
4052 Basel
Switzerland
Tel. +41 61 683 77 34
Fax +41 61 302 89 18
www.mdpi.com

Electronics Editorial Office
E-mail: electronics@mdpi.com
www.mdpi.com/journal/electronics

www.ingramcontent.com/pod-product-compliance
Lightning Source LLC
LaVergne TN
LVHW071945080526
838202LV00064B/6682